The Earth's Biosphere

Also by Vaclav Smil

China's Energy
Energy in the Developing World
Energy Analysis in Agriculture
Biomass Energies
The Bad Earth
Carbon Nitrogen Sulfur
Energy Food Environment
Energy in China's Modernization
General Energetics
China's Environmental Crisis
Global Ecology
Energy in World History
Cycles of Life
Energies
Feeding the World
Enriching the Earth

VACLAV SMIL

THE EARTH'S BIOSPHERE

Evolution, Dynamics, and Change

The MIT Press
Cambridge, Massachusetts
London, England

This book was set in Galliard, Bodoni, and Futura by Graphic Composition, Inc.
Printed on recycled paper and bound in the United States of America.

Library of Congress Cataloging-in-Publication Data

Smil, Vaclav.
The earth's biosphere : evolution, dynamics, and change / Vaclav Smil.
p. cm.
Includes bibliographical references (p.) and index.
ISBN 0-262-19472-4 (hc. : alk. paper)
1. Biosphere. I. Title.
QH343.4.S655 2002
577—dc21 2001058705

Contents

Preface

A random perusal of recent publications in such disparate fields as bacterial genomics and Internet encryption or organic chemistry and astrophysics shows that scientific reductionism, the practice I liken to the drilling of ever deeper holes, is thriving more than ever. Synthetic endeavors, such as scanning both nearby terrain and distant horizons, preoccupy only a small minority of scientists but their frequency and scope are increasing. One of the principal reasons for the rejuvenation of scientific syntheses (which used to make up such a large part of nineteenth century science) is the growing realization that the very survival of modern civilization is inextricably tied to the fate of our environment, to changes in the Earth's biosphere. And anybody with even the most basic understanding of the environment realizes that in order to minimize the impact on the biosphere modern societies need integrated, multidisciplinary studies that inform by the breadth of their syntheses.

I wrote *The Earth's Biosphere* to further this fundamental understanding. I have tried to do that by (1) offering a new synthesis that is informed by the latest advances in a multitude of disciplines, ranging from geochemistry and geophysics to ecology and energetics, (2) explaining both well-known and largely unknown fundamentals of the biosphere's structure and dynamics, and (3) trying to convey the many amazing attributes of the Earth's realm of life. These remarkable characteristics are met at every scale, from sub-cellular to planetary—the elaborate organization of protein machines within every living cell, the incredible abilities of microbial metabolism in both mundane and extreme environments, the grand-scale transfers of heat in the Earth's ocean and atmosphere, and the life-sustaining complexities of global biogeochemical cycles.

I have also tried to use some of my professional advantages and personal predilections—bridging the natural and the social sciences—in order to write a multifaceted book that does not fit easily into any simple categories. I know that some readers will find this approach too broad or too taxing. Their patience may be particularly tried as they encounter many pages devoted to the intricacies of microbial life—but this more-detailed focus on bacteria and archaea is merely an acknowledgment of the still underappreciated dominance and ubiquity of these organisms in the biosphere and of their irreplaceable roles in the cycles of life.

I am also sure that some specialists will feel that many topics have not been given enough attention or that explanations of some particulars would have benefited from a greater detail or from a more subtle understanding. I plead guilty on all counts. Syntheses cannot be written as lengthy encyclopedias and, no matter how hard one tries, subtle, confident understanding of myriads of details making up the mosaic of life is beyond anyone's abilities. Such are the perils of this nonreductionist intellectual adventure; its rewards, I hope, are some new perspectives.

As with any book of this kind, I am indebted to thousands of scientists whose work I cite, admire, or criticize and without whose findings and ideas I would not have been able to scan the global horizon and deliver this new synthesis. My particular thanks go to Marty Hoffert, John Katzenberger, and David Schwartzman, who read the entire typescript, corrected some lapses, and suggested numerous additions and improvements.

Any remaining mistakes and misinterpretations are mine—and only mine. As any true interdisciplinarian I have been always aware of numerous weaknesses in my quest for grand syntheses, and I have never had any illusions about the impact of my books. I just do my best. But I am also always mindful of Seneca's sobering verdict regarding our understanding of the world in general:

Veniet tempus quo posteri nostri tam aperta nos nescisse mirentur. (The day will come when posterity will be amazed that we remained ignorant of matters that will to them seem so plain.)

The Earth's Biosphere

1

Evolution of the Idea

From Vernadsky to a Science of the Global Environment

A new character is imparted to the planet by this powerful cosmic force. The radiations that pour upon the Earth cause the biosphere to take on properties unknown to lifeless planetary surfaces, and thus transform the face of the Earth.
Vladimir Ivanovich Vernadsky, *Biosfera*

The Earth is not just an ordinary planet!
Antoine de Saint Exupéry, *The Little Prince*

Unlike so many ideas that have unclear, or contested, origins, there is no dispute about the first use of the term "biosphere." Eduard Suess (1831–1914)—a famous Austrian geologist who, together with Charles Lyell (1797–1875) and Louis Agassiz (1807–1873), was one of the three greatest nineteenth-century synthesizers of the rising discipline of Earth science—coined the term (fig. 1.1). Suess became eventually best known for his monumental, although now quite outdated, three-volume set *Das Antlitz der Erde* (Suess 1885–1904), which summarized the contemporary understanding of all major geological features of the Earth.

In fact, Suess literally tossed the new term away, just once and without an explicit definition, in his pioneering book on the genesis of the Alps (Suess 1875). Suess noted that

one thing seems to be foreign on this large celestial body consisting of spheres, namely, organic life. But this life is limited to a determined zone at the surface of the lithosphere. The plant, whose deep roots plunge into the soil to feed, and which at the same time rises into the air to breathe, is a good illustration of organic life in the region of interaction between the upper sphere and the lithosphere, and on the surface of continents it is possible to single out an independent biosphere. (p. 3)

This is a peculiarly limited, and flawed, definition: it appears that Suess did not even consider microorganisms that are abundant both in the lower layer of the atmosphere and in the ocean. And perhaps only Suess's professional

1.1 Eduard Suess (1831–1914), Austrian geologist who coined the term "biosphere" in 1875. Portrait courtesy of Dr. Fritz Popp, Institut für Geologie, Wien Universität.

preoccupation with crustal forms and orogenesis explains his glaring exclusion of marine life from his concept of the biosphere. And seeing the biosphere as independent (*selbständig* in the original) is an inexplicable denial of myriads of links between organisms and their environment. In any case, the new term remained an oddity, and it took a long time before it entered the scientific vocabulary. Five decades had to pass before a Russian scientist reintroduced the concept, defined it in great detail and put it at the core of his interdisciplinary examination of life on the Earth. And incredibly, only more than four decades after that rigorous reintroduction did the concept become widely known outside Russia and begin providing stimulation for long overdue fundamental research as well as for raising public awareness of the need to protect the Earth's environment.

Two developments pushed the concept of the biosphere to the center stage of scientific attention during the last generation of the twentieth century: concerns about unprecedented rates and scales of anthropogenic environmental change and the global monitoring of the Earth's environment from increasingly sophisticated satellites. As a result, few challenges facing civilization during the twenty-first-century will be as daunting, and as critical, as the preservation of the biosphere's integrity.

Vernadsky's Biosphere

Vladimir Ivanovich Vernadsky (1863–1945) was the scientist who elaborated the concept of the biosphere and who is now generally acknowledged as the originator of a new paradigm in life studies (fig. 1.2).[1] Vernadsky belonged to that remarkable group of Russian researchers and thinkers who flourished during the last decades of the nineteenth and the first decades of the twentieth century and whose contributions proved so important for the progress of many disciplines because of their bold departures in new directions. The group's most illustrious names include Dimitri Ivanovich Mendeleev (1834–1907), the author of the periodic table of elements; Vasili Vasil'evich Dokuchaev (1846–1903), the founder of modern soil sci-

1. In references to his work published in French and English I am using the common transcription "Vernadsky"; in references to his Russian-language publications or books about Vernadsky I am using the correct transcription "Vernadskii." Perhaps the most useful book dealing with Vernadsky's life and scientific accomplishments in a wider setting of Russia's history, including reminiscences of his contemporaries, is the volume edited by Aksenov (1993). Other useful English-language books on Vernadsky's life and work include Tumilevskii (1967) and Sokolov and Ianshin (1986). Two additional English-language books are a translation of Balandin's (1982) work and Bailes (1990).

1.2 Vladimir Ivanovich Vernadsky (1863–1945). Photo courtesy of Eric Galimov, V. I. Vernadsky Institute of Geochemistry and Analytical Chemistry of the Russian Academy of Sciences, Moscow.

ence; Ivan Petrovich Pavlov (1849–1936) and Ilya Il'ich Mechnikov (1845–1916), Nobelians (1904 and 1908) in medicine and physiology; Konstantin Eduardovich Tsiolkovsky (1857–1935), the visionary pioneer of space flight; and Sergei Winogradsky (1856–1946), discoverer of chemotrophic metabolism in bacteria and one of the creators of modern microbiology.

Vernadsky's scientific career advanced smoothly. Born into a well-off Ukrainian family (his father was a professor of political economy in Kiev and in Moscow and was also engaged in liberal politics), he studied natural sciences at the Faculty of Physics and Mathematics of St. Petersburg University, where both Mendeleev and Dokuchaev were among his professors; mineralogy was his speciality. In 1888–1889 Vernadsky spent two years studying in Munich and in Paris, where he worked with Henri Louis Le Châtelier and with Pierre Curie. Le Châtelier (1850–1936) was an expert in high-temperature studies and in behavior of gas mixtures, and he remains best known for his eponymous principle: "Every change in one of the factors of an equilibrium occasions a rearrangement of the system in such a direction that the factor in question experiences a change in the sense opposite to the original change." Pierre Curie (1859–1906) was a corecipient (with his wife, Marie Curie-Skłodowska, and Henri Becquerel) of 1903 Nobel Prize in physics for their work on radiation phenomena.

Vernadsky's doctoral thesis was submitted to the Moscow University in 1897. A year later he became an extraordinary, and in 1902 an ordinary professor. His first book, *Osnovi kristalografii* (*Fundamentals of Crystallography*), came out two years later (Vernadskii 1904). In 1905, the year of Russia's first democratic revolution, Vernadsky became a founding member of the Constitutional Democratic Party, and between 1908 and 1918 a member of its central committee. Members of the party were known as *kadets,* the name derived from the first letters of the party's Russian name. In 1909 Vernadsky turned from crystallography to geochemistry, and three years later he was elected a full member of the Russian Academy of Sciences. Shortly after the beginning of World War I, he founded and chaired the Commission for the Study of Natural Productive Sources (Russian acronym KEPS), whose goal was to assist the country's military effort. Prominent scientists also participated in the war effort in Germany, France, and the United Kingdom, and later in the United States, foreshadowing the much greater impact of science during World War II.

Beginnings and Interruptions

In the spring of 1916 Vernadsky went to Crimea to the *dacha* of Mikhail Bakunin, brother of the famous anarchist. Then, in July, he prospected, with his favorite student Aleksandr E. Fersman (1883–1945), for bauxite in the Altai Mountains. Afterward he spent several weeks at his own comfortable *dacha* in Shishaki that he built in 1911 on the high left shore of the Psel River, halfway between Poltava and Mirogorod. These were the months and the places where he began to think systematically and to make notes—"with exceptionally broad intentions" in mind—about living matter as the transformer of solar energy and about its planetary importance.

His intentions were to go beyond the unsatisfactory ways in which contemporary biology was dealing with life: examining it either without any references to its environment or merely as adapting to diverse environmental conditions. Fossil fuels, carbonate and phosphate deposits, coral reefs, soils, and atmospheric gases were obvious manifestations of life's importance in actively shaping and transforming the Earth. Vernadsky intended to answer a critical question that he jotted down on a small piece of paper preserved in his archive: "What importance has the whole organic world in the general scheme of chemical reactions on the Earth?" But his answers had to wait: soon after he posed that question the old world of the Russian Empire fell apart. The revolutionary regime that overthrew the czar in February 1917 was short-lived: it was itself overthrown, with German help, by Lenin's Bolsheviks in October 1917, and after years of civil war a new Communist state emerged victorious and the Union of Soviet Socialist Republics was officially established in November 1921 (Walsh 1968; Bradley 1975).

The year 1917, so fateful for Russia, was personally eventful and tragic for Vernadsky. His daughter Nyuta died of tuberculosis, which he, too, contracted. As a prominent *kadet,* he became a deputy to S. F. Oldenburg, the minister of education in the provisional government. On November 6, shortly after the Communist takeover, he wrote in his diary: "Very sad and apprehensive about the future." Yet he was also lucky. When the Bolshevik revolution came in 1917 *kadets* were among the principal enemies to be eliminated by the new regime, and thousands of them, and hundreds of thousands of others, perished in the violence, anarchy, and hunger of the subsequent civil war (for the history of Russia's civil war and political parties see Bradley 1975 and Brovkin 1994).

Three years of disjointed, perilous, and uncertain life were ahead of him. In 1918 Vernadsky fled Moscow, organized the Academy of Sciences in a temporarily free Ukraine, and became its first president. Later he was separated and reunited with his family, got ill with typhoid in February and March of 1920, and two months later became a deputy rector, and in September a rector, of the Tauride University in Simferopol. He was joined in this Crimean refuge, protected by General Wrangel's army (the reactionary Whites, in the Communist parlance), by such outstanding scientists as Ioffe and Tamm. Abram Fedorovich Ioffe (1880–1960), whose main achievements were in crystal physics, was the founder of one of the USSR's principal schools of physics, and Igor Evgenievich Tamm (1895–1971) was a corecipient of the 1958 Nobel Prize in physics. The Communist reconquest of the Crimea in November 1920 ended abruptly that episode of Vernadsky's career.

His son Georgi evacuated with Wrangel's retreating forces, but Vernadsky remained. A letter from N. A. Semashko, Lenin's commissar for health care, spared Vernadsky any persecution. Moreover, he, with a few others, was assigned a special railway car in which to return to

Moscow. Soon after his arrival in Moscow, the family moved back to Petrograd, where Vernadsky tried to resume his work and accepted Fersman's invitation to do research in Russia's north, on the Kola Peninsula near Murmansk. But on July 14, 1921, Vernadsky was arrested and brought to the city's *cheka* headquarters. Without an intervention at the highest level, there would have been no hope: "No release could happen automatically, only shooting happened automatically" (Aksenov 1993).

After spending a day in prison he was, to everybody's surprise, released following two hours of interrogation: A. P. Karpinsky, the president of the Academy, sent telegrams to Lenin, Semashko, and Lunacharsky (the commissar for education) stressing that Vernadsky never fought against the Soviet power. Within a day after his release he left with his daughter Nina, a medical student, for Kola. In May 1922 Vernadsky, his wife, and Nina got passports to travel, via Prague, where his son Georgi settled as an emigrant, to Paris. They arrived on July 8, 1922, and their stay was to be just for one academic year. Vernadsky's Sorbonne lectures on geochemistry were published as a book (Vernadsky 1924), and his stay was extended, with the university's and the Rosenthal Foundation's help, as he searched for funds in the United States to establish a biogeochemical laboratory (Vernadsky 1923). Would he have joined thousands of Russian intellectuals in exile if this laboratory idea were realized?

From Paris, in a letter to his old friend Ivan Il'ich Petrunkevich (1843–1928), one of the leaders of the prerevolutionary democratic opposition to the czarist regime living in exile in Prague, he made his position, and his sense of duty, quite clear:[2]

If I were much younger—I would emigrate. My universal feelings are much stronger than my nationalist feelings. But now it is difficult and impossible, as one always needs to lose several years on securing a position. I have no illusions—to live in Russia is extraordinarily difficult. . . . Even if my attempts of moving to America were successful, I would still feel obliged to return and then to leave.

But his last year spent in France was hardly wasted in waiting. He nearly completed a new book whose beginnings went to those prerevolutionary wartime thoughts and notes. The family left Paris in December 1925, and in February 1926, staying again with his son in Prague, Vernadsky penned the preface to his new book. In its very first sentence he made clear what makes it unique (Vernadsky 1998): "Among numerous works on geology, none has adequately treated the biosphere as a whole, and none has viewed it, as it will be viewed here, as a single orderly manifestation of the mechanism of the uppermost region of the planet—the Earth's crust" (Vernadsky 1998, p. 39). He stressed that he did not want to construct new hypotheses but "strive to remain on the solid ground of empirical generalization." Vernadsky and his wife returned home to a renamed city, Leningrad, in March 1926 (daughter Nina remained abroad). *Biosfera* was published just three months later, and the book's printing of 2,000 copies sold out fast (Vernadskii 1926).

Biosfera

Biosfera was made up of two lengthy scientific essays, the first entitled "Biosphere in the Cosmos," containing 67 sections (each one having typically between two and five paragraphs); the other, "The Domain of Life," containing 160 sections. Reading *Biosfera* at the beginning of the twenty-first century is a very interesting experience. Vernadsky's predilection for grand but concise generalizations

2. Vernadsky's letters to Petrunkevich are archived at the Columbia University; extensive excerpts are printed in Aksenov (1993), where the quoted passage (my translation) is on pp. 287–288.

continues to evoke frequent admiration at how well he set out many concepts, both fundamental and intricate ones, and how crisply he stated his conclusions. Inevitably, there are also generalizations and conclusions that have not withstood the test of time; indeed, some of them were arguable even at the time when they were written.

In addition, the meaning of some statements and sentences remains opaque or altogether obscure. And there are also signs of Vernadsky the mystic. But this is not a critical exposé of *Biosfera* aimed at singling out and discussing the book's factual errors and arguable opinions. My goal here is to review, with some brief comments, those fundamental generalizations and conclusions that have remained unassailable; only secondarily will I note some of those assertions or hypotheses that have been invalidated by subsequent research—or opinions that are distinctly unfashionable today. The latter include, above all, Vernadsky's progressivism, which is the very opposite of the idea of evolution driven by random mutations.

The term "biosphere" is mentioned for the first time in the book's second sentence—but without any definition:

The face of the Earth viewed from celestial space presents a unique appearance, different from all other heavenly bodies. The surface that separates the planet from the cosmic medium is the biosphere, *visible principally because of light from the sun, although it also receives an infinite number of other radiations from space, of which only a small fraction are visible to us. (Vernadsky 1998, p. 43)*

The third section would have made, I believe, a better opening:

Activated by radiation, the matter of the biosphere collects and redistributes solar energy, and converts it ultimately into free energy capable of doing work on Earth. . . . This biosphere plays an extraordinary planetary role. The biosphere is at least as much creation of the sun *as a result of terrestrial processes. (p. 44)*

And a few paragraphs later Vernadsky stressed that "we can gain insight into the biosphere only by considering the obvious bond that unites it to the entire cosmic mechanism" (p. 47).

The biosphere is thus seen as a region of transformation of cosmic energy, specifically of the solar radiation. Its major segments (ultraviolet, visible, and infrared) are transformed in different regions and by different means, and photosynthetic conversion of visible wavelengths producing innumerable compounds rich in free energy extends the biosphere "as a thick layer of new molecular systems" (p. 50). Diffusion of living matter by multiplication creates the ubiquity of life: as "organisms have developed and adapted to conditions which were initially fatal to them. . . . [L]ife tended to take possession of, and utilize, all possible space" (p. 60).

Living matter is thus "spread over the entire surface of the Earth in a manner analogous to gas" (p. 59) as a continuous envelope, and its most characteristic and essential trait is its uninterrupted movement, proceeding with "an inexorable and astonishing mathematical regularity" (p. 61). The result is life that occurs on land, penetrates all of the hydrosphere, and can be observed throughout the troposphere. It even penetrates the interior of living matter itself in the form of parasites. The section on growth contains a number of theoretical examples of the maximum possible reproductive capacities of arthropods and bacteria. And although properly stressing the importance of gaseous exchange involving living organisms, Vernadsky greatly underestimated the total mass of atmospheric oxygen, concluding that it is "of the same order as the existing quantity of living matter" (p. 70). In reality, the atmosphere contains about 1.1 Et (10^{18} t) of oxygen, whereas even the most liberal estimate of the Earth's biomass is no larger than 10 Tt of fresh matter (which is mostly water; see chapter 7). Consequently, oxygen is at least five orders of magnitude more abundant than biomass.

The remainder of the first essay is taken up largely by a discussion of photosynthesis. Here Vernadsky also erred by concluding that "the hydrosphere, a majority of the planetary surface, is always suffused with an unbroken layer of green energy transformers" (p. 80) and in maintaining that "the total mass of green life in the ocean exceeds that on land because of the larger size of the ocean itself" (p. 73). The latter claim was a common misconception during the late nineteenth century and the early decades of the twentieth. In reality, the standing terrestrial phytomass is at least 200 times as large as the biomass of marine phytoplankton and macrophyta (for details see chapter 7).

But Vernadsky presented accurate estimates of typical conversion efficiencies of solar radiation into new plant mass: their large-scale averages are mostly less than 1 percent. The cyclical link between the living matter and the atmosphere is also rightly emphasized: "The gases of the biosphere are generatively linked with living matter which, in turn, determines the essential composition of the atmosphere. . . . The gases of the entire atmosphere are in an equilibrium state of dynamic and perpetual exchange with living matter. Gases freed by living matter promptly return to it" (p. 87).

Finally, Vernadsky pointed out the distinction between rapid and slow cycling of dead organic matter: although most of it is recycled rather rapidly into new living tissues, a small share leaves the biosphere for extended periods and it "returns to living matter by another path, thousands or millions of years afterwards" (p. 87). Vernadsky called the generation of this enormous mass of minerals unique to life "the slow penetration into the Earth of radiant energy from the sun" (p. 88), and adding it to his vastly exaggerated estimated of photosynthesizers, he estimated the total weight of the biosphere at 10^{24} g. And he concluded the first part's last paragraph by maintaining that "although we do not understand the origin of the matter of the biosphere, it is clear that it has been functioning in the same way for billions of years" (p. 89).

The book's second, and longer, part deals mostly with the spatial extent of the biosphere. In its first sentence Vernadsky acknowledged Suess's authorship of the term "biosphere as a specific, life-saturated envelope of the Earth's crust" (p. 91). Then a rather detailed discussion of the Earth's various spheres begins by summarizing the contemporary ideas about the planet's core, the overlaying region ("mantle" in today's terminology), and the crust. Vernadsky also introduced five separate classifications of the Earth's envelopes: thermodynamic, gaseous, chemical, paragenetic, and radiation-based. Overlaps and duplications do not make this approach very clear.

As far as the organisms are concerned, Vernadsky followed the tripartite division—autotrophs, heterotrophs, and mixotrophs—proposed by W. Pfeffer (1881). Autotrophs use only inorganic matter to build their bodies, transforming raw materials into complex organic compounds. Heterotrophs must use ready-made organic compounds in their metabolism; mixotrophs combine organic and inorganic sources of nutrients. Autotrophs include photosynthesizing organisms—not just green plants but also many species of bacteria—and autotrophic bacteria lacking light-sensitive pigments and producing new living matter independent of solar radiation (for details see chapter 3).

Vernadsky rightly stressed the importance of single-celled organisms (Monera) in the biosphere: "Monera are ubiquitous, existing throughout the ocean to depths far beyond the penetration of solar radiation, and they are diverse enough to include nitrogen, sulfur, and iron bacteria. . . . One is led to conclude that bacterial abundance is a ubiquitous and constant feature of the Earth's surface." This led him to conclude "that we should therefore expect that the bacterial mass in the biosphere would far exceed the mass of green eukaryotes" (p. 109). Today's best appraisals confirm this beyond any doubt (see chapter 7).

Vernadsky also marveled at "curious secondary equilibria between sulfate-reducing bacteria and autotrophic organisms that oxidize sulfides" (p. 110), and at an analogical exchange between autotrophic bacteria that oxidize nitrogen and heterotrophic organisms that deoxidize the nitrates. But he was mistaken in concluding that during the Archean era "the quantity of living green matter, and the energy of solar radiation that gave it birth, could not have been perceptibly different in that strange and distant time from what they are today" (p. 108). We now know that the solar output was actually weaker at that time, and the total mass of "green matter" was smaller, than they are today (see chapters 3 and 4).

The remainder of the book is devoted to a fairly thorough exploration of the limits of life. Vernadsky only cited examples of short-term toleration of extreme pressures, temperatures, and radiation exposures rather than exploring the extremes of sustainable metabolism (see chapter 6). Some of his conclusions that are permanently valid (unless, of course, one subscribes to panspermia: see chapter 2) include the fact that "by all appearances the natural forms of life cannot pass beyond the upper stratosphere" (p. 119) and that the shortest wavelengths of electromagnetic radiation are deleterious to life. Vernadsky also estimated fairly correctly the maximum expected depth of the subterranean biosphere, but other conclusions, including the maximum depth of life in the ocean, have been altered considerably with the research advances of the last two generations.

After delimiting the biosphere Vernadsky turned to biogeochemical cycles in the hydrosphere, stressing the differences between planktonic and littoral organisms (or "living films," in his terminology), the reducing environment in the marine mud (the realm of anaerobic bacteria), the action of living organisms that "separates calcium from the sodium, magnesium, potassium and iron of the biosphere, even though it is similar in abundance to these elements" (p. 137), formation of biogenic phosphorite deposits, and releases of hydrogen sulfide by bacteria-reducing sulfates, polythionates, and complex organic compounds.

On land he assigned all living matter to just one living film of the soil and its fauna and flora and stressed the terrestrial life's dependence on water. The closing segment, on the relationship between the living films and concentrations of the hydrosphere and those of land, is very brief. Vernadsky reiterated that "life presents an indivisible and indissoluble whole, in which all parts are interconnected both among themselves and with the inert medium of the biosphere," and that "the biosphere *has existed throughout all geological periods,* from the most ancient indications of the Archaean" (p. 148).

And, immediately, he restated this conclusion once again in a slightly different form: "*In its essential traits, the biosphere has always been constituted in the same way.* One and the same chemical apparatus, created and kept alive by living matter, has been functioning continuously in the biosphere throughout geologic times, driven by the uninterrupted current of radiant solar energy." Still he was not satisfied and a few paragraphs later carried the conclusion too far. First he claimed that "all the vital films (plankton, bottom, and soil) and all vital concentrations (littoral, sargassic, and fresh water) have always existed." Then he reiterated that the changes in the total mass of living matter "could not have been large, because the energy input from the sun has been constant, or nearly so, throughout geological time." (pp. 148–149). We know now that both of these conclusions are wrong.

At the very end of the book Vernadsky returned to an idea that he explicitly stated at its very beginning and that sets him directly against the modern worshipers of blind randomness and selfish genes (see chapter 3). In the third section of the first essay he wrote:

> *Ancient religious institutions that considered terrestrial creatures, especially man, to be* children of the sun *were far nearer the truth than is thought by those who see earthly beings simply as ephemeral creations arising from blind and accidental interplay of matter and forces. Creatures on Earth are the fruit of extended, complex processes, and are an essential part of a harmonious cosmic mechanism, in which it is known that fixed laws apply and chance does not exist. (p. 44)*

Then he simply, but emphatically, concluded: "*But living matter is not an accidental creation*" (p. 149).

Later Elaborations

Biosfera and a number of Vernadsky's subsequent writings on this topic entered the canon of Russian science almost immediately. Vernadsky's name, mainly because of his political activities, was well known to Russia's prerevolutionary intelligentsia. When he returned from Paris as one of the doyens of Russian science, he was 63, but his white beard and hair made him look older (fig. 1.3). Although he withdrew from any political activity and refused to join the Communist Party, Vernadsky became known not only to a new generation of scientists brought up by a new state, but also to many ordinary educated Russians and Ukrainians. Knowledge and acceptance of Vernadsky's ideas abroad was another matter.

1.3 Vernadsky during the last decade of his life. Photo courtesy of Eric Galimov, V. I. Vernadsky Institute of Geochemistry and Analytical Chemistry of the Russian Academy of Sciences, Moscow.

The first French translation of *Biosfera* was published by Félix Alcan as early as 1929 (Vernadsky 1929). In France, Vernadsky's ideas were already fairly well known to fellow specialists in geology and geochemistry, most notably to Pierre Teilhard de Chardin (1881–1955), a Jesuit geologist and paleontologist who later became world known for

1.4 Pierre Teilhard de Chardin (1881–1955), photographed in Clermont-Ferrand in 1923. Photo used with the permission of the Fondation Teilhard de Chardin, Paris.

his unorthodox theological views and for his more than two decades of work in China (Teilhard de Chardin 1956, 1966; fig. 1.4). Vernadsky's ideas were also appreciated by the famous French philosopher Henri Bergson (1859–1941) and by the mathematician and philosopher Édouard Le Roy (1870–1954). Outside France and Germany, and particularly in the English-speaking world, Vernadsky's work remained largely unknown. An abridged English version of *Biosfera* was finally published in Arizona 60 years after the book's original appearance, in 1986 (Vernadsky 1986), and English readers had to wait another 12 years for a complete translation of *Biosfera* (Vernadsky 1998).

After the exiling of Trotsky in 1929, Stalin further tightened his grip on the country, and rigid ideological toeing of the Party line began affecting, though not yet totally controlling, both the Academy of Sciences and the universities. In 1930 Vernadsky was replaced as the Director of the Commission on the History of Science by Nikolai Bukharin (1888–1938), who before the decade's end became one of the most prominent victims of Stalin's terror. In 1934 Boris L. Lichkov (1888–1996), Vernadsky's friend and collaborator, was arrested and deported. KEPS, Vernadsky's creation going back to the war years, was reorganized and attached to the Gosplan, the state planning agency whose chief interest lay in maximum plunder rather than in rational use of the country's vast natural resources.

But Vernadsky's scientific stature made it possible for him not only to publish papers in France, but also to attend some scientific meetings. He traveled to France in 1932 (also visiting Germany on this trip), in 1933, and, for the last time, in 1935. Throughout the 1930s and the early 1940s Vernadsky was engaged in a remarkable range of projects. He wrote on the problems of time in modern science, on radiogeology, on Goethe as a naturalist, and on scientific thought, but he kept elaborating his ideas about the biosphere, about its boundaries, and its composition and structure. The two most notable extensions of his original work were a detailed listing of the biosphere's biogeochemical functions (Vernadskii 1931) and a systematic contrast between living and inert matter of the biosphere that he expressed in an extensive table (Vernadskii 1939). Vernadsky's list of biospheric functions reads like many recent enumerations of environmental (or nature's) services (Daily 1998).

Vernadsky distinguished nine principal biochemical functions of the biosphere, beginning with three concerning the atmosphere: the gas function, that is, biogenic formation of atmospheric gases; the oxygen function, the formation of free O_2; and the oxygenating function, producing many inorganic compounds. The fourth function was the binding of calcium, in both simple and complex compounds, by numerous marine organisms, and the fifth was the formation of sulfides (most commonly hydrogen sulfide [H_2S] by sulfate-reducing bacteria and iron sulfide [FeS_2]). All organisms, albeit to different extent, take part in the ubiquitous sixth function of concentrating many elements from their dilute environmental concentrations. This process of bioaccumulation is, of course, most obvious in the case of carbon, life's principal building block, but the biomass of many organisms either contains unusually high levels of rare elements or at least binds them in concentrations well above the biospheric mean. The long list of such elements includes, most notably, silicon, calcium, potassium, sodium, nitrogen, iron, manganese, and copper. In contrast, the seventh function, the decomposition of organic compounds, accompanied by releases of carbon dioxide (CO_2) and other gases, is carried out overwhelmingly by bacteria and fungi, and only bacteria perform the eighth function, the decomposition producing such gases as hydrogen (H_2), H_2S, and methane (CH_4). And all aerobic organisms perform the last of Vernadsky's nine functions, metabolism and oxygen-consuming respiration resulting in CO_2 generation and biosynthesis and migration of organic compounds.

1.5 Evelyn Hutchinson (1903–1991), one of the twentieth century's most influential ecologists and an avid promoter of Vernadsky's work. Photo courtesy of William Sacco, Yale University.

Vernadsky's sixteen-point contrast of living and nonliving matter in the biosphere, formulated in June 1938, is noteworthy for a number of reasons: its systematic sweep, its brevity, and its accurate formulation of several fundamental realities that have since become the universally acknowledged attributes of life (Vernadskii 1939). Here I will note just the most important conclusions, quoting them from the only two English-language publications of Vernadsky's work prepared while he was still alive (Vernadsky 1944; 1945). Evelyn Hutchinson (1903–1991) of the Department of Zoology at Yale was responsible for their introduction of Vernadsky's work to the English-speaking world (fig. 1.5). Hutchinson was the twentieth century's most eminent limnologist and the founder of the Yale school of ecology, whose influence is still felt in disciplines ranging from systems ecology to biogeochemistry. Vernadsky's son George, who had been a professor of Russian history at Yale since 1927, translated the work.

To begin with, Vernadsky was skeptical about life's extraterrestrial origins: "Living natural bodies exist only in the biosphere and only as discrete bodies. . . . Their entry into the biosphere from cosmic space is hypothetical and has never been proven" His second point—the unity of life—has received incontrovertible confirmation by modern sequencing of genomes: "Living natural bodies, in their cellular morphology, protoplasmic nature and reproductive capacity have a unity, which must be connected with their genetic connection with each other in the course of geological time." In contrast, "inert natural bodies are extremely diverse and have no common structural or genetic connections" (Vernadsky 1945, p. 2).

Vernadsky's fourth point puts humanity's role in transforming the Earth into a wider evolutionary context: "The rise of the central nervous system has increased the geological role of living matter, notably since the end of the Pleistocene." His sixth point is an excellent definition of what modern general system studies came to call open systems: "There is a continual stream of atoms passing to and from living organisms from and into the biosphere. Within the organisms a vast and changing number of molecules are produced by processes not otherwise known in the biosphere. Inert natural bodies change only from outside causes, with the exception of radioactive materials" (Vernadsky 1945, p. 3).

In his twelfth point, Vernadsky expressed this essential quality of life in thermodynamic terms: "The processes in living matter tend to increase the free energy of the biosphere." In contrast, "all inert processes, save radioactive disintegration, decrease the free energy of the biosphere" (Vernadsky, pp. 3–4). This entropic distinction became a key feature of von Bertalanffy's (1940) and Schrödinger's (1944) definitions of life. Von Bertalanffy's (1940) definition reads as follows: "The organism . . . is an open system in a (quasi) steady state, maintained constant in its mass relations in a continuous change of component material and energies, in which material continually enters from, and leaves into, the outside environment" (p. 521).

In the next point Vernadsky notes that, although extremely complicated, the chemical composition of living bodies is definitely "more constant than the isomorphous mixtures constituting natural minerals." The fourteenth point notes that "isotopic ratios may be markedly changed by the processes in living matter." This property makes it possible, among other things, to establish the extent and the intensity of biogenic contributions to sedimentary rocks throughout the biosphere's evolution. In the closing, sixteenth point, Vernadsky returns to a thermodynamic formulation: "The processes of living natural bodies are not reversible in time" (Vernadsky 1945, pp. 3–4).

In his eighth point Vernadsky repeated his earlier conclusion that the mass of living organisms "has remained fairly unchanged in the course of historical time" (Vernadsky 1944, p. 504) and added that apparently this mass increases in the course of geological time toward a limit and that "the process of the occupation of the terrestrial crust by living matter is not yet completed" (p. 504). Obviously, it is now most unlikely that such a process could be completed on a planet that will soon support close to ten billion people (see chapter 9). In the subsequent point he delimited the minimum size of living bodies to be "of the order of 10^{-6} cm. The maximum size has never exceeded $n \cdot 10^4$. The range, 10^{10}, is not great" (p. 504). We now know that the some cells are an order of magnitude smaller (see chapter 2).

Besides these elaborations, a large part of Vernadsky's writings during the last years of his life was devoted to the idea of the noosphere, literally the envelope of the mind. As with the biosphere, Vernadsky was not the originator of the idea, and he readily acknowledged its beginnings in the writings of Yale geologist James Dwight Dana (1813–

1895) and physiologist and geologist Joseph LeConte (1823–1901), who believed that the evolution of living matter is proceeding in a definite direction. Dana (1863) called the process "cephalization," irregular growth and perfection of the central nervous system culminating in the human brain. LeConte (1891) wrote about the psychozoic era. Édouard Le Roy, building on Vernadsky's ideas and on discussions with Teilhard de Chardin, came up with the term "noosphere," which he introduced in his lectures at the College de France in 1927 (Le Roy 1927). Vernadsky saw the concept as a natural extension of his own ideas predating Le Roy's choice of the term. In one of his French lectures in 1925, identifying humanity as a new geological, perhaps even cosmic, force resulting from human intelligence. To illustrate the impact of this new geological force that had emerged quite imperceptibly and over a relatively short (in evolutionary terms) period of time, he noted that "that mineralogical rarity, native iron, is now being produced by the billions of tons. Native aluminum, which never before existed on our planet, is now produced in any quantity. The same is true with regard to the countless number of artificial chemical combinations newly created on our planet" (Vernadsky 1945, p. 9).

Vernadsky was aware of the price paid for this progress as "the aerial envelope of the land as well as all its natural waters are changed both physically and chemically." During the twentieth century humanity, for the first time ever, permeated the whole biosphere, and instant communications enveloped the entire planet. Vernadsky perceived man to be "striving to emerge beyond the boundaries of his planet into cosmic space. And," he added, "he probably will do so" (Vernadsky 1945, p. 9).

Vernadsky the empiricist thus saw the idea of the noosphere as an accurate description of existing, and even more so of coming, realities. Humanity adds up to an insignificantly small share of the Earth's living matter, and its environmental, indeed planetary, impact "is derived not from its matter, but from its brain. If man . . . does not use his brain and his work for self-destruction, an immense future is open before him." Significantly, Vernadsky also put a great stress on the unity of mankind: "One cannot oppose with impunity the principle of the unity of all men as a law of nature. I use here the phrase 'law of nature' as this term is used more and more in the physical and chemical sciences, in the sense of an empirical generalization established with precision" (1945, p. 8). The conclusion was then clear: "The noosphere is the last of many stages in the evolution of the biosphere in geological history" (p. 10).

Diffusion of the Idea

After Vernadsky's death on January 6, 1945, a number of his pupils and followers in Russia continued work based on his ideas, and they prepared his collected writings for publication (Vernadskii 1954–1960). The Institute of Geochemistry and Analytical Chemistry of the Academy of Sciences, a research vessel, and a Ukrainian research station in Antarctica, as well as streets and squares in Russia and Ukraine, carry Vernadsky's name. But neither his name nor his idea of the biosphere became widely known abroad. In 1948 Evelyn Hutchinson, who, as noted above, introduced Vernadsky's work to the American scientists in 1945, published a short article, "On Living in the Biosphere," in *Scientific Monthly*. After offering a variant on Vernadsky's definition of the biosphere, Hutchinson noted that besides the customary division of resources (raw materials and energy) used within the biosphere

> *there is also a third very important though inseparable aspect, namely, the pattern of distribution. Most fossil sunlight, or chemical energy of carbonaceous matter, is diffused through*

sedimentary rocks in such a way as to be useless to us. . . . I am almost tempted to regard pattern as being as fundamental a gift of nature as sunlight or the chemical elements.

Hutchinson also stressed his concern about the wastefulness of modern society, in which laboriously collected materials are, after a brief use, discarded and most noncombustible waste is eventually carried, in solution or as sediment, to the ocean:

Modern man, then, is a very effective agent of zoogenous erosion . . . affecting most powerfully arable soils, forests, accessible mineral deposits, and other parts of the biosphere which provide the things that Homo sapiens *as a mammal and as an educatable social organisms needs or thinks he needs. The process is continuously increasing in intensity, as populations expand. (Hutchinson 1948, p. 393)*

But it took another two decades before concerns about the state of the Earth's environment made biosphere an indispensable term of modern scientific understanding.

Widespread adoption of Vernadsky's idea of the biosphere and common use of the term as one of the key operative words of modern scientific thought came from a combination of diverse developments that focused attention on the state and fragility of the Earth's environment. Concerns about environmental pollution—on scales ranging from local to global—led to new legislative initiatives and international treaties, and scientific advances in a number of disciplines demonstrated clearly the unity of the global environment and the necessity of its protection.

Environmental Consciousness

The first major shift of attitudes toward environmental pollution resulted from the dependence of modern high-energy civilization on the combustion of coal. Episodes of severe air pollution created by high concentrations of particulate matter and SO_2 caused premature deaths among infants and the elderly and led to the passing of relatively strict antipollution laws: the British Clean Air Act in 1956 and the corresponding U.S. legislation of the same name in 1963 (Brimblecombe 1987; Smil 1993). Radioactive fallout from atmospheric testing of increasingly powerful nuclear weapons and the resulting global health hazard, especially to the unborn and babies, was terminated when the United States, the Soviet Union, and the United Kingdom (but not France and China) signed the Test Ban Treaty in August 1963. On the Soviet side the movement toward the treaty started in 1957 when Andrei Dmitrievich Sakharov (1921–1989), the creator of the country's hydrogen bomb, wrote an article on the effects of low-level radiation and then persisted in lobbying the Communist Party leadership to conclude the treaty.

Ubiquitous use of pesticides and particularly the concerns such use raised about long-term and long-range effects of DDT on biota was a major reason for Rachel Carson's writing *Silent Spring* (1962). These concerns were soon followed by additional concerns about other persistent organic pollutants (Eckley 2001) and by Patterson's (1965) findings of very high lead contamination in humans stemming from the presence of this heavy metal in gasoline, solder, paint, and pesticides. Concern about the effects of pesticides is still very much with us (Harrad 2001), whereas exposure to lead has been greatly reduced through its elimination from gasoline and paints.

By far the most notable international scientific event of the 1950s was the International Geophysical Year, conceived by Lloyd Berkner in 1950: it took place between July 1957 and December 1958 to coincide with the high point of the eleven-year cycle of solar activity. Measurements taken during that period ranged from seismological investigations and soundings of the ocean floor to observations of stratospheric dynamics; they spanned literally the entire planet (including the establishment of permanent U.S. South Pole station) and involved the participation of sixty-seven countries (Chapman 1959). Of course, 1957 was also the year of the Earth's first artificial satellite, and the small beeping Sputnik was soon followed by the first weather-observing platform. TIROS I (for "television infrared observing satellite") sent its first fuzzy image on April 1, 1960, and TIROS satellites have monitored the Earth's clouds ever since (fig. 1.6). And 1957 was also the year when a paper by Roger Revelle and Hans Suess launched the sustained investigation of the greenhouse gas phenomena (Revelle and Suess 1957). Revelle and Suess concluded that

human beings are now carrying out a large scale geophysical experiment of a kind that could not have happened in the past nor be reproduced in the future. Within a few centuries we are returning to the atmosphere and oceans the concentrated organic carbon stored in sedimentary rocks over hundreds of millions of years. (p. 19)

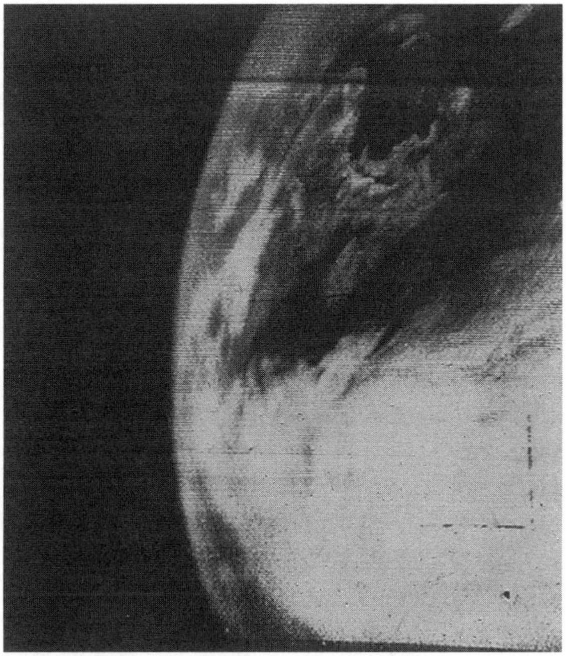

1.6 The first low-resolution television image transmitted from space by the TIROS I satellite on April 1, 1960. NASA stores the image at <http://antwrp.gsfc.nasa.gov/apod/image/0004/first_tiros1_big.gif>

The first systematic measurements of rising background CO_2 levels were organized by Charles Keeling in 1958 at Mauna Loa in Hawai'i, and at the South Pole (Keeling 1998; fig. 1.7).

Planning for the International Biological Programme began in 1959, and its decade of work took place between 1964 and 1974 (Worthington 1975). The program's principal goals were to investigate the productivity of terrestrial, freshwater, and marine ecosystems, particular production processes, human adaptability, and use and management of (mainly genetic) resources. Several synthesis volumes resulting from the program introduced unprecedented global perspectives, particularly to the study of major terrestrial ecosystems, including the two final syntheses: Breymeyer and Van Dyne (1980) for grassland ecosystems and Reichle (1981) for forests.

In 1968 the United Nations Educational, Scientific, and Cultural Organisation (UNESCO) convened an intergovernmental conference of experts on "Use and Conservation of the Biosphere" in Paris (UNESCO 1968). In the introduction to the conference's published proceedings, Kovda et al. (1968) summarized Vernadsky's key

1.7 Mauna Loa Observatory is situated 3,400 m above sea level on the northeastern slope of Hawaii's largest volcano. Clouds cover part of the southern slope of Mauna Kea in the background. Image available at <http://www/cmdl.noaa.gov/obop/mlo/mlo2b.gif>

ideas and presented the latest quantitative estimates of major biogeochemical flows, the Earth's total biomass, and the productivity of its main ecosystems. This conference led to the establishment of UNESCO's Man and the Biosphere (MAB) Programme three years later; two decades after that, this program continues interdisciplinary research and training, now with particular emphasis on more than 300 biosphere reserves in eighty-five countries (UNESCO 2000).

The first detailed maps of the ocean floor showing the global system of undersea ridges were also published during the 1960s, and they were followed by a rapid acceptance of the idea of plate tectonics. This new geological paradigm made it possible to understand the global nature of the grand processes that formed the continents and ocean floor (Wilson 1972; Kahle 1974). The late 1960s and the early 1970s brought a number of inquiries into the interactions of population, resources, and the environment, most notably a sweeping survey by Ehrlich et al. (1970). Two new large-scale environmental worries emerged during that same period, as Swedish scientists identified the process of ecosystemic acidification (Royal Ministry for Foreign Affairs 1971) and Barry Commoner wrote about the consequences of human interference in the nitrogen cycle through rising applications of synthetic fertilizers (Commoner 1971; Smil 1985).

The same period also saw the first extended space flights. Gemini's ten missions between March 1965 and November 1966 were a transitional step between the pioneering Mercury flights and eleven Apollo launches (October 1968 to December 1972) that culminated in six Moon landings. Gemini and Apollo images showed the thinness of the Earth's blue atmosphere above the planet's curved surface. And nothing could illustrate better the unity and the fragility of life than the first photographs of the entire Earth taken by Apollo crews, those unforgettable mixtures of the planet's whites, blues, and browns against the blackness of the cosmic void in the images taken during the journey to the Moon (fig. 1.8) and from its surface (fig. 1.9).

1.8 Earth photographed on July 16, 1969, from a distance of 158,000 km by Apollo 11 on its journey to Moon. *Source:* NASA ID AS 11-36-5355.

All of these varied strands contributed to a new wave of rising environmental consciousness, and the publication of a special issue of *Scientific American* in September 1970 was undoubtedly a key event articulating these concerns. Entitled simply *The Biosphere,* with a detail from *The Dream,* Henri Rousseau's famous oil painting of tangled tropical rain forest plants and lurking beasts, on its cover, the issue finally made the term widely known even among nonbiologists. Evelyn Hutchinson introduced *The Biosphere* in a masterly overview of the subject, explaining how the

1.9 Earth-rise above the Moon horizon, photographed by Apollo 11 astronauts in July 1969.
Source: NASA ID AS 11,44,6552.

Earth's thin film of life is sustained by energy flows and by grand biogeochemical cycles (Hutchinson 1970).

Hutchinson defined three ingredients that make the biosphere special as a terrestrial envelope: existence of liquid water in large quantities, an ample supply of external energy, and the interfaces between the liquid, solid, and gaseous states of matter. He stressed the fragility and inefficiency of the photosynthetic process that is responsible for creating all but a tiny fraction of living matter and its dependence on movement and cycling of numerous elements. Hutchinson recognized that "if we want to continue living in the biosphere we must also introduce unprecedented processes," but, in conclusion, he was extremely pessimistic about the prospect of extended human life on Earth:

> *Many people, however, are concluding on the basis of mounting and reasonably objective evidence that the length of life of the biosphere as an inhabitable region for organisms is to be measured in decades rather than in hundreds of millions of years. This is entirely the fault of our own species. It would seem not unlikely that we are approaching a crisis that is comparable to the one that occurred when free oxygen began to accumulate in the atmosphere. (p. 53)*

But the last three decades of the twentieth century brought some signs of hope. Although most of the human onslaughts that Hutchinson lamented have continued and some have even intensified, there has been a rise of unprecedented awareness of the biosphere's unique qualities—of its longevity, complexity, diversity, resilience, and fragility—and a growing realization of the necessity to limit its degradation and to protect its integrity. Satellite observations have played a particularly important role in fostering the idea of the unique and unified biosphere and its vulnerability.

Satellite Monitoring

The ability to observe natural processes on a planetary scale from the Earth's satellites advanced rapidly during the last forty years of the twentieth century (Slater 1986; Szekielda 1988; Rao et al. 1990; Danson and Plummer 1995). After the pioneering TIROS weather satellites came the Nimbus series. Launched into a sun-synchronous orbit—a polar orbit (in this case with 99° inclination) in which the satellite crosses the Equator at the same local solar time—these satellites could scan the entire Earth. In 1966, applications technology satellites (ATSs) began to be placed in geostationary orbit near the Equator, 36,000 km above the Earth, where they match its rotation speed and remain over a fixed point on the surface. Six ATS satellites were launched between 1966 and 1974. They could scan the entire visible hemisphere and send the image in less than half an hour, and their successive scans could be used to generate moving sequences of cloud systems and pressure cells. For the first time ever we could see the Earth's entire atmosphere in motion.

As revealing as these satellites were for studying the Earth's dynamic atmosphere, their resolution was of no help in investigating the biosphere's most obvious manifestation, the annual ebb and flow of photosynthesis by forests, grasslands, croplands, and wetlands, and the anthropogenic change of the areas covered by natural ecosystems. That changed with the launching of the first earth resources technology satellite (ERTS), which was sent into polar orbit in July 23, 1971, making fourteen revolutions a day at an altitude of 900 km and passing over the same spot on the Earth every 18 days (EROS Data Center 1978–1986).

Although its three-color TV cameras failed shortly after launch, the satellite's multispectral scanning systems (MSSs) recorded reflected radiation in four different spectral bands between 0.5 and 1.1 μm with a resolution of 80 m (fig. 1.10). Chlorophyll reflects less than 20% of the longest wavelengths of visible light but about 60% of near-infrared radiation. These differences in reflectance can be used to distinguish vegetated and barren areas, to perform relatively detailed ecosystemic mapping, and, when backed up by good ground observations, to estimate total phytomass. Landsat, as the satellite was soon renamed, produced land use data with a degree of resolution sufficient for a fairly reliable mapping of ecosystems and their changes. A second Landsat followed in January 1975, the third in March 1978, and Landsat D, launched in 1982, carried not just an improved MSS but also a thematic mapper (TM) with resolution of 30 m. The latest

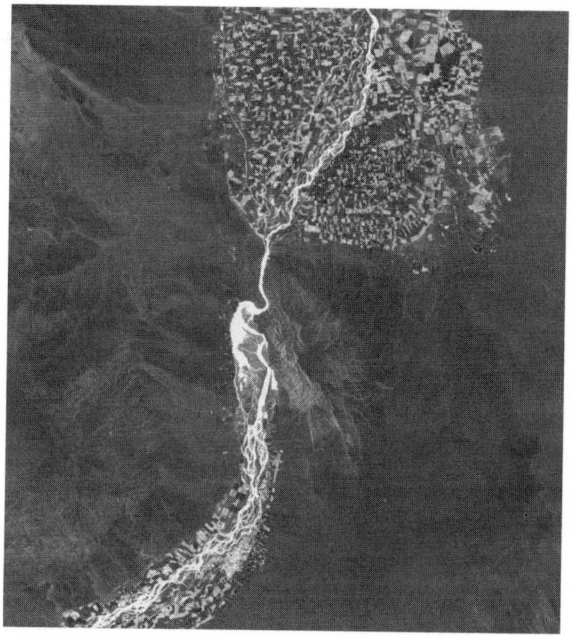

1.10 The first LANDSAT images had a relatively coarse resolution of 80 m, but they were very useful for tasks ranging from uncovering previously unrecognized geological features to mapping land use. In this image the intensively cultivated fields in eastern Ningxia in northern China, irrigated by water from the Huanghe, contrast with their desert surroundings.

Landsat, Landsat 7, was launched in April 1999. Its multispectral scanner has, once again, a resolution of 30 m, but its new, panchromatic images have a resolution of 15 m, and it is equipped with a thermal infrared sensor with a resolution of 60 m.

Standardized false-color images derived from MSSs, with hues ranging from red-magenta for dense vegetation to the blue-grey of populated areas, have provided a view of the Earth that is more than just visually stunning (the first two worldwide selections of these images are Short et al. 1976 and Sheffield 1981; these were followed by a number of national Landsat atlases). Landsat imagery has made it possible to monitor shifts of desert margins in the Sahel and snow and ice cover in the Arctic, to determine the abundance of icebergs in Antarctica, to identify new seismic faults in California and new crustal fractures in South Africa, to map tropical deforestation throughout Amazonia, to chronicle the expansion of urban areas, and to estimate crop yields in China and Russia.

The French System Probatoire pour l'Observation de la Terre (SPOT), flying since 1985, revealed details at a resolution of 10–20 m, and U.S. intelligence satellites have for many years relayed images with resolution below 15 cm (Richelson 1998). These images remain classified, but multispectral four-meter resolution color images and panchromatic one-meter images from the Ikonos 2 satellite, launched on September 24, 1999, are now commercially available. Obviously, this imagery is too detailed for any global observation, and monitoring with resolution far lower than that of even the Landsat's MSS has brought the most convenient, and now the most widely used, method of charting the ebb and flow of global photosynthesis.

An advanced very-high-resolution radiometer (AVHRR) was installed for the first time in 1978 on the National Oceanic and Atmospheric Administration's (NOAA) TIROS-N, and it has been put on board the series' polar-orbiting satellites ever since, with the latest NOAA-K, L, and M carrying an enhanced version (Rao et al. 1990; D'Souza et al. 1996). AVHRR senses in wavelengths ranging from visible (0.58–0.68 μm) to far infrared (11.5–12.5 μm) to map day and night clouds and surfaces, delineate surface water, ice and snow, and measure sea surface temperatures. Its maximum resolution of 1.1 km is sufficient to appraise large-scale patterns of vegetation coverage and to detect seasonal and statistically significant year-to-year variability in global plant conditions (Gutman and Ignatov 1995; fig. 1.11).

The Coastal Zone Color Scanner (CZCS), launched on Nimbus 7 in October 1978, was the first sensor to monitor the intensity of phytoplankton production (by measuring pigment concentration) in coastal seas. The second CZCS followed in 1986, and the Sea-Viewing Wide-Field-of-View Sensor (SeaWIFS), launched in 1997, does the same job even better. The U.S. Navy's Remote Ocean Sensing System (NROSS) provides information on ocean waves and eddies, and a joint U.S.-French TOPEX/POSEIDON project (launched in 1992) is equipped with two high-precision altimeters to measure changes resulting from winds, currents, and gravity. Altimeter measurements, corrected for the total water vapor content in the atmosphere, make it possible to chart the height of the seas with an accuracy of less than 10 cm. The forthcoming Jason 1 project will achieve accuracy better than 2.5 cm (for details see the Web site in appendix H). Combined information from these two missions revealed a detailed global pattern of ocean circulations. Other monitored variables have ranged from solar irradiance and surface shortwave radiation to tropospheric temperatures and stratospheric ozone.

Terra, launched on December 18, 1999 and designed for six years of service, is now the flagship of the National Aeronautic and Space Administration's (NASA's) Earth Observation System. This 5.19-t satellite is equipped with five advanced sensors that enable it to assess the state of the biosphere in unprecedented detail (King and Herring 2000). Its Advanced Spaceborne Thermal Emission and Reflection Radiometer (ASTER, with high resolution of 15–90 m) monitors the surfaces in the visible and infrared spectrum. Clouds and the Earth's Radiant Energy

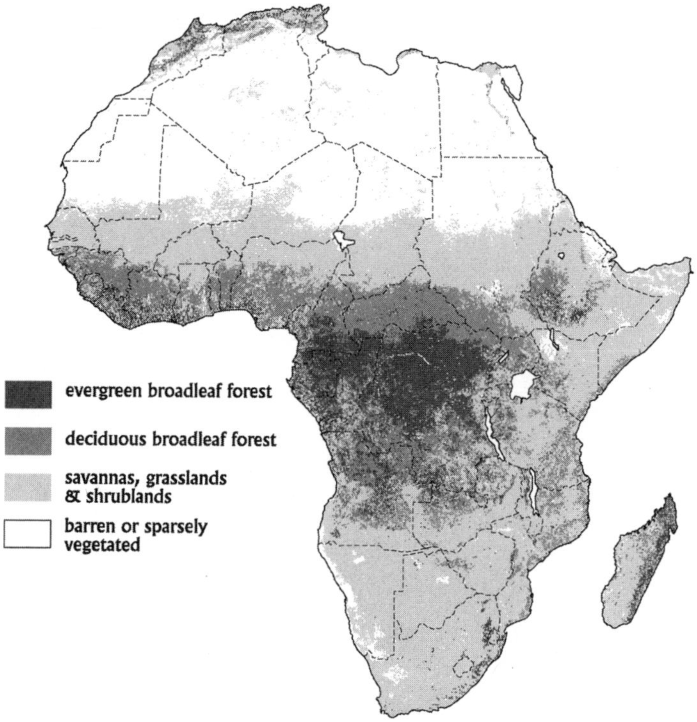

1.11 The first AVHRR was mounted on the TIROS-N satellite in 1978. AVHRR images, acquired with resolution of 1–4 km, made it easy to map large-scale patterns of the Earth's vegetation. This image shows Africa's main biomes: evergreen (tropical rain) and seasonally green forests, savannas, and shrublands. Derived from a color image available at <http://edcdaac.usgs.gov/glcc/gif/gigbp_10.gif>

System (CERES; resolution 20 km) records radiation fluxes throughout the atmosphere. Measurement of Pollution in The Troposphere (MOPITT; resolution 22 km) traces air pollutants to their sources. Multiangle Imaging SpectroRadiometer (MISR, resolution 275 m–1.1 km) generates stereoscopic images of clouds and smoke plumes. And MODerate-Resolution Imaging Spectroradiometer (MODIS, resolution 250 m–1 km) monitors the Earth in thirty-six discrete spectral bands, making it possible to track clouds as well as ocean phytoplankton (see appendix H for Terra's URLs).

Perhaps the most unusual addition to the growing number of Earth-observation machines is Triana, a satellite stationed at the LaGrange neutral gravity point between the Earth and the Sun, far enough (1.6 Gm) from the Earth to allow the transmission of full-color images of the planet's entire sunlit side. Besides this obviously widely appealing planetary view, the mission will also measure directly the Earth's albedo, cloud coverage, and aerosol optical thickness and improve our understanding of solar winds (NASA 2001).

Global Environmental Change

The Massachusetts Institute of Technology sponsored the first interdisciplinary attempt at a systematic evaluation of global environmental problems in the summer of 1970 (Study of Critical Environmental Problems 1970). The study's list of key problems included (in no rank order) emissions of carbon dioxide from fossil fuel combustion, particulate matter in the atmosphere, cirrus clouds from jet aircraft, the effects of supersonic planes on the stratospheric chemistry, thermal pollution of waters, DDT and related pesticides, mercury and other toxic heavy metals, oil on the ocean, and nutrient enrichment of coastal waters. Only the effects of supersonic planes would not make today's list, as the expensive and wasteful Concorde remains the only supersonic transport in even a limited operation: the U.S. Boeing project to develop a supersonic aircraft was canceled by Congress shortly after the SCEP meeting. The other concerns, however, are still very much with us.

The first-ever global meeting devoted solely to the environment was the U.N.-organized Conference on the Human Environment in Stockholm in 1972, dominated by national concerns and political posturing, rather than any cooperative attempts at environmental protection. Although the host country's report focused on acid rain, Brazil asserted its right to engage in extensive tropical deforestation, and Maoist China claimed to be free of any environmental problems (U.N. Conference on the Human Environment 1972). Establishment of the United Nations Environmental Programme (UNEP) was one of the few tangible outcomes of the meeting (Sandbrook 1999).

OPEC's quintupling of oil prices in 1973–1974 and the additional quadrupling in 1979–1980 diverted the world's attention from environmental concerns to matters of energy supply, inflation, and economic stagnation, but as those worries eased, environmental awareness intensified during the 1980s, particularly after 1986 (Smil 1987, 1993). Throughout the decade a wide range of traditional concerns—air and water pollution, hazardous and radioactive wastes, pesticide residues, soil erosion—were receiving more research and legislative attention, some of them because of major environmental mishaps. These included the dioxin release in Seveso near Milan in 1976, massive methyl isocyanate poisoning in Bhopal in 1984, and the Chernobyl nuclear disaster in 1986, and they brought more attention to the lasting problems of toxic wastes and safety of nuclear energy generation (Smil 1993). Higher concentrations and more frequent occurrence of photochemical smog in large cities of every continent and rising demand for municipal and industrial water supplies made it clear that most of the effort to reduce air and water pollution still lay ahead. But a new class of truly global environmental concerns rapidly became prominent.

Concerns about the Biosphere's Integrity

Although many other factors played their part, increasing recognition of three kinds of global environment change was largely responsible for the emergence of concerns on the biospheric scale. The first was an increasing realization that conversions of natural ecosystems in general, and extensive tropical deforestation in particular, are causing an unprecedented loss of biodiversity (Wilson 1986). The second key impulse was the discovery of a substantial seasonal weakening of the ozone layer above the Antarctica (Farman et al. 1985; Butler et al. 1999; fig. 1.12). And the third main cause was a revival of concerns about global climate change driven by increasing emissions of greenhouse gases that might lead to a rapid rate of planetary warming, with both its predictable and its unknown consequences (Carbon Dioxide Assessment Committee 1983).

Concatenation of these worries elevated global environmental change to a leading item of international policymaking. The Bruntland Report of 1987 (World Commission on Environment and Development 1987) and the Earth Summit in Rio de Janeiro in 1992 were the two most talked-about political responses, though each had, not surprisingly, very limited practical consequences. By far the most effective practical response was a rapid conclusion of an international treaty to phase out production and use of chlorofluorocarbons responsible for the loss of stratospheric ozone. The Vienna Convention for the Protection of the Ozone Layer was concluded in March 1985, and the Montreal Protocol designed to phase out ozone-depleting substances was signed in 1987, just two years after the discovery of the seasonal ozone loss above Antarctica (Ko et al. 1994).

Institutionalized scientific response to global environmental change included the establishment of numerous large-scale research programs that have become known by their acronyms. The International Geosphere-Biosphere Programme (IGBP) was set up in 1986 with the mission of describing and understanding the interactive processes and changes that regulate the Earth's environment and make it suitable for life, as well as the human impacts on this complex planetary system. The acronymic maze of IGBP's core projects includes BAHC (Biospheric Aspects of the Hydrological Cycle), GCTE (Global Change and Terrestrial Ecosystems), IGAC (International Global Atmospheric Chemistry Project), JGOFS (Joint Global Ocean Flux Study), LOICZ (Land-Ocean Interactions in the Coastal Zone), and PAGES (Past Global Changes). For details, see OECD (1994) and IGBP's Web site (see appendix H). Besides research reports, the IGBP also publishes a *Global Change Newsletter* available at its Web site.

From the start the IBGP collaborated closely with the World Climate Research Programme (WCRP), set up in 1980 by the World Meteorological Organization (WMO), and with the International Human Dimensions of Global Environmental Change Programme (IHDP). But the cooperative effort that has received perhaps the greatest attention during the 1990s has been the Intergovernmental Programme on Climatic Change (IPCC), set up by the WMO and UNEP to prepare periodic reviews of the state of the art of climatic change research and to evaluate a wide range of possible environmental, economic, and social consequences of changing climate. Its reports (Houghton et al. 1990; 1996; 2001), resulting from extensive international exchanges between hundreds of natural and social

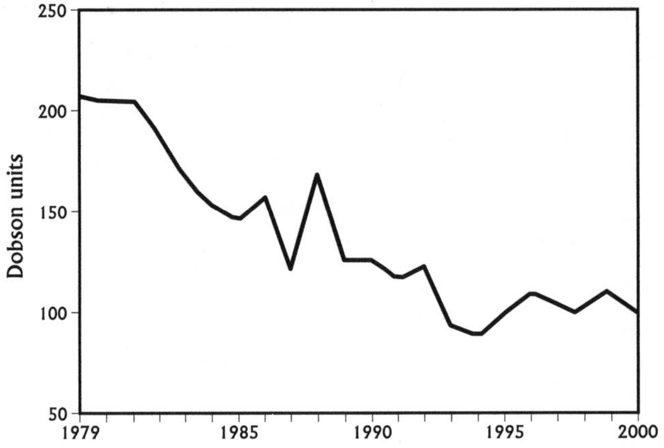

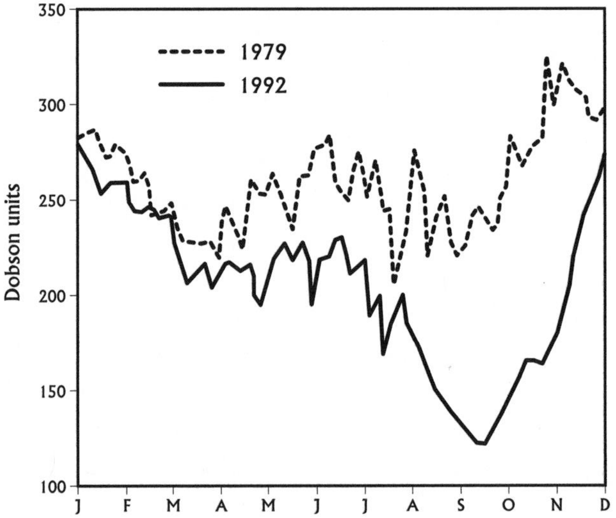

1.12 Evolution of the Antarctic ozone hole, shown here in minimum values of Dobson units between 1979 and 2000 and as a comparison of monthly ozone minima in 1979 and 1992. Based on data available at <http://www.epa.gov/ozone/science/hole/size.html> and <http://www.cpc.ncep.noaa.gov/products/stratosphere/>

scientists, have been widely quoted, and also intensely debated, and they form the principal input into lengthy, continuing negotiations about limits on greenhouse gas emissions.

All major (and a large number of smaller) countries now participate in various components of global environmental research. Besides climatic change, the principal topics of such research now embrace global energy and biogeochemical cycles and budgets, photosynthetic productivity and biomass storage of major ecosystems, changes of stratospheric ozone and effects of ultraviolet-B (UVB) radiation on biota, various atmospheric aerosols (of volcanic, vegetation, and combustion origins), mapping of global land use and its changes, inventories of biodiversity, status and degradation (particularly erosion) of croplands and grassland, and land-ocean interactions in coastal zones.

Another manifestation of this globalization of environmental studies has been the emergence of new journals explicitly devoted to the Earth's entire environment and to its links with other events and processes. *Global Biogeochemical Cycles* began publication in 1987, *Global Environmental Change* in 1991, *Global Change Biology* in 1995, and *Global Change & Human Health* in 2000. The first three-volume *Encyclopedia of Global Environmental Change* appeared in 2002. But scientific understanding will have to be translated into universal and effective action if we are to preserve the biosphere's integrity: the idea of planetary management may seem preposterous to many, but at this time in history there is no rational alternative.

Managing the Biosphere

In closing his introduction to *The Biosphere* in 1970, Evelyn Hutchinson cited from Vernadsky's letter written shortly before his death to Alexander Petrunkevich: "I look forward with great optimism. I think that we are experiencing not only an historical change, but a planetary one as well. We live in a transition to the noosphere." Unfortunately, Hutchinson commented, "the quarter-century since those words were written has shown how mindless most of the changes wrought by man on the biosphere have been. Nonetheless, Vernadsky's transition in its deepest sense is the only alternative to man's cutting his lifetime short by millions of years." (p. 53)

I agree with Hutchinson's assessment. Although the task is such that even adjectives like "awesome" and "immense" are completely inadequate for describing its complexity and duration, we have no choice unless we are willing to accept an early demise of our species. We have already altered the biosphere to such an extent that the only rational way out is to understand as best possible its intricate functions—and then to make sure that the future changes we inflict on the global environment will remain within tolerable limits. Obviously, the collective role of human consciousness will be essential if this unprecedented process of planetary management is to succeed (I will return to this topic in chapter 9).

Although it would be easy to be highly pessimistic about the prospect of success of such a monumental undertaking, it is also necessary to recognize the far from insignificant progress we have already made in understanding some of the essentials. We now have both global and virtually instant coverage of just about every essential biospheric variable and a large part of this fascinating information is now easily accessible on the World Wide Web (appendix H). In the case of some variables it is now possible to survey a generation (20–25 years) of changes. New perspectives afforded by this global and repetitive coverage have brought both good and bad news. They have made it possible to refute claims about the southward march of the Sahara (Tucker and Nicholson 1999), to quantify for the first time with reasonable accuracy annual losses of Brazil's unique tropical rain forests (Skole and Tucker 1993), and to show that many populous low-income countries, and China in particular, has substantially more farmland than the total given by official statistics (Smil 1999a).

The combination of global monitoring and expanding computer capabilities has also resulted in a variety of increasingly more realistic simulations of complex biospheric realities. Three-dimensional models of global atmospheric circulation are used not only in routine weather forecasting but also in modeling long-term shifts in the Southern Oscillation (El Niño phenomenon) and forecasting changes arising from different concentrations of greenhouse gases. The ultimate, and still rather distant, goal is a coupling of terrestrial, hydrospheric, and atmospheric models in realistic simulations of the biospheric behavior. As imperfect as our understanding remains, science is well ahead of effective action. The rapid conclusion and subsequent successful implementation of the Montreal treaty banning the production and use of chlorofluorocarbons (CFCs) is an example—unfortunately, an atypical one—of what can be done.

The unprecedented international agreement represented by the Montreal treaty was so swift and so effective because the action was guided by solid scientific understanding and because commercial alternatives to CFCs were readily available (Rowland 1989; Manzer 1990). Neither is true in the case of reducing greenhouse gas emissions. Our understanding of the inherent complexities of the biospheric carbon cycle (for details see chapter 5) and of factors involved in global climatic change remains inadequate: too often we can do no better than narrow down the numbers within a factor of two. And there are no ready, large-scale substitutes either for fossil fuels, the largest anthropogenic sources of both CO_2 and CH_4, or for nitrogen fertilizers (the leading source of nitrous oxide [N_2O]). Consequently, we have been unable to offer any firm guidance for effective management on the global scale. I will return to these challenges in more detail in the closing chapter of this book. Now I will begin a systematic coverage of the biosphere, probing first the probability of life in the universe, and then looking at life's evolution and metabolism before exploring the biosphere's extent, mass, productivity, and grand-scale organization.

2

Life in the Universe

Attributes, Constraints, and Probabilities

Living matter gives the biosphere an extraordinary character, unique in the universe.
Vladimir Ivanovich Vernadsky, *Biosfera*

A full realization of the near impossibility of an origin of life brings home the point how improbable this event was. This is why so many biologists believe that the origin of life was a unique event . . . no matter how many millions of planets in the universe.
Ernst Mayr, *The Growth of Biological Thought*

As this is not a science fiction book, the only kind of life considered here is the only one we know, the Earth's life made up of carbon-based macromolecules forming cells whose fresh mass is largely water. I will explain the astonishingly complex, yet in so many ways so impressively simple, underlying genetic and metabolic unity of carbon-in-water life; its basic cellular organization, including the fundamental split into prokaryotic and eukaryotic organisms; and its two no less fundamentally different ways of securing energy and carbon for biosynthesis through auto- and heterotrophy. For many more details from different perspectives—cellular, evolutionary, paleobiographical—a curious reader should consult such recent books as Rensberger (1996) or Fortey (1998).

I will not begin the story of life by tracing the evolution of the universe according to the standard cosmological model, the prevailing scientific paradigm whose explanations abound in recent publications (Peebles 1993; Coles and Lucchin 1995). I will go back "only" some five billion years (Ga). At that time, when the universe was about two-thirds its present size, a new planetary system began forming a galaxy we now call the Milky Way. Given the known requirements and tolerances of carbon-in-water life, I will outline the basic constraints on stars and planets needed to create the environment conducive to life's eventual emergence and evolution. This examination will be pursued in greater detail by looking at specific conditions on the Earth: at its surfaces and their transformation, and its changing atmosphere, particularly during the Hadean (4.3–3.8 Ga bp) and Archean (3.8–2.5 Ga bp) eras (appendix A lists detailed subdivisions of evolutionary eras).

Only then will I address the elusive topic of the origin of life and the less elusive, but still highly uncertain, matter of the biosphere's earliest evolution during the Archean era. Has this all been — as the logic of large galactic and stellar numbers and the deterministic nature of chemical reactions would seem to dictate — an inevitable development, just one of many demonstrations of a universal process that is replicated throughout myriads of galaxies? Or is the Earth's biosphere an incredibly improbable singularity? I will contrast arguments for both sides, but although I favor the latter possibility, I will conclude the chapter as an agnostic stressing our profound ignorance in these matters.

Carbon-Based Life

Biochemical reactions are carried out in water solutions, or at water interfaces, and water is both an essential metabolite and an excellent solvent. Usually more than half of all living biomass is the triatomic molecule of hydrogen and oxygen, with shares ranging from just over 35% in some insects to more than 95% in many fresh leaves and shoots (for details see chapter 5). But once dehydrated, the biomass of bacteria or butterflies has elemental composition very much like the biomass of tulips or tunas, with carbon, because of its abundance and chemistry, being always the single most important presence, accounting typically for about 45% of the total.

Hydrogen, helium, carbon, oxygen, and nitrogen are, in that order, the five most common elements in the universe. Helium is chemically inert, as is N_2, the most common, hard-to-cleave, form of nitrogen. Hydrogen has only one, and oxygen two electrons to share, whereas carbon has four. Carbon is thus able to form covalent bonds (the strongest interatomic links with shared electrons) with most nonmetals, and it joins readily with many radicals (atoms or molecules with unpaired electrons). Carbon atoms can also join together to make hexagonal rings or to form long, linear spines of macromolecules (fig. 2.1). Arthur Needham (1965) described carbon's chemistry as the joining of the opposites, as a rare combination of momentum with inertia: carbon compounds are formed relatively easily but are both fairly stable and far from unreactive. And Primo Levi (1984) extolled carbon's uniqueness this way: "Carbon, in fact, is a singular element: it is the only element that can bind itself in long stable chains without a great expense of energy, and for life on earth (the only one we know so far) precisely long chains are required" (pp. 226–227).

Carbon, hydrogen, and oxygen dominate biomass composition, but neither proteins nor nucleic acids (details on both follow in the next section) could be built with just those three elements. Ultimate analysis of proteins shows 50–55% carbon, 21–24% oxygen, and 6.5–7.5% hydrogen, but also 15–18% nitrogen (16% is used as the mean), and 0.5–2% sulfur. Nucleic acids cannot be formed without phosphate groups, and phosphorus is also essential for intracellular energy conversion via adenosine triphosphate (ATP; more on it later in this section). Biomass is thus overwhelmingly composed of covalently bonded atoms of carbon, hydrogen, oxygen, and nitrogen.

Six other elements — calcium, potassium, magnesium, iron, sodium, and chlorine — are relatively abundant in organisms, and more than half a dozen metals have irreplaceable functions in many enzymes. These metallic elements include cobalt, copper, manganese, molybdenum, nickel, selenium, tungsten, and zinc, but often just a single atom per molecule is needed. Altogether more than 30 elements have been identified as recurrent constituents of biomass, but more than 95% of life's mass is made of carbon, hydrogen, oxygen, and nitrogen, and the share is

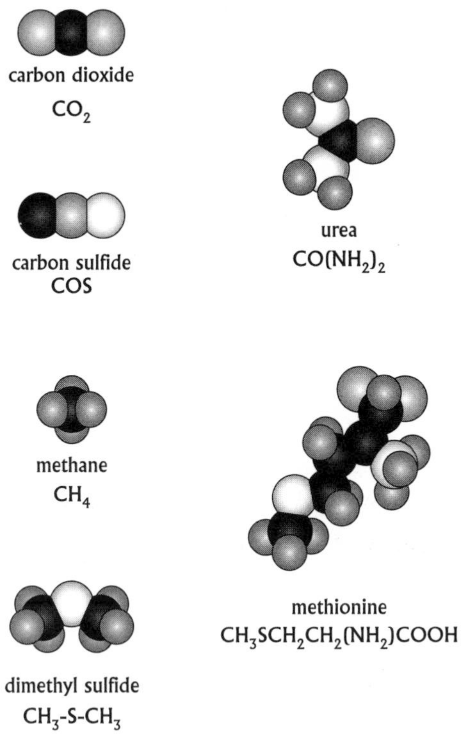

2.1 Carbon compounds common in the biosphere range from triatomic CO_2, the source of the element for photosynthesis, to amino acids that make up proteins (methionine shown here).

99.9% when sulfur and phosphorus are added. Listings of these elements do not, obviously, define life.

Concise yet all-encompassing definitions of life are not easy to produce, and perhaps they are impossible within the framework of physics and chemistry (Chyba and McDonald 1995). Many published attempts have been inadequate, inelegant, or imprecise, with different factual emphases and biases of individual scientists tilting the definitions in many different directions. According to Dawkins's "gene's eye-view" perspective, life is nothing but the maximization of deoxyribonucleic acid (DNA) survival, a merciless contest among genes where "DNA neither cares nor knows. DNA just is. And we dance to its music" (Dawkins 1995, p. 85). Schrödinger's (1944) famous thermodynamic definition sees life as "order based on order," and hence the only way for an organism to be alive is "by continually drawing from its environment negative entropy" (p. 72).

Various more or less comprehensive listings of life's attributes almost always include reproduction (self-replication), mutation and evolution by natural selection, metabolism, homeostasis, biosynthesis of complex compounds, and sometimes, pace Schrödinger, the ability to exist in improbable states, that is, not in equilibrium with the surrounding environment (Horowitz 1986; Margulis and Sagan 2000; Chao 2000). Such definitions eliminate candidates that may fulfill many or most of the listed requirements. Viruses are certainly the most peculiar forms of almost-life: they evolve as a result of mutation and natural selection, posing a tremendous challenge in developing effective vaccines.

But they are alive only inside the infected cells, never on their own: they are mere fragments of ribonucleic acid (RNA) or DNA (but not both) enclosed in a protein shell, with no enzymes and hence no independent metabolism (Levine 1992). Although animal cells are the most common hosts, since the early 1970s viruses or virus-like particles—some of them large, 140–190 nm in diameter, icosahedrons—have been found in ten out of fourteen classes of eukaryotic algae, including *Chlorella,* perhaps the most widely distributed algal genus in the biosphere (Van Etten and Meints 1999).

Self-replication is assured through storage and transmission of genetic information by nucleic acids. The process must be stable enough to ensure that the progeny itself will be able to replicate, yet sufficiently flexible to cope with a changing environment and to allow for the emergence of adaptations and eventually new species. Biosynthesis (anabolism) of an immense variety of carbon-based polymers organized in cells, tissues, and organs is the most obvious manifestation of life's intricate metabolism (Cartledge et al. 1992). Catabolism, the intracellular conversion and excretion of metabolized compounds, is metabolism's opposite. The resulting growth negates the entropic slide and keeps organisms in states far removed from equilibria with their surroundings.

Life's fundamental functional and structural unity has been revealed only by scientific advances that have made it possible to understand organisms at cellular, molecular, and submolecular levels. The most impressive achievements of this revealing descent in scale — including the discoveries of structural complexities of nucleic acids and proteins, storage and replication of genetic information, and the ability to sequence complete genomes — have come only during the second half of the twentieth century. Hunter's (2000) recounting of this scientific progress is an excellent guide through these complexities.

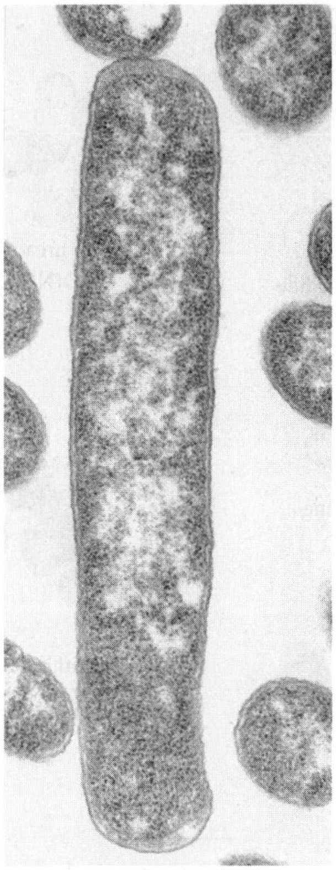

2.2 *Escherichia coli* is one of most abundant and perhaps the best-known and most-studied bacteria in the biosphere. Transmission electron microscope image by K. A. Sjollema, Laboratory for Electron Microscopy, Biologisch Centrum, Rijksuniversiteit Groningen.

Life's Biochemical Unity

Life's enormous variety is just a matter of adaptive expressions of internal unity of organisms. The most notable functional commonalities of living organisms include their shared machinery of genetic coding and protein production and the reversible intracellular energy storage used to power all metabolic processes mediated by enzymes. Life's structural unity is manifested by the biosynthesis of carbon-based macromolecular polymers constructed from simpler subunits (monomers). The bulk of any dry biomass is made up of varying mixtures of just four kinds of polymers: proteins (dominant in animals), carbohydrates (making up most of the plant mass), lipids, and nucleic acids (with particularly high shares in some bacteria and fungi). Three-µm-long *Escherichia coli,* perhaps the best known of all bacteria, contains more than a thousand different proteins and hundreds of nucleic acids, as well as scores of lipids and carbohydrates, whereas human cells produce more than fifty times as many kinds of different molecules (fig. 2.2).

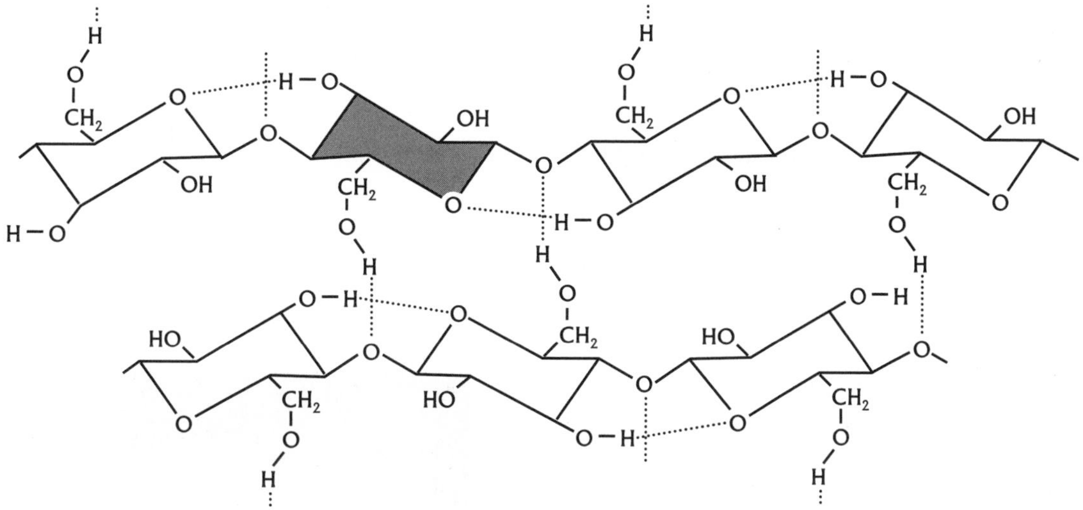

2.3 A small segment of cellulose, the biosphere's most abundant macromolecule, made up of about 3,000 units of a simple sugar, glucose (Smil 2000a).

All of these polymers are composed of a surprisingly limited number of monomers: only about 50 different compounds—five nucleotides, twenty-one amino acids, fifteen monosaccharides, and ten fatty acids—make up more than 90% of biomass. These compounds make up both the enormously complex intracellular machinery needed to perpetuate life as well as an astonishing variety of specialized tissues and organs that has evolved to cope with environmental opportunities and constraints. Whereas aperiodic polymers—nucleic acids, linear unbranched assemblies of nucleotides and proteins made of amino acids—store, transmit, and process genetic information and catalyze biochemical reactions, periodic polymers are indispensable as long-term energy stores and as structural components. Simple sugars, above all glucose, are polymerized to produce carbohydrates. Long-chained cellulose, consisting of about 3,000 units of glucose, is the most abundant structural macromolecule in the biosphere (fig. 2.3; for further discussion see chapter 7).

All polymers of nucleotides, amino acids, and sugars are produced by the process of dehydration condensation, as water is removed to form bonds between monomers. Lipids, used as building blocks of cellular membranes and internal organs, are composed of long chains of fatty acids (fig. 2.4). In Crick's (1966) summation, "molecules in cells are either large polymers, or the small molecules used to form these polymers. . . . It is not worth the cell's while to build up complicated molecules except by simple repetitive processes. Thus medium-size molecules are rarely produced" (p. 35).

The instructions for making all the proteins necessary for the growth of every organism, as well as the regulatory

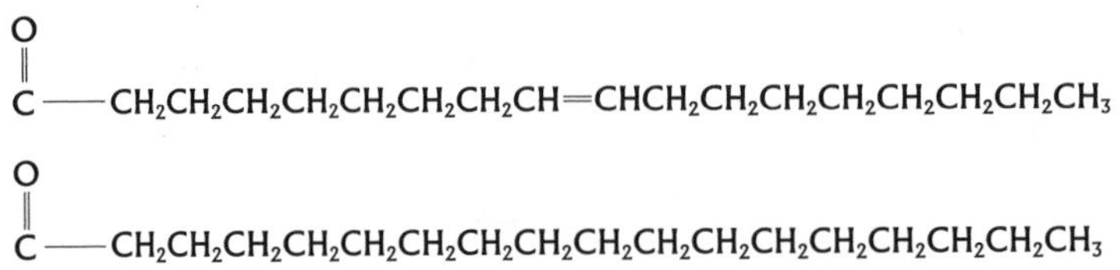

2.4 Two examples of fatty acids, whose hydrocarbon chains are essential building blocks of lipids.

sequences that determine the timing and quantity of protein synthesis, are encoded in the form of a right-handed, double-stranded helix of DNA, whose structure was first decoded by Watson and Crick (1953; fig. 2.5). The two strands, running antiparallel to each other, are made of polymerized nucleotide chains, with each monomer composed of three parts: a five-carbon sugar (deoxyribose in DNA, ribose in RNA) is the backbone bound to the phosphate group, which is on the outside of the helix, and to a base, a nitrogen-containing structure on the inside (fig. 2.6). The nucleotides are named after one of the four possible bases: two purines, adenine (A) and guanine (G), and three pyrimidines, thymine (T), cytosine (C), and (only in RNA) uracil (U). The strands are held together by hydrogen bonds formed between complementary bases (A-T and C-G). Mutations change a DNA sequence by adding, subtracting, or altering one or more bases.

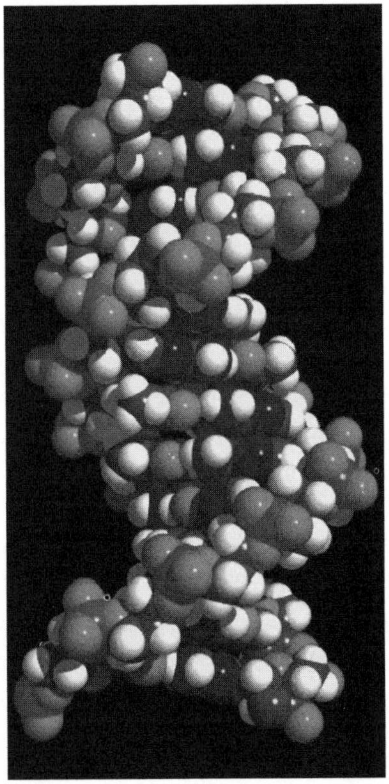

2.5 Space-filling model of the DNA helix.

Most genes have between 10,000 and 100,000 nucleotides, and we are now able to determine entire nucleotide sequences from one end of a chromosome to the other. The techniques needed to do so were first introduced in 1977 (Maxam and Gilbert 1977; Sanger et al. 1977), and by the 1990s they were perfected to rapid, efficient, and fully automated commercial applications (Alphey 1997; Suhai 2000). The first complete genome—

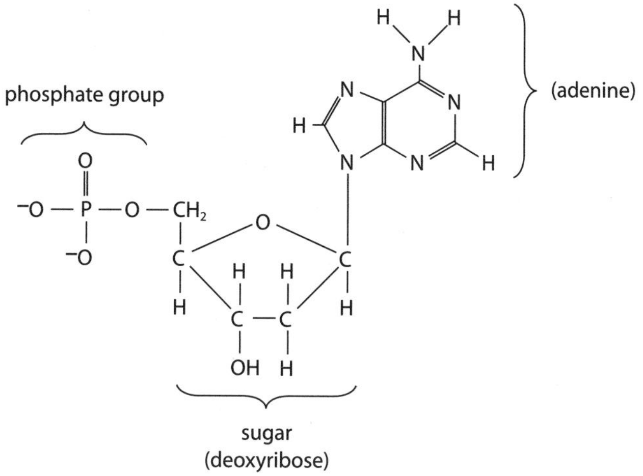

2.6 Structure of adenine (A) nucleotide, one of the four constituents of DNA.

for *Mycoplasma genitalium,* a parasitic bacterium coding for just 482 genes—was deciphered eighteen years after the techniques were introduced (Fleischmann et al. 1995). By the beginning of 2000, complete sequences became available for more than twenty microbial species (TIGR 2000), and the entire human genome sequence of about three billion base pairs was completed by the end of that year (IHGSC 2001). Functional genomics has a rich and bright future (Dhand 2000).

The twenty amino acids whose combinations make proteins are all variations of a single basic structure. A central carbon atom is bonded to four distinct groups: a hydrogen atom, a triatomic amino group (-NH_2), a carboxylic acid group (-COOH), and a variable group, a so-called side chain, which accounts for the differences among amino acids and is specified by genetic coding. In glycine, the simplest amino acid, the side chain is just a single hydrogen atom; in contrast, arginine and lysine have long, linear side chains, and the chains of phenylalanine, tryptophan, and tyrosine include hexagonal rings (fig. 2.7).

Justus von Liebig's work laid the foundations for amino acid chemistry: by treating milk solids with acid, he released, purified, and analyzed the molecules of leucine and tyrosine; but it was only in 1935, nearly seventy-five years after Liebig's pioneering work, that the present list of twenty amino acids was completed (Dressler and Potter 1991). After the discovery of DNA's structure, the biggest remaining question to be resolved in order to understand the chemical fundamentals of life was how DNA directs the synthesis of proteins. This puzzle—in Crick's (1966) words, the need "to translate from the four-letter language of a nucleic acid into the twenty-letter language of the protein" was answered by the early 1960s (p. 43). Proteins are synthesized by assembling constituent amino acids according to a nucleotide sequence specified in the cell's DNA. First, the enzyme RNA polymerase transcribes the

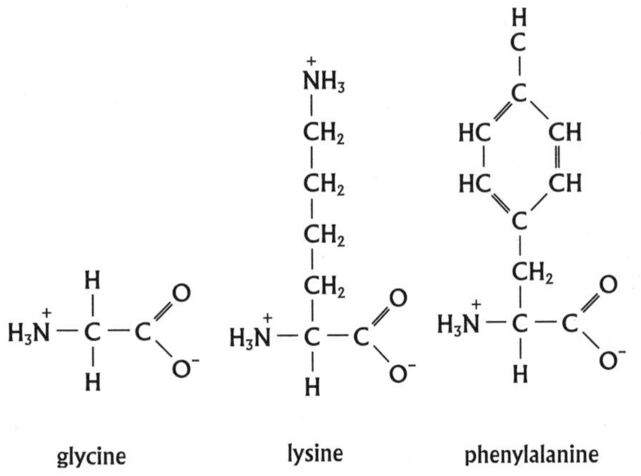

2.7 Three representative amino acids: glycine has a single hydrogen atom as its side chain; lysine has a straight nitrogen-containing hydrocarbon side chain; and phenylalanine's side chain is a planar hydrocarbon ring.

genetic information from the chromosomal DNA to an intermediary molecule, messenger RNA (mRNA). That molecule carries the blueprint to ribosomes, complex catalytic conglomerates serving as the sites of polypeptide production in cells (RNA differs from DNA by using ribose instead of deoxyribose and uridine instead of thymidine as one of its bases). Ribosomes are 15–25 nm across, made of two pieces of unequal size (small and large subunits, SSUs and LSUs) and consisting of RNA (about 55% of total weight) and fifty-three to eighty different proteins (de Duve 1984; Hill et al. 1990; Frank 1998).

The SSU is the site of translation initiation, accomplished with the help of transfer RNA (tRNA), which latches onto amino acids and attaches them to the growing polypeptide chain (the third type of RNA in ribosomes, rRNA, serves as a structural component; for more on it, see chapter 3). All amino acids, except glycine, are chiral molecules, existing as two mirror images, but protein synthesis uses only the L (laevorotatory, or left-handed) form. This handedness of life may have its origins in space billion years ago (Clark 1999). Studies of amino acids in the Murchison meteorite showed L-alanine to be twice and L-glutamic acid three times as abundant as their counterparts. This could have been caused by a narrow range of circularly polarized light of one handedness formed by scattering of dust particles spinning around the direction of the local magnetic field and destroying just one enantiomer.

As the mRNA is threaded through ribosomes, adjacent groups of three nucleotides (codons) determine the next amino acid to be added to the increasing polypeptide chain. Added amino acids are joined, end to end, by peptide bonds formed by combining the carboxy group of the first amino acid with the amino group of the next one (and eliminating water in the process). Every cell contains genetic instructions to produce 10^4–10^5 different proteins, but actual output is only a fraction of that total. Polypeptide chains produced by ribosomes are folded spontaneously into protein molecules. When adjacent in a three-dimensional structure, two cysteine residues in different parts of the polypeptide chain can be oxidized; the resulting covalent bonds of disulfide bridges stabilize the three-dimensional structures by providing structural stiffness, particularly to extracellular proteins secreted by cells (fig. 2.8).

Whereas the information carried by nucleic acids resides in their one-dimensional sequence of bases, specific properties (and hence activities) of individual protein molecules are determined by their three-dimensional structure (Gierasch and King 1990; Branden and Tooze 1991). It was disappointing when x-ray crystallography revealed protein molecules, in contrast to the neat packing of

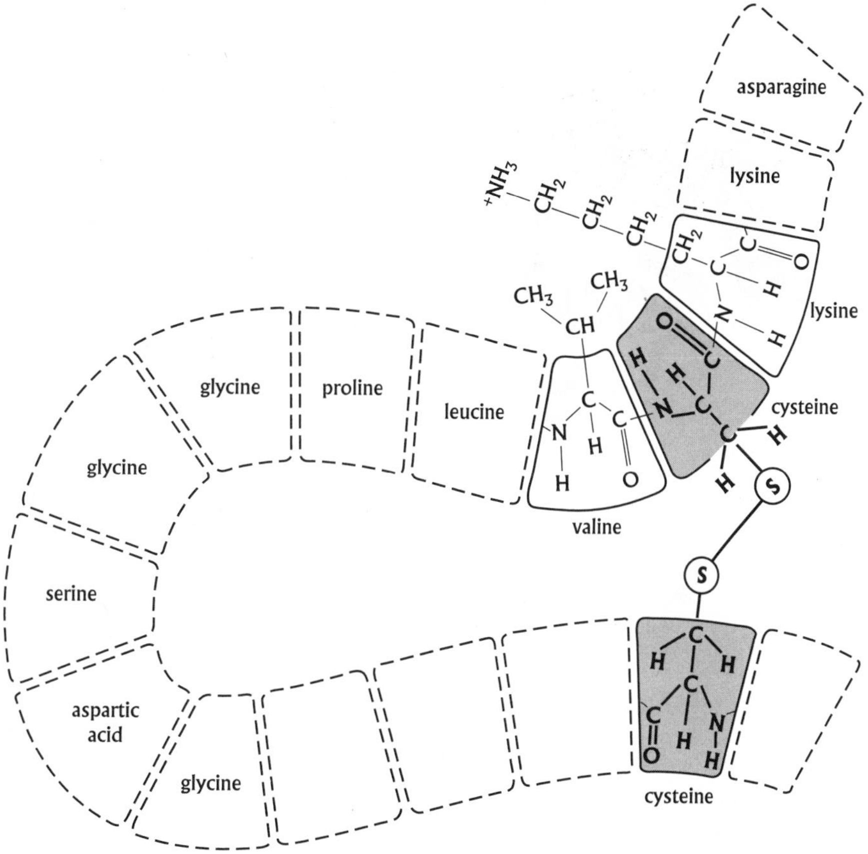

2.8 The three-dimensional structure of long polypeptide chains is fixed by covalent disulfide bonds formed between two adjacent cysteines.

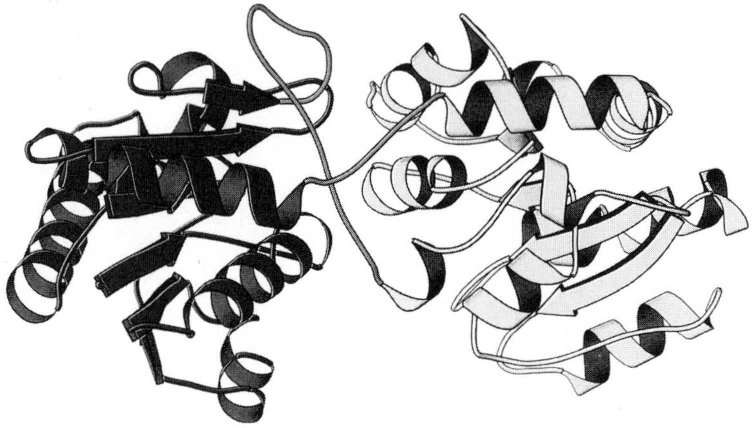

2.9 The most popular way of depicting the complexities of protein folding is by using schematic diagrams, introduced by Jane Richardson (1985), in which β sheets are represented simply by arrows showing both the direction of each strand and the connections of strands to one another along the polypeptide chain.

double-stranded DNA, to be bewilderingly complex, asymmetrical, and irregular. Kendrew et al. (1958) commented on the lack of symmetry in his low-resolution model of myoglobin and noted that the entire arrangement "seems to be almost completely lacking in the kind of regularities which one instinctively anticipates, and is more complicated than has been predicted by any theory of protein structure" (p. 665).

However, proteins share such regular secondary structures (a specific amino acid sequence is their primary structure) as α helices and β sheets—both described first by Pauling and Corey (1951)—which are connected by loop regions and form the hydrophobic core and hydrophilic surface of the molecule. In turn, simple combinations of a few secondary structures are repeatedly encountered as structural motifs, some of which are associated with particular functions. The presence of repeated structural motifs suggests that these proteins arose from gene duplication and fusion of small subdomains (Miles and Davies 2000). Hairpin β motif, Greek key motif, and β-α-β motif are common supersecondary structures.

Finally, domains—polypeptide chains, or their parts, which can fold independently to form stable three-dimensional structures—are built from structural motifs (fig. 2.9). The complete information needed to specify a protein's three-dimensional structure is contained within its amino acid sequence, but the ability to use this information to predict actual protein folding has proven elusive. It now appears, however, that the fundamental physics underlying folding may be much simpler than previously believed and that folding rates and mechanisms are largely determined by the topology of the spontaneously folded state (Baker 2000).

Functionally the most important intracellular proteins are enzymes, catalytic compounds indispensable in cell growth and differentiation and development of organisms as well as in all basic metabolic processes. They belong to one of six functional classes, whose names describe their catalytic activity—oxidoreductases, transferases, hydrolases, lyases (bond breakers), isomerases, and ligases (bond formers)—and they work with astonishing efficiency and accuracy, speeding up chemical reactions and producing extremely high yields of correct isomers in cells. Some enzyme complexes function literally as machines equipped with springs, levers, and rotary joints. H^+-ATP synthase is the world's smallest rotary motor (Block 1997). But not all enzymes are proteins: RNA molecules are also able to catalyze some chemical transformations, namely, the excisions of material from the RNA chain (Cech 1986).

Besides enzymes, other functional proteins include carriers (such as hemoglobin), defensive molecules of the immune system, and hormones. A great variety of structural proteins are encountered in specialized tissues and organs ranging from bones and tendons to beaks and nails. "Proteome," a term coined in 1994, describes the complete set of proteins that is expressed and modified by the entire genome during a cell's lifetime, and proteomics, following in the footsteps of genomics, strives to separate and identify all of these constituents (Abbott 1999; Pandey and Mann 2000).

The overall biochemical plan of the propagation of life is thus both immensely complex and conceptually fairly simple. Consequently, Alfred Hershey (1970) felt that it should be possible to express it in a few words, and his attempt is worth quoting as a fundamental summary of our understanding:

> *First, the genotype resides in DNA—more importantly, in the linear sequence of the four nucleotides in single DNA strands. Second, nucleotide sequences . . . are transcribable into complementary sequences according to simple one-to-one rules: the four bases form only two interstrand pairs. . . . This code or its equivalent is used for DNA replication, gene transcription, and synthesis of ribosomal and transfer RNAs. . . . Third, sequences in one of the two complementary strands transcribed into messenger RNA, are translatable into amino-acid sequences in proteins according to a second code unrelated to the first. This is a non-overlapping triplet code (three bases per amino acid) usually called the genetic code. (p. 698)*

A skeletal formulation going beyond the genetic coding would be as follows: DNA makes RNA, RNA makes proteins, proteins (enzymes) govern metabolism, metabolism produces monomers, monomers are biosynthesized into polymers, polymers build tissues, tissues constitute organs and organisms, and organisms are open systems, constantly importing energy and exchanging water, gases, and solids with their environment to biosynthesize, replicate, and mutate.

Most organisms store energy over longer periods of time (this may be days to decades, depending on their life spans) either as carbohydrates or lipids, but all need a more general form of intracellular energy to power directly a vast variety of enzymatic reactions. This energy at the last step comes from the hydrolysis of the phosphate bond in ATP; this compound, present in every metabolizing cell, acts as the principal energy carrier in the biosphere (fig. 2.10). Evolution selected this phosphate as the energy

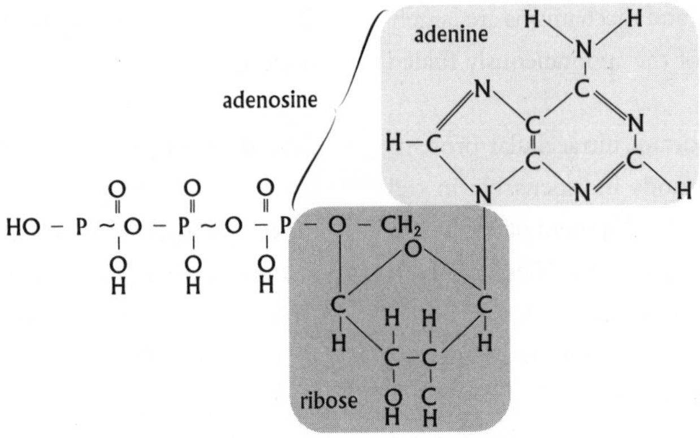

2.10 ATP is composed of adenine (6-aminopurine), ribose, and a triphosphate tail (Smil 1999d).

carrier because its enzymatic hydrolysis (loss of the endmost phosphate group) releases 31 kJ/mole, a relatively large amount of energy (Westheimer 1987).

The reverse reaction, catalyzed by ATP synthase, takes place when energy is not immediately needed, and it reforms the compound by adding a third phosphate molecule to adenosine diphosphate (ADP). ATP is not the cellular fuel per se: other energy sources (solar radiation in photosynthesizers, chemical energy released from digested biomass in microorganisms and animals) are used to maintain a high ATP-to-ADP ratio, and hence ATP can be hydrolyzed again and again for energy release. The intensity of ATP generation is thus stunning (Broda 1975). A 60-kg man consuming daily about 700 g of food in carbohydrates would make and use no less than 70 kg of ATP, roughly 3 g of ATP for every dry gram of body. This rate is minuscule compared to intensities achieved by respiring bacteria. *Azotobacter* breaking down carbohydrates while fixing large amounts of dinitrogen produces 7000 g of ATP for each gram of its dry mass!

Cells and Organisms

Membrane-bound cells, providing optimum conditions for biochemical processes, are the smallest functioning units of life. Robert Hooke drew them first as he observed them in a section of cork in 1665, but only in 1839 did Matthias Schleiden and Theodor Schwann state unequivocally that "all organisms are composed of essentially like parts, namely of cells" (Nurse 2000). Today we know that even the smallest organisms require at least 300 nonribosomal proteins, and these proteins, together with at least 250–400 genes and ribosomes needed for their synthesis, can fit into a sphere no less than 250–300 nm across, including the bounding membrane (Steering Group for the Workshop on Size Limits of Very Small Microorganisms 1999; appendix B).

Indeed, free-living bacteria with diameters of 300–500 nm are common in oligotrophic environments. The best-known bacterial species, *Escherichia coli,* measures 500–700 nm across, whereas the smallest parasitic organism, *Mycoplasma genitalium,* is just 150 nm across. A recent claim that free-living nanobacteria with diameters of mere 50 nm were isolated from blood (Kajander et al. 1997) has been met with much skepticism about the viability of such isolates, much like the already clearly refuted claim of nanobacteria in a Martian meteorite (McKay et al. 1996; Abbott 1999). Underneath their stiffened walls cells have membranes acting as selective barriers: impermeable to many organic molecules synthesized in the cell, but allowing the flux of various chemicals in and out of the container not just by passive permeability, but also by active pumping. Consequently, intracellular concentrations of some compounds may be orders of magnitude above the levels in the surrounding environment. The internal organization of cells was unraveled gradually through increasing microscopic powers, and then by more discerning techniques.

Light microscopes can show details of about 0.2 μm, sufficient to resolve cellular organelles as mitochondria; electron microscopes can distinguish such large molecular assemblies as ribosomes and viruses, but the cells have to be dried and metal-plated and hence cannot be observed alive using such microscopes. The internal structure of purified and crystallized compounds can be investigated through X-ray crystallography; in contrast, nuclear magnetic resonance (NMR) spectroscopy can work with molecules in solution to reveal the distance between a specified pair of atoms. An enormous literature describes cellular organization, function, and development (de Duve 1984; Rensberger 1996) and a recent discovery, apoptosis, the process of programmed cell suicide activated by a cascade of enzymes (Kumar 1998; Hengartner 2000).

Proteins, organized in functional assemblies of ten or more molecules that interact with other large complexes, form the bulk of any cell's dry mass. Alberts's (1998) analogy is fitting:

Indeed, the entire cell can be viewed as a factory that contains an elaborate network of interlocking assembly lines, each of which is composed of a set of large protein machines . . . like the machines invented by humans to deal efficiently with the macroscopic world, these protein assemblies contain highly coordinated moving parts. Within each protein assembly, intermolecular collisions are not only restricted to a small set of possibilities, but reaction C depends on reaction B, which in turn depends on reaction A—just as it would in a machine of our common experience. (p. 291)

These protein machines—as well as storage, transfer, and translation of genetic information, enzymatically driven and ATP-mediated biosyntheses, and the use of structural polymers—are common to all life forms. But there are two different plans of cellular organization, both obviously very successful in evolutionary terms: cells lacking a nucleus and nucleated cells. Haeckel (1894), in one of the first comprehensive classifications of life, assigned organisms with the first kind of cellular organization—including bacteria and blue-green algae, as cyanobacteria used to be known—to single-celled Monera. This distinction between cells without and with a nucleus eventually led Edouard Chatton (1937) to combine two qualifiers (before and good) with another Greek word, nut, to coin the terms *procariotique* and *eucariotique* ("prokaryotic" and "eukaryotic," respectively, in their common English spellings).

When examined using an electronic microscope, prokaryotic cells appear as tiny rigid vessels filled with amorphous cytoplasm lacking internal membranous organelles (mitochondria, lysosomes, peroxisomes), with all genetic information stored in a single DNA double helix (fig. 2.11). However, prokaryotic metabolism is not a disorganized, inefficient affair: even without membrane-bound specialized compartments, prokaryotic cells display

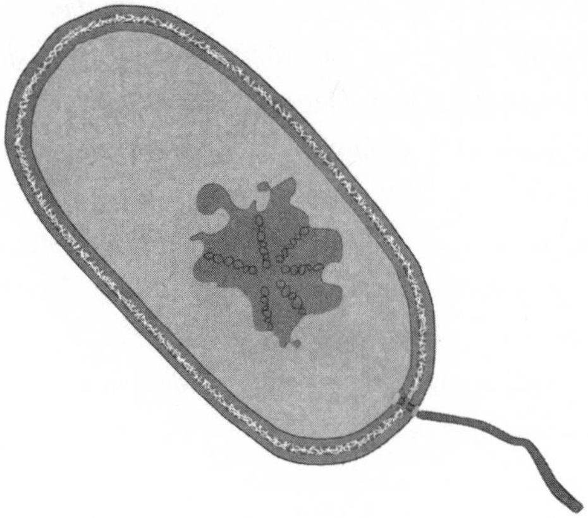

2.11 Prokaryotic cells are filled with cytoplasm and their DNA is concentrated in poorly defined nucleoids that are not separated by a membrane from the rest of the interior. Based on Hoppert and Mayer (1999).

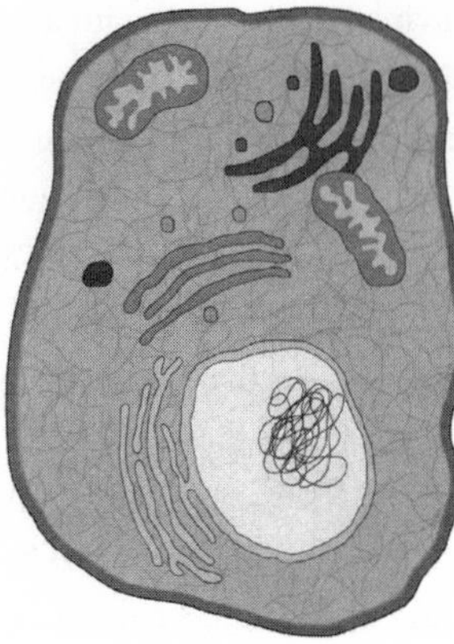

2.12 The nucleus, a membrane-delimited compartment containing DNA, is the key distinguishing feature of all eukaryotic cells. Other specialized compartments include mitochondria, sites of energy conversion; Golgi apparatus, devoted to processing proteins synthesized by the endoplasmic reticulum; lysosomes, sites of intracellular digestion; and peroxisomes, the most abundant kind of cellular microbodies that oxidize enzymes. Based on Hoppert and Mayer (1999).

a high degree of subcellular organization and, as attested by the rapid spread of many bacterial infections, are able to metabolize with stunning efficiency (Hoppert and Mayer 1999).

In contrast, eukaryotic (fungal, plant, protist, and animal) cells have a nucleus containing chromosomes, each one with many double helices of DNA encoding thousands of genes (fig. 2.12). Discovered in 1831, the cell nucleus has been one of the most studied and best known, but least understood, cellular organelles (Lamond and Earnshaw 1998). Eukaryotes also have distinct intracellular compartmentalization, and specialized organelles, such as mitochondria for energy production and chloroplasts for solar energy capture in plants. There are other biochemical and structural differences between eukaryotic and prokaryotic cells. A few exceptions aside, prokaryotic cells are much smaller than the eukaryotic ones, typically less than 5 μm, exceptionally above 10 μm across, compared to 10^1, 10^2, and even 10^3 μm across for nucleated cells. Prokaryotic cell walls contain peptidoglycan, a cross-linked polymer made up of glucosamine chains and amino acid bridges, rather than cellulose, as the strengthening ingredient; and they reproduce asexually, relying on nuclear binary fission rather than on mitosis, the process of cell division producing two identical daughter cells.

This fundamental divide between eukaryotes and prokaryotes was later confirmed by molecular sequences (Gray and Doolittle 1982), and additional distinguishing features added by later research include differences in ribosomes and in the presence of RNA polymerase (prokaryotes have only one, eukaryotes have three polymerases) and its makeup. The two kinds of cells also have a fundamentally different logic of gene regulation (Struhl 1999). All aerobic eukaryotes generate ATP through the oxidation of reduced carbon compounds inside mitochondria—membrane-enclosed cell organelles introduced into cells billions of years ago through fusion with a microorganism—and through excretion of water (Brown 2000). Anaerobic eukaryotes (including some fungi and ciliates and diplo- and trichomonads) must obviously use compounds other than O_2 (nitrate, for example) as electron acceptors, or they simply produce H_2 (Embley and Martin 1998).

Cells arise from self-assembly of their components from molecular building blocks and, in turn, they self-assemble into tissues that then form organs: the final outcome is a hierarchical organization of less or more complex organisms. In spite of their enormous diversity, all organisms can be assigned to one of two basic classes determined by their feeding strategies. The terms "autotrophy" (self-nourishment) and "heterotrophy" (feeding on other organisms) were formally introduced by a German plant physiologist, Wilhelm Pfeffer (1845–1920), who defined autotrophs as organisms able to synthesize their cells from CO_2 as their main source of carbon (Pfeffer 1897). Heterotrophs must secure their carbon either directly (herbivores, detritivores, saprovores) or indirectly (carnivores) from autotrophs: they obtain their monomers from food by digesting proteins, carbohydrates, and lipids.

This fundamental division cuts across the cellular divide: autotrophs include not only all plants (obviously eukaryotic) but also a large number of prokaryotic bacteria. And, as I will show in some detail in the next chapter, autotrophy is not solely the ability to photosynthesize, that is, to use solar radiation as the energy source for the conversion of CO_2, water (H_2O), and mineral nutrients into new biomass: there are numerous microbial autotrophs thriving in the total absence of sunlight. And there are also facultative autotrophs that can grow heterotrophically when appropriate organic compounds are available. Methanogenic microbes can do so (for more on them, see chapter 3) and recently discovered marine photoheterotrophic bacteria, which metabolize organic carbon when available but can switch to photosynthesis when such carbon is scarce, make up more than 10% of the total microbial biomass in the upper open ocean (Kolber et al. 2001).

This is also an apposite place to note that many unicellular organisms—be they bacteria or eukaryotes, such as yeasts—may frequently behave as well-organized multicellular entities. Many bacterial colonies show intriguing spatial organization, and most, if not all, individual cells can communicate with their neighbors via lactones and peptides (Kaiser and Losick 1993). For example, different yeast colonies (*Candida, Saccharomyces*) not only are capable of short-range intracolony cell-to-cell communication but also can send long-distance (relatively speaking) signals between neighboring colonies by using pulses of volatile ammonia (Palkova et al. 1997). These transmissions—the first type nondirected, the second enhanced and directed toward a neighbor colony—result in growth inhibition of the facing parts of both colonies. Consequently, some microbiologists are favoring the concept of bacteria as multicellular organisms (Shapiro and Dworkin 1997).

Universal Prerequisites

In considering physical prerequisites of long-lasting biospheres we do not have to be troubled by contentious problems of life's origin: it does not matter if the first organisms arose de novo on a planet or if one believes that they arrived from elsewhere in the galaxy (more on this later in this chapter). In any case, the necessities for supporting a long-lived and hence relatively stable biosphere are obvious: a suitable star must hold an appropriate planet in a correct orbit for a sufficiently long period of time. Quantifying these prerequisites is not that easy.

Candidate stars do not include supergiants, dwarfs, or most of the binaries and multiple stars. Orbits, too, are restricted to relatively narrow habitable zones around a star, but these zones may change with the star's age, and in the presence of radiation-absorbing gases their extent can be modified by dynamic atmospheric feedbacks: the three terrestrial planets (Venus, Earth, and Mars) offer excellent illustrations of these realities. Even a basic list of suitable planetary attributes must contain more than a dozen items—from mass to atmospheric composition—with some of these variables locked in compensatory reinforcing feedbacks. For example, a slower rotation (say 72 hours) of an Earth-like body would have much milder consequences for surface temperature if 80% of such a planet were covered by ocean, with land evenly interspersed in it. In contrast, extreme heating and cooling would take place under such slower rotation on a planet with a small ocean restricted to a single longitudinal hemisphere.

In addition, extreme values of most of these variables will vary depending on what level of biospheric complexity and longevity we specify: a planet habitable for what kind of life? A more accurate general term would be biocompatible planets, or planets with biospheres—but it would be also a much more restrictive definition. A planet could have only a subterranean biosphere inhabited by slowly metabolizing chemolithotrophic microbes (for details on these organisms, see chapters 3 and 6), or its benign environments could be limited to just a small share of its surface, and/or to brief periods of time. Consequently, a planet's biosphere could be restricted to patches similar to those inhabited by the Antarctic bacteria embedded in ice, which come to life only for short spells of time when temperatures rise to near 0°C—making it merely temporarily biocompatible rather than permanently habitable.

Star Candidates

Galactic environment and a star's intragalactic motion matter a great deal, with locations in quieter regions and with nonextreme orbits obviously preferable. In both respects the Sun is an excellent star (for more on it, see chapter 4). Its galactic neighborhood is one characterized by an extremely low presence of interstellar gas: there are no massive clouds within 100 light-years of the star, and this so called Local Bubble is such an extreme void that even the best laboratory vacuums are about ten million times denser (Frisch 2000). Although the Sun's trajectory is taking the solar system away from the bubble, there is no imminent danger of encountering any large, dense interstellar clouds.

The Sun's orbit around the galactic center is less elliptical than are the orbits of similar nearby stars: this prevents the solar system's passage through the inner galaxy, where destructive supernova explosions are more common. In addition, the small inclination of its orbit prevents the Sun from abruptly crossing the galactic plane, a trajectory that would stir up the Oort cometary cloud. And the Sun's closeness to the galaxy's corotation radius allows it to avoid crossing the galactic spiral arms, which would also increase the exposure to supernovas. Moreover, the Sun, unlike most of the stars, is solitary, and hence much more likely to generate suitable planetary orbits. Binaries and multiple star systems can have planets orbiting within habitable zones (more on these zones in the next section), but stable orbits around two stars are restricted to radii between 0.3 and 3.5 times the stellar separation (Donnison and Mikulskis 1992), and those around multiple stars are even more unlikely.

Stars release enormous quanta of energy by fusing hydrogen. The energy output (luminosity) of young stars slowly increases, and during most of their life, called the main sequence, the fusing goes on near the star's center. But eventually the core hydrogen is used up, and stars of the Sun's size and larger are transformed rather rapidly into red giants, whereas the largest stars (with the initial mass greater than about eight solar masses) are destroyed in supernova explosions. A star's mass thus has obvious implications for its longevity and gravitational pull: stars must be sufficiently, but not excessively, massive in order not to age too rapidly and to hold a suitable planet in just the right orbit.

Stars more than 1.5 times as massive as the Sun (spectral types O, B, and A) experience very high rates of thermonuclear reactions and hence spend relatively short spans on the main sequence before undergoing drastic transformations (fig. 2.13). Even the largest suitable candidates, stars in the F2 class, which are 1.4 times the Sun's mass, consume their hydrogen too rapidly and spend no more than 3 Ga on the main sequence. Consequently, any complex life could evolve on such planets only if its evolution would proceed much faster than it did on Earth. In contrast, small and long-lasting stars (abundant and cool M dwarfs having less than half of the Sun's mass) produce too little energy to illuminate more distant planets, and the faintest among them radiate it almost totally in infrared wavelengths and with a large (tens of percent) variability. Moreover, the tidally synchronized rotation of planets orbiting in habitable zones around such stars (much like the Moon's rotation relative to the Earth) would cast one hemisphere into perpetual darkness, where its thin atmosphere would freeze, clearly not a condition compatible with long-term habitability.

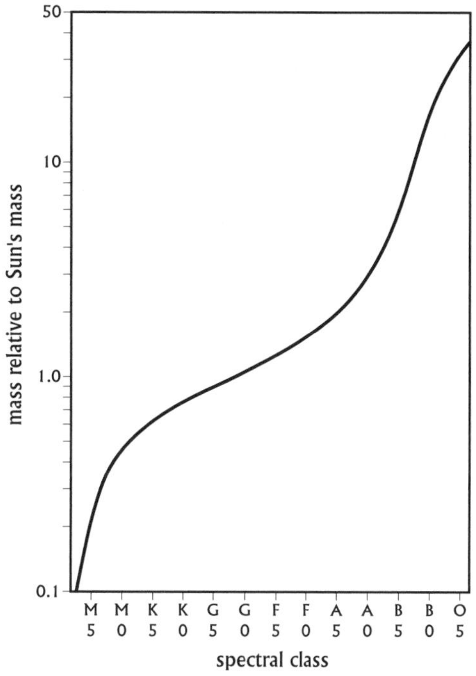

2.13 The mass of main-sequence stars as a function of spectral class. Based on Dole (1964).

Habitable Planets

I will use the term "habitable planets" only for those bodies in space whose attributes can permit the emergence of complex eukaryotic, carbon-in-water life whose evolution eventually proceeds to a widespread diffusion of highly encephalized organisms. Dole (1964) and Shklovskii and Sagan (1966) published the first systematic examinations of the prerequisites of complex life in the universe; recent publications on habitability include works by Aczel (1998), Lewis (1999), Dole (1996), Kasting (1993, 1997) and Lissauer (1999).

Planetary mass is obviously the key determinant not just of surface gravity and atmospheric pressure, but also of atmospheric composition, water retention, surface features, thermal flux from the interior, and the rate of tectonic rearrangement of continents and oceans. Small planets would have insufficient gravity to hold breathable atmospheres, and they would be also cooler, as they could not retain large volumes of greenhouse gases. Because of their larger surface area–to–mass ratio, they would cool faster, and hence their cycle of geotectonic renewal would end sooner. The smallest planets (there is only an arbitrary line between them and large asteroids) would not be able even to assume spherical shapes and would not allow for differentiation of planetary environments. Moreover, anything airborne would remain aloft for very long periods of time.

On the other hand, the high gravity, hotter planetary interior, and less rigid crust of very massive planets would produce flat surfaces and deeper oceans. Very dense atmospheres that could not be penetrated by sunlight and gravity impeding or precluding movement of all larger organisms would be among the greatest obstacles to complex life on such planets. The upper limit on gravity for a habitable planet is 1.5 g, which corresponds to 2.35 Earth mass and 1.25 Earth radius. Dole (1964) set the lower limit, using a relatively complex consideration of atmospheric characteristics at between 0.4–0.6 Earth mass; the lower value implies surface gravity of 0.68 g and radius 0.78 times the Earth's value.

Whereas it is easy to point out intolerable rates of planetary rotation, it is much more difficult to specify an acceptable range. Faster rotation rates would result in angular velocities that would destabilize planet's shape; very slow rates of rotation would separate the planet into excessively heated and intolerably cooled segments. An easy range would be between eight and forty-eight hours, whereas the extremes suggested by Dole (1964)–two to three hours and ninety-six hours—would be harder to cope with. Changes in one of the key orbital variables—distance from a star, orbital eccentricity, average illuminance, and inclination of the equator—have obvious consequences for the other parameters.

The simplest definition of the habitable zone around a star is the space within which a planet can support liquid water. This zone lies between two spherical shells centered on the star. Orbiting closer to the star than the inner shell would produce intolerably high inputs of radiation, resulting in photolysis of water followed by escape of hydrogen to space. A conservative estimate is that a solar flux about 1.1 times the present insolation rate (1370 W/m^2; for details, see chapter 4) would trigger this process, and so a habitable planet could not be closer than 0.95 astronomical units, or AU (the Earth's mean orbital distance from the sun). But with the right kind of clouds cooling the surface, this zone might move a bit closer to the Sun: Dole (1964) calculated 0.725 AU for a planet with no equatorial inclination. Venus, at 0.7 AU, is too close, and it also has the wrong atmosphere, more than ninety times as dense as the Earth's and replete with CO_2, and so it has a surface temperature of 750 K and a massive dissociation of water.

The outer edge of the habitable zone lies at the distance where greenhouse gases could not compensate for the diminished flux of solar radiation to keep water liquid. Kasting et al. (1993) put this distance at just 1.37 AU on the basis of volcanic CO_2 forming dry ice and shutting off the temperature-stabilizing carbonate-silicate cycle (see chapters 4 and 5). But radiative warming by clouds could, on a

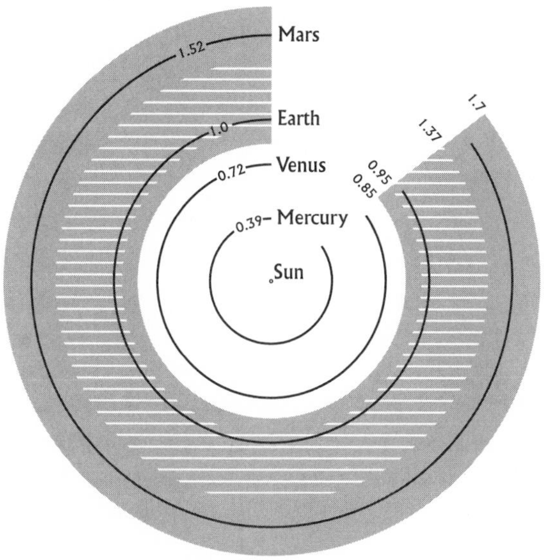

2.14 Conservatively (darker shade) and more liberally (lighter shade) defined habitable zones for an Earth-like planet in the solar system. All values are in astronomical units (1 AU = 149.6 Gm, the Earth's mean orbital distance).

planet sufficiently large to retain its volatiles, move the edge all the way to 2 AU: this means that the more massive Mars (with orbit at 1.52 AU) would be habitable (Dole's maximum was close to 1.4 AU with only 10% of the surface habitable).

A conservatively delimited habitable zone for the solar system ranges from 0.95 to 1.37 AU, and a more liberal range is twice as large, 0.85–1.7 AU. In the first instance a habitable planet could orbit within a zone reaching two-thirds of the way to the orbit of Mars but not quite to the orbit of Venus (at 0.72 AU; fig. 2.14). Main-sequence stars brighten as they age, and hence their ecospheres shift outward. A continuously habitable zone that takes into account the lower luminosity of young main-sequence stars — the young Sun was about 30% less luminous than it is now — spans an overlap of habitable zones at two widely separated points in time. For the Sun it would be, conservatively defined, just 0.2 AU wide (0.95–1.15 AU), but it could range from 0.95 up to 1.7 AU, including both the Earth and Mars.

Moderate orbital eccentricities (0.2–0.3 AU) would not create any intolerable problems with seasonal temperature ranges. Acceptable axial inclination of a planet's equator to the plane of its orbit could be, everything else being as it is on the Earth, as much as 80° and still leave at least 10% of the surface habitable, and up to 60° and still leave one-half of its surface suitable for complex life (fig. 2.15). Venus, the Earth's close planetary twin — its radius is 95% and mass 82% of the Earth's totals, and its orbit is close to the edge of the Sun's ecosphere — is the best example of the paramount importance of atmospheric composition. Its massive CO_2 atmosphere is more than ninety times as dense and more than 2.5 times as warm (750 vs. 288 K) as the one on the Earth.

This list far from exhausts the planetary prerequisites for supporting life. Certainly the most prominent among many other essential necessities are the planet's sufficient age and relatively stable climate. The latter requirement presupposes the long-term maintenance of many tolerable rates, ranging from wind speeds and terrigenic dust generation to the intensity and frequency of volcanic eruptions, atmospheric electricity discharges, and constant meteoroid infall. Excessive rates of any of these processes could, singly or in combination, make complex life impossible, even if all other conditions were met. The no less obvious importance of atmospheric composition will be reviewed later in this chapter.

Judging by the Earth's example, an evolutionary span of at least 4 Ga may be needed for the emergence of highly

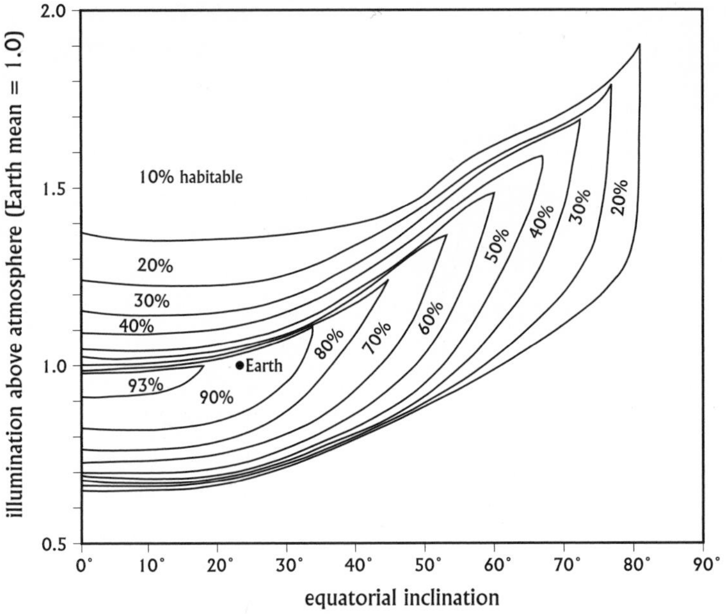

2.15 Depending on a planet's equatorial inclination, habitable zones would range between less than 10% and more than 90% of the planet's surface. Based on Dole (1964).

encephalized organisms, and so a planet lasting just a fraction of that time may never get a chance to progress beyond prokaryotic life. In spite of the Earth's past climatic oscillations, none of those events was severe enough to shut down an unfinished evolutionary experiment, but at least two episodes of extreme cooling may have come very close (for more on these episodes, see the next chapter), and so it is highly realistic to conclude that irreversible climatic shifts could have made other planets uninhabitable after even relatively long spans of biocompatible climates. And the importance of avoiding, or surviving, encounters with other space bodies—particularly with massive comets and asteroids—is self-evident (see the next chapter).

Life on the Earth

The Earth's formation is now understood to be a part of a sequential process of the emergence and consolidation of the solar system whose details have been elaborated by interdisciplinary research of the past two generations (Wetherill 1990; Taylor 1992). I will note just the basic highlights of this standard model. Measuring decay rates of long-lived radioactive nuclides (most often that of ^{238}U decaying to ^{206}Pb with the half-life of 4.5 Ga) has made it possible to establish the time of the planet's formation with an error of just 50 Ma, but the events of the subsequent 200 Ma, as well as the changes during the Hadean era, are open to speculation, as there is very little direct physico-chemical evidence from this earliest and very turbulent age of the Earth.

Two global processes have been particularly important for life's evolution on Earth: gradual growth of the continental crust, which has provided a variety of terrestrial environments needed for the subsequent evolution of plants and animals, and the emergence and the persistence of the atmosphere, whose greenhouse gases have kept most of the planet above freezing most of the time (snowball Earth phenomenon is discussed in chapter 3) and whose changing composition led eventually to the rise of oxygen-breathing organisms.

My review of the current understanding of the origins of the Earth's life will note the major points of broad consensus, and it will also stress many profound uncertainties. These uncertainties concern not just the developments during the precellular era (was it an RNA world?), but also the genetic search for a common ancestor (rooting of the universal tree of life), the origin of eukaryotic cells, and the rise of the Earth's oxygenated atmosphere. Afterward I will move to less-disputed ground and describe the milestones of life's evolution that we have been able to identify from the sparse remnants of Proterozoic life.

Formation and Early Evolution of the Planet

The Sun was formed by gravitational collapse of a slowly rotating, dense, interstellar molecular cloud, and planets grew by agglomeration of planetary embryos that formed in the solar nebula. Interstellar grains borne by the infalling gas were originally very small (< 0.1 μm), and their accumulation was determined by nongravitational forces. Once the planetesimals (tiny planets) became larger than 10 km in diameter, gravitational effects became dominant. Collisions of about 10^{10} such bodies were needed to assemble the Earth, and the planet may have attained virtually all of its present mass in less than 80 Ma after the beginning of the final accumulation of larger planetesimals (Wetherill 1990). The Earth's growth was dominated by a number of giant impacts involving bodies whose mass was between that of Mercury and Mars or even larger.

As a result of these impacts the forming Earth would have been extremely hot, with a massive solar-composition (H_2 and He) protoatmosphere blanketing a global magma ocean through which metal rained out to form the protocore (Harper and Jacobsen 1996). Decay of ^{238}U to ^{206}Pb puts the time of initial accretion at 4.55 Ga ago, and the planet's growth through planetesimal impacts continued for another 120–150 Ma (see appendix C for the chronology of major milestones in the Earth's and the biosphere's evolution). By 4.44–4.41 Ga ago, the Earth had a differentiated interior, and it retained its first atmosphere. But as Wetherill (1990) conceded, it is possible that the standard model of the Earth's formation is incorrect. Cold accretion from dust particles cannot be excluded, and the processes and sequences that governed many subsequent events, including the planet's differentiation into a heavy, iron-dominated core, silicate mantle, and light crust, are no less contentious (Jacobs 1992).

In any case, the crust amounts to a mere 0.6% of the Earth's volume, and its thickness ranges from just around 5 km below the oceans to more than 30 km below the continents, and up to 80 km below the Himalayas. Most of the Earth's mass (82%) is in the mantle, which goes to a depth of about 2,900 km; the underlying outer core is fluid, and the inner one is solid (Jacobs 1992). Major oceans may have appeared during the early Hadean era, as soon as 200 Ma after the Earth formed, in a hot (about 100 °C) and reducing environment, and their pH of 5.8 gradually rose to near neutral (Morse and Mackenzie 1998). Early Archean oceans were strongly stratified, with a deep anoxic bottom

(Lowe 1994). The first oceanic crust appeared as early as 4.5 Ga ago, formed most likely at ocean ridges (for more on this see chapter 4), and composed mostly of komatiite (magnesium-rich volcanic rock), with basalt becoming more important later (Condie 1989).

The continental crust came shortly after. No crustal rocks have survived from the first half billion years after the Earth's formation 4.55 Ga ago. The oldest preserved terrestrial rocks—metamorphosed Acasta gneiss found southeast of Great Bear Lake in Canada's Northwest Territories—was initially dated to 3.96 Ga ago, but a later examination with the super high-resolution ion micro-probe (SHRIMP) concluded that the zircon grains in the rock crystallized at about 4.03 Ga ago (York 1993; Stern and Bleeker 1998). Rocks aged 3.7–3.9 Ga ago were also identified in Australia, Greenland, and Antarctica. But the oldest crust, now destroyed, certainly predated these rocks. In 1983 examination by SHRIMP revealed that zircon crystals—eroded from their parental rock and later incorporated in quartzite (metamorphosed sandstone) from Mount Narryer in Western Australia—were 4.1–4.2 Ga old (Froude et al. 1983). Wilde et al. (2001) pushed the date of the earliest Australian zircon grains to 4.404 Ga ago. This means that some patches of continental crust existed within 150 Ma of the Earth's accretion.

Because of relatively rapid rate of crustal recycling by more intensive plate-tectonic processes (see chapter 4), the subsequent growth of the continental crust proceeded slowly until about 3 Ga ago, when the Archean Earth was dominated by oceanic lithosphere (Lowe 1994). The most important spurt of crustal growth took place between 3 and 2.6 Ga ago, during the transition from the Archean to the Proterozoic era (Taylor and McLennan 1985, 1995). The other notable episodes of growth were between 1.7 and 2 Ga ago and between 1.1 and 1 Ga ago, by which time the Earth had acquired at least 90% of today's continental crust (fig. 2.16). The cooling of the Earth's surface, the emergence of oceans, and the growth of the continental crust provided a variety of niches for prokaryotic life—but life's evolution was always influenced by the planet's changing atmosphere.

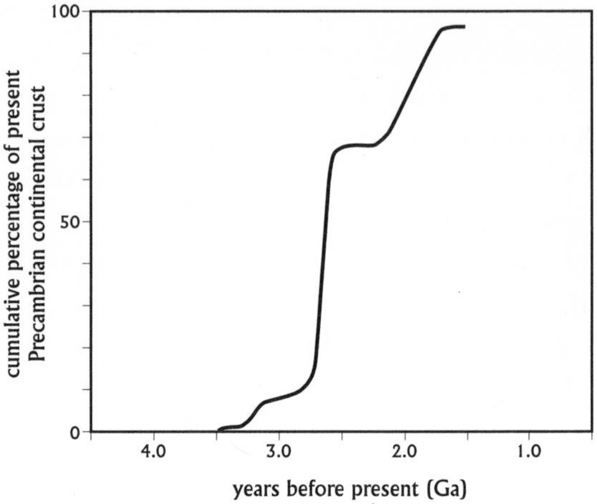

2.16 Approximate growth of the Earth's Precambrian continental crust, derived from the present age distribution of Precambrian rocks. Based on Taylor and McLennan (1985).

Changing Atmosphere

The impacts of large space objects would have created a sequence of peculiar atmospheres during the first 600–700 Ma of the Archean era. An impactor exceeding 440 km in diameter could have vaporized the planet's entire ocean and could have temporarily (for several months) saturated the air with hot rock vapor. A millennium of a dense hot steam atmosphere would follow, and its eventual condensation and rain out would recreate a new ocean (Sleep et al. 1989; fig. 2.17). There may have been only one such event

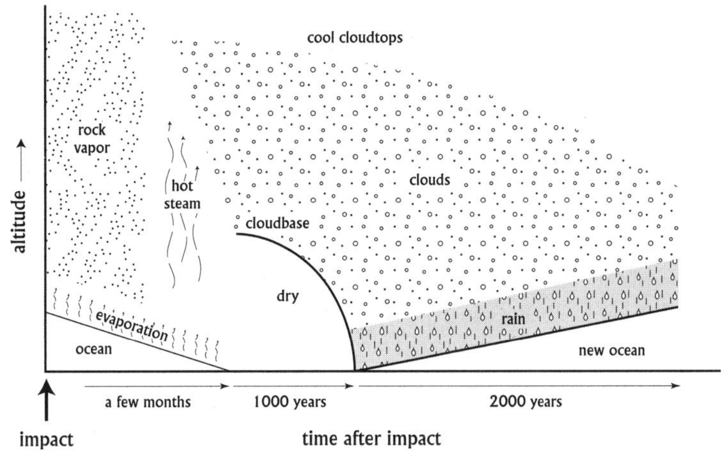

2.17 Consequences of an ocean-vaporizing impact following a collision with a huge (500-km-diameter) object during the Archean era. Based on Sleep et al. (1989).

before 4 Ga—the one that formed the Moon—but collisions with smaller bodies could have repeatedly vaporized a layer of surface water. A body with radius over 70 km could have vaporized the upper 100 m of the ocean, and such incidents could have persisted until 3.8 Ga ago (Lissauer 1999).

The atmosphere that emerged after these large impacts ceased was originally thought to be a highly reducing one, containing CH_4 ammonia (NH_3), H_2, and H_2O, and reactions among these compounds, energized by lightning, could have produced the molecules of life (see the next section). The current consensus is that methane and ammonia, neither of which is released by volcanoes, were not present in any significant amount, and that the first persistent atmosphere was dominated by CO_2, H_2O, and N_2 (Kasting 1993). Such a neutral or weakly reducing composition would make it difficult to explain how life's molecules could have been synthesized from atmospheric gases, but it would not have been inimical to the emergence of life. Levels of CO_2 in this early Archean atmosphere had to be much higher than those today in order to generate the stronger greenhouse gas effect needed to compensate for the lower luminosity of the young Sun.

Negative feedback among atmospheric CO_2, surface temperature, and silicate weathering acted (everything else being equal) as a planetary thermostat. Cooler temperatures would reduce the rates of weathering, and CO_2 emitted by volcanoes would accumulate in the atmosphere and increase its temperature; the resulting higher rates of weathering would then lower the atmospheric CO_2 levels. Although CO_2 levels were roughly 1,000 times the present concentration at the beginning of Archean era, they may have eventually declined to about 100 times the present concentration by 2.5 Ga ago and to roughly ten times by

600 Ma ago. Methane, a more potent greenhouse gas than CO_2, could have been the additional cause of surface warming once it was emitted in substantial quantities by the anaerobic decay of biomass or by methanogenic bacteria (Pavlov et al. 2000). As speculative and quantitatively uncertain as this grand scenario is, it is less contentious than the problem of atmospheric oxygen and ozone.

Cloud's (1968) conclusion that atmospheric O_2 levels rose appreciably only around 2 Ga has been widely accepted on the basis of three separate lines of geologic evidence. Banded iron formations (BIFs), chemical sediments containing alternating layers of oxidized ferrous (FeO) and ferric (Fe_2O_3, hematite) iron and iron-bearing silica, are abundant in successions older than 1.9 Ga but are rare in younger formations.[1] BIFs required only very low levels of oxygen, most likely provided by photosynthesis in shallow coastal basins to which the dissolved Fe^{+2}, which originated from midocean ridges, was transported through the anoxic ocean. But some Archean microenvironments, above all cyanonbacterial mats, could have had rather high O_2 levels.

In contrast, redbeds formed by subaerial oxidation are common in deposits younger than 1.9 Ga. Finally, abundant deposits of pyrite (FeS_2) and uraninite (UO_2), minerals that are unstable in oxidizing environments, overwhelmingly predate that time. Rye and Holland (1998) support this conclusion through their examination of paleosols, placing the onset of the dramatic rise of O_2 levels at about 2.25 Ga ago.

1. BIFs are the world's largest source of iron ores. Origins of the richest BIFs—Labrador Trough and Lake Superior District in North America, Minas Gerais in Brazil, Krivoy Rog in Ukraine, and Transvaal in South Africa—date to 2.7–1.9 Ga ago. BIFs contain about 1×10^{14} t Fe, enough for 10^5 years at the current rate of extraction.

The isotopic record of sedimentary sulfides bolsters this scenario, suggesting low concentrations of seawater sulfate and atmospheric oxygen in the early Archean 3.4–2.8 Ga ago, with the earliest O_2 increase starting in the early Proterozoic, 2.5 Ga ago (Canfield et al. 2000). And Canfield and Teske (1996) used phylogenetic and sulfur isotope studies to infer a significant evolutionary radiation of nonphotosynthetic marine sulfide-oxidizing bacteria between 1.05 and 0.64 Ga ago. These events were driven by a rise in atmospheric O_2 to 5–18% of today's level, a change that eventually triggered the evolution of animals. But scenarios involving relatively late oxygenation of the biosphere have not been universally accepted. The presence of superoxide dismutases, O_2-protective enzymes, in early prokaryotes indicates that some prephotosynthetic oxygen existed even before 3.5 Ga ago (Pierson 1994). Low concentrations of free oxygen were thus almost certainly a factor in early Archean evolution, and the low $^{13}C/^{12}C$ ratios that have prevailed since at least 3.5 Ga ago, as well as microfossils and stromatolites from the same period, confirm the great antiquity of early oxygenic photosynthesis (Towe 1994; Schopf 1999; Hofman et al. 1999).

The presence of steranes—sedimentary molecules derived from sterols whose biosynthesis requires molecular oxygen—in 2.7-Ga-old rocks indicates that by the late Archean there was enough atmospheric oxygen to support aerobic respiration by single cells (Knoll 1999). Geochemical evidence of microbial mats developed on the soil surface between 2.7 and 2.6 Ga ago and preserved in unusually ancient carbonaceous soils in South Africa (Watanabe et al. 2000) further strengthens the argument that an ozone (O_3) shield sufficiently effective to prevent damage by ultraviolet radiation was present at that time. Ohmoto et al. (1993) initially suggested that atmospheric O_2 levels have been within 50% of the present concentration since the earliest Archean, 3.8 Ga ago.

Ohmoto's later (1996) study of paleosols led him to conclude that the minimum O_2 pressure for the 3.0–2.2 Ga atmosphere was the same as for the post–1.9 Ga atmosphere. More recently Ohmoto also identified 2.3-Ga-old laterites, iron-rich deposits formed by the leaching of iron from upper layers of rock by organic acids and the depositing of the leachate as oxides in a deeper layer. There is also an alternate explanation for the formation of uraninite and pyrite, minerals unstable in the oxic atmosphere, whose presence in pre–2.2 Ga alluvial conglomerates has been offered as a proof of low O_2 levels (Ohmoto et al. 1993).

Suggested explanations for the timing of the oxygen rise at about 2 Ga ago include increased burial of organic carbon caused by intensified continental rifting and orogeny (see the previous section and chapter 5) and increased generation of O_2 by eukaryotic phytoplankton. Marine photosynthesis could become highly productive only with the emergence of a sufficiently effective ultraviolet screen (eliminating wavelengths below 290 nm), whose existence may have required just 1% of the present atmospheric O_2 level. The total mass of O_2 generated from the burial of carbon sediments is more than twenty-five times larger than the current atmospheric oxygen content, and most of the released gas had to be thus consumed by oxidation of hydrogen, carbon monoxide (CO), and, above all, H_2S and FeO (Warneck 2000). This was not a smooth process. The most marked jump in O_2 levels took place during the Permo-Carboniferous era as a consequence of the diffusion of vascular plants and enhanced burial of organic matter in swamps (Beerling and Berner 2000).

Uncertainties concerning atmospheric oxygen extend into the Phanerozoic era, as we lack a completely satisfactory explanation of the mechanism that has kept post-Carboniferous atmospheric oxygen levels within a remarkably narrow range. Concentrations below 10% would cause extinctions of large animals, those above 30% would ignite global wildfires. In a recent model Van Cappellen and Ingall (1996) propose the supply of phosphorus in oceanic photosynthesis as the key regulator. Because the burial of organic matter results in net oxygen production, the P availability for oceanic photosynthesis limits overall O_2 generation; should the atmospheric O_2 level fall, the retention of P by sediments would diminish, and a large flux of the element from the sediments to the ocean could intensify photosynthesis and hence the burial of organic matter—and restore the atmospheric O_2 level. But the model, as pointed out by Kump and Mackenzie (1996), has several weaknesses, and the actual regulation is much more likely to be a matter of multiple feedbacks.

Origin of Life

A disclaimer first: I will leave aside in my discussion of the origin of life both the possibility of instantaneous spontaneous generation of life from nonliving matter and accidental or directed panspermia (transmission of life from elsewhere in the universe). Those readers who want to amuse themselves with the first possibility (they are at least a century and a half too late to espouse it in earnest, as Redi and Pasteur refuted it convincingly) should consult Farley's (1977) historical review. Those who are inclined to believe the second possibility should read its original exposé by Svante Arrhenius (1908), a great Swedish chemist and one of the first Nobelians (1903), or its elaboration by Francis Crick and Leslie Orgel (1973, 1981).

The original theory credited the interstellar transfer of extraordinarily resistant bacterial spores—perhaps such as those produced by *Bacillus subtilis* (fig. 2.18)—with the implantation of life on the Earth. Crick and Orgel introduced the idea of directed panspermia, that is, a deliberate transfer of organisms by means of interstellar spaceship travel. Of course, if the first theory were true, there would be no discernible boundaries to life, as cryptobiotic

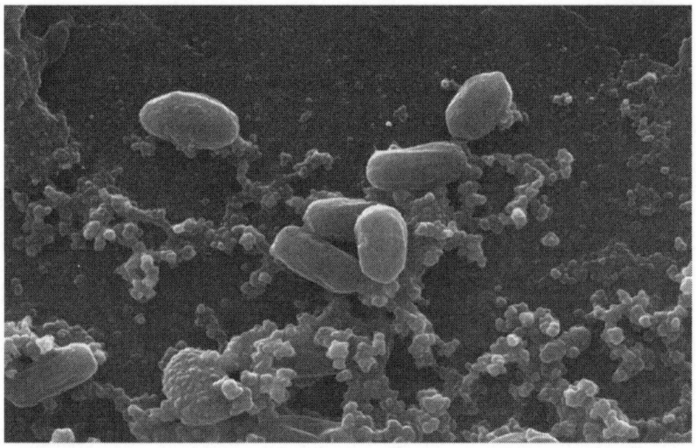

2.18 Scanning electron microscope image of *Bacillus subtilis* spores, perhaps the best candidates for interplanetary transmission of life. A detail from the image by Mathew Larson, Department of Civil and Environmental Engineering, University of Illinois, available at <http://www.itg.uiuc.edu/exhibits/iotw/archive/20010510-original.jpg>.

microorganisms would be drifting, albeit with exceedingly low densities, through the cosmic void. They would, after eons of travel through empty space and upon randomly encountering suitable planets, establish new harbors of concentrated life.

Several years later Hoyle and Wickramasinghe(1978) became enthusiastic advocates of the origin of life in molecules arriving from interstellar clouds. A key quote conveys their conviction: "The essential biochemical requirements of life exist in very large quantities within the dense interstellar clouds of gas. . . . We might speak of the Earth as having become 'infected' with life-forming materials . . . the picture is of a vast quantity of the right kind of molecules simply looking for suitable homes, and of there being very many suitable homes" (Hoyle and Wickramasinghe 1978, p. 157). There is no doubt about the universal presence of a variety of carbon-based polyatomic molecules and other compounds required for life (H_2O, NH_3) in space. Organic compounds have been repeatedly detected in interplanetary dust particles, meteorites, and comets. Delsemme (2001) argues that a rain of comets hitting the early Earth brought not only all of the planet's water but also organic molecules to jump-start its life. In Lovelock's (1988) words, a galaxy is "a giant warehouse containing the spare parts needed for life." Laboratory simulations of shocks created with high-energy lasers show that hydrogen cyanide, acetylene, and amine groups are produced by repeated shocking of methane-rich gas mixtures resembling cometary material (McKay and Borucki 1997).

And examination of the magnetic field in the Martian meteorite ALH84001 (the same one that was said to contain microfossils: see later in this chapter) showed that the temperature in its interior did not get hot enough to destroy bacteria (Weiss et al. 2000). Consequently I am not risibly dismissing the possibility of imported life, but given our ignorance, I am not going to add any speculative elaborations. As for the testaments to directed panspermia, they remain limited to UFO sightings and ET movies. But speculations cannot be left behind when dealing with a variety of generally favored theories concerning life's origin. Numerous writings review this fascinating, but also frustrating, field (Miller and Orgel 1974; de Duve 1991, 1995a; Deamer and Fleischacker 1994; Chyba and McDonald 1995; Davies 1998; Dyson 1999a; Woolfson 2000). As its theories are based on selective choices of realities and on credulity-stretching assumptions and extrapolations, it has not been difficult to point out their weaknesses, inconsistencies, and unwarranted conclusions.

The very basis of Oparin's (1924, 1938) durable theory of the heterotrophic origin of life in a prebiotic soup, certainly the most influential and the most cited proposal of its kind, was toppled by the realization that a highly reducing atmosphere, needed to produce the requisite organic compounds, did not exist on the Hadean Earth (see the previous section). Miller's (1953) classical experiment of producing several simple amino acids (mostly glycine and alanine), other organic acids (glycolic, lactic, acetic), and urea by electrical discharges in a mixture of methane, ammonia, hydrogen and water (claimed to replicate the Earth's early atmosphere) suffers from the same problem, and in addition his products were a mixture of R and L amino acids.

Thomas Gold's advocacy of the origin of life deep underground, where it could be protected from asteroid impacts and radiation and could draw on sulfur compounds and methane for its metabolism, has been given some support by recent discoveries of subterranean microbial life (for much more on these organisms, see chapters 3 and 6). But his claim that the deep, hot biosphere may extend down to 10 km and to temperatures between 150 and 300°C does not have many convinced adherents (Gold 1992, 1999). Similarly, high temperatures are a key obstacle to the acceptance of a theory placing the origin of life in hydrothermal vents active along ocean ridges (Corliss et al. 1981; for more on these vents and associated life, see chapters 3 and 6). Organic compounds are decomposed, rather than synthesized, and nucleic acids are rapidly hydrolyzed at vent temperatures (Miller and Bada 1988).

Christian de Duve (1991, 1995a, 1995b), who shared the 1974 Nobel Prize in medicine for his work on cell structures, believes that life began with a "protometabolism." Short polypeptids were formed from thioesters of amino acids in water solution in strongly acidic, sulfurous hot springs, volcanic lakes, or hydrothermal vent. A high-energy thioester bond played the role later assumed by ATP, and primitive chemical reactions eventually led to the

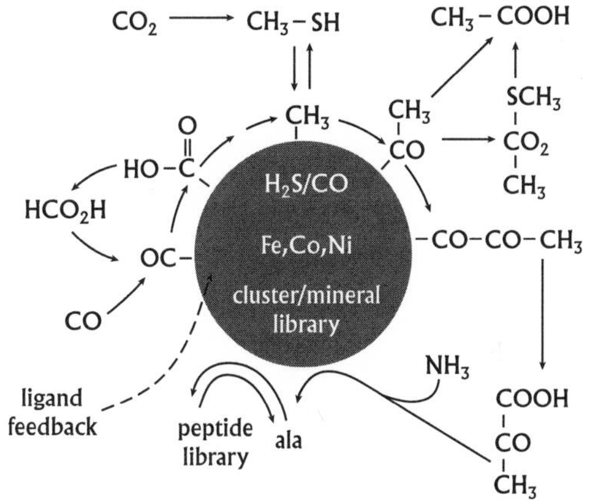

2.19 Reactions in the iron-sulfur world postulated by Wächtershäuser (1992).

synthesis of RNA and formation of cells. A similarly radical proposal has come from Wächtershäuser (1992), who believes that the first biosyntheses and polymerizations took place on the surface of pyrite. Simple organic compounds can become bonded to positively charged FeS_2 surfaces, and continuous production of pyrite by the reaction of FeS with H_2S provides the energy required for the eventual emergence of enzymes and nucleic acids from this archaic metabolism (fig. 2.19).

A key weakness of this intriguing theory is that the FeS/H_2S system does not reduce CO_2 to amino acids. Perhaps nothing summarizes the contentious nature of all of these theories better than Miller's reaction (recall that the relevance of his own pioneering work is now very much in doubt): he called the panspermia concept "a loser," the vent hypothesis "garbage," and dismissed the pyrite theory as "paper chemistry" (Horgan 1991). As in other cases of profound uncertainty, I find it counterproductive either to champion or to denigrate any particular theories of life's origin and remain content with pointing out their most obvious weaknesses.

And an ingenious solution to the most intractable of all chicken-egg problems — how to get proteins formed without DNA when DNA cannot be formed without proteins — is much less appealing on a closer examination than its supporters claim. The idea of the so-called RNA world has its origin in the discovery that some types of RNA can act as their own enzymes (Cech 1986). After RNA formed from ribose and other organic compounds, according to this line of reasoning, its molecules eventually became capable of self-replication and began synthesizing enzymes that helped the RNA to make double-stranded versions of itself, leading to the evolution of DNA (Gesteland and Atkins 1993; Woolfson 2000).

But how did RNA, a compound difficult to synthesize in a controlled environment, arise in the prebiotic world, and how could this now rather inept molecule, which copies itself only with much sophisticated laboratory help, reproduce itself so easily and so accurately (Shapiro 1988)? It also appears very unlikely that the components needed for the synthesis of ribonucleotides existed in more than trace amounts on the early Earth. In addition, ribose and other sugars have surprisingly short half-lives for decomposition at neutral pH (Larralde et al. 1995), and polyphosphates and activated phosphates were not abundant prebiotic compounds and hence would have been an unlikely prebiotic free energy source. For these and other reasons, Lazcano and Miller (1996) conclude that it is unlikely that RNA was the magical prebiotic molecule that single-handedly ushered in the DNA-RNA-protein world.

And in spite of remarkable progress in tracing the genetic history of life, we cannot be certain if its origins were hot or cold. RNA-based molecular phylogenies (for details see the next chapter) indicate that all thermophilic prokaryotes belong to the earliest branches of the evolutionary tree, and this fact has been used to argue that the common prokaryotic ancestor was hyperthermophilic, presumably a chemolithotroph growing on CO_2 (Woese 1987; Holm 1992; Szathmary 1997). Other molecular-phylogenetic studies point strongly to a heterotrophic origin of life based on inorganic nutrition, with photosynthesis and use of organic matter coming later (Pace 1997).

Another support for a thermophilic origin of life comes from the just-mentioned theory of a pressurized iron-sulfur world as well as from new experiments modeling volcanic or hydrothermal conditions. Wächtershäuser (1992) suggested that autotrophic metabolism of low-molecular-weight constituents could have developed fairly rapidly in an environment of FeS and hot magmatic exhalations. Cody et al. (2000) provided experimental support for this theory by showing that the synthesis of pyruvic acid, a key compound present in metabolic pathways producing amino acids and sugars, takes place in the presence of organometallic phases under high temperature (250°C) and pressure (50–200 Mpa).

But a cold origin for the earliest ancestor seems no less compelling, because the building blocks of those organisms would have been unstable at high temperatures: only later did extreme thermophiles evolve biochemical adaptations to protect their amino acids and nucleic acids (Forterre 1996). Galtier et al. (1999) challenged the thermophilic hypothesis of life's origin by deducing the G + C nucleotide content of the common ancestor and found it incompatible with survival at high temperature. But their conclusions were, in turn, challenged by Di Giulio (2000), who concluded that the late stage of genetic-code structuring took place at high temperatures.

As for the search for the universal ancestor of all existing life, Woese (1998a) argues that it could not have been a discrete entity, a single organismal lineage, but rather

a loosely knit, diverse conglomeration of primitive cells that evolved as a unit, and it eventually developed to a stage where it broke into several distinct communities, which in turn become the three primary lines of descent. . . . The universal ancestor is not an entity, not a thing. It is a process characteristic of a particular evolutionary stage.

There are not many solid facts to work with. We do not know what the features of the pre-DNA world were, how the nucleic acids got organized, how RNA begat DNA, how the first cells arose, whether they originated in hot or cold environments, and whether they were auto- or heterotrophic. Nor can we exclude the possibility that the first life forms were entirely destroyed by a catastrophic collision with a space body and that evolution had a second (even a third?) start. Yockey's (1995) observation that the current origin-of-life research floats improbably in midair like the roof of a house built by an architect of the Grand Academy of Lagado, whom Gulliver encountered during his travels, remains apposite. Starting with the roof has the advantage of quick and uninterrupted construction sheltered from weather, but the faith in the floating roof must come first, as the builders strive to extend walls toward the foundation in the real world of critical acceptance.

At least we know that all of the earliest organisms were prokaryotic, that they were anaerobic, that they had to cope with a high ultraviolet flux penetrating through the atmosphere devoid of O_3, and that their existence was repeatedly imperiled by the Earth's collisions with relatively large space bodies. And we get on a firmer ground when we begin dealing with the oldest preserved organisms of the Archean era — but even here there is no shortage of uncertainties.

Precambrian Organisms

Once we move past the Earth's violent Hadean era into the early Archean time we encounter the oldest preserved rocks — and hence a possibility of finding fossilized organisms that inhabited the earliest relatively stable biosphere. In spite of the obvious fact that the Precambrian (anything older than 544 Ma ago) fossil record is poor compared to the riches preserved from the Phanerozoic era (from 543 Ma ago to present), advances in fairly accurate rock dating and in isotopic, microscopic, and phylogenetic studies have greatly enhanced our knowledge of the earliest biosphere. Schopf (1983 and 1999), Schopf and Klein (1992), and Bengtson (1994) summarize this new understanding.

Not surprisingly, there is not any clear answer to the question about the oldest preserved evidence of life. I use "evidence" rather than "organism," because the latest claimant of the age record is an isotopic anomaly rather than a cell. Crustal rocks older than 3.5 Ga have undergone intense metamorphosis, and hence the only way to find geochemical evidence of the earliest life is to look within minerals that are resistant to metamorphic change. Mojzis et al. (1996) were the first to do so, using ion-microprobe measurements of isotope composition of elemental carbon inclusions within grains of apatite from the remains of 3.8-Ga-old metamorphosed sediments of the BIF from the Isua supracrustal belt in West Greenland and from a similar formation from nearby Akilia Island that may be older than 3.85 Ga.

The difference between the ratio of carbon's two stable isotopes in a tested sample, the less abundant ^{13}C and the dominant ^{12}C, and their ratio in a standard carbonate, $\delta^{13}C$ are used to trace carbon's origin. The element entering the biosphere from the mantle has $\delta^{13}C$ around –5.5, but because of the preferential photosynthetic uptake of ^{12}C, phytomass $\delta^{13}C$ is low, between –20 and –30. As these isotopic fingerprints are largely preserved in sediments, we can trace the origins of autotrophic carbon assimilation in graphitized organic matter rather reliably to at least 3.5 Ga ago, but concerns about the accuracy of dating the parental rock and the faintness of the isotopic signature make some researchers question the reliability of the 3.8-Ga dating.

The oldest identified microfossils belong to eleven taxa of cellular filaments (genus *Primaevifilum* being dominant) embedded in the early Archean chert (microcrystalline quartz) in northwestern Australia whose age was rather reliably dated to 3.465 Ga (Schopf 1993, 1999). Individual cells can be clearly seen in these 10–100 μm-long and about 1 to nearly 20 μm-wide filaments resembling today's Oscillatoriaceae, ancient and still common family of cyanobacteria. Their already relatively complex structure indicates a considerable period of previous evolution. Photoautotrophs, both anaerobic and oxygen-producing, were thus early colonizers of shallow seas, and they had to have evolved rapidly sometime between 3.9 Ga (the end of intense space encounters) and 3.6 Ga ago.

The rapidity of their evolution was dictated by two key factors. All of the ocean's water had to be recirculated through submarine vents once every 10 Ma, and any organic compounds would be altered or destroyed by temperatures surpassing 300 °C (Lazcano and Miller 1994). The limited stability of initial monomers and synthesized polymers was another obvious factor. Lazcano and Miller (1996) estimated that the development of a 100-kb genome of a DNA/protein primitive heterotroph into a 7,000-gene filamentous cyanobacteria would have required only 7 Ma.

The subsequent isotopic record indicates a remarkable constancy of photosynthetic CO_2 reduction. Long-term means of $\delta^{13}C$ at –25 for organic (reduced) carbon and at about 0 for inorganic (oxidized) carbon indicate that during the past 3.5 Ga about one-fifth of the element in near-surface sediments has come from organic matter (for more detail see chapter 5). Between 3.5 and 2 Ga ago, the bulk of this carbon fixation was accomplished by cyanobacteria-like organisms whose extremely conservative evolution led Schopf (1999) to coin the term "hypobradytely," an exceptionally slow rate of morphological evolutionary change characteristic of these prokaryotes.

New molecular evidence—namely, the extraction of biomarkers including 2α-methylhopanes derived from membrane lipids synthesized in large amounts only by cyanobacteria—pushes the existence of these autotrophs to at least to 2.7 Ga ago (Brocks et al. 1999; Summons et al. 1999). By 2.5 Ga ago cyanobacteria dominated the biosphere, and their oxygen production has been explained not just as a search for a ubiquitous electron donor, but as a way to poison the environment for competitors. By far the most notable fossilized accumulations of cyanobacteria, found throughout the long Precambrian era, appear as stromatolites. These accretions are organosedimentary structures produced in shallow seas by finely layered, mucilage-secreting, mat-building cyanobacteria, usually associated with photosynthesizing bacteria, which are lithified by grains of calcite precipitated onto the growth surfaces of the microorganisms (Grotzinger and Knoll 1999).

Stromatolites can still be found in shallow waters of the Persian Gulf, along the western coast of Australia, in Bahamian reef zones, and where their predators are limited, but their extent is minuscule compared to their Precambrian dominance. Moundlike accretions come in different

2.20 Stromatolites along the shore-ocean interface in Hamelin Rock Marine Nature Reserve, Shark Bay, Western Australia. Original color photograph available at <http://www.fotopositiv.com/photopositiv-pics/photopositiv-pics/australien.../c%20aus58-12.jp>

shapes, with the domical *Cryptozoon* from Australia's Shark Bay being the most often photographed stromatolite assemblage (fig. 2.20). Studies of modern stromatolites show growth patterns that may be applicable to ancient forms: periods of rapid sediment accretion, when gliding filamentous cyanobacteria are dominant, alternating with prolonged hiatal periods whose climax communities are dominated by endolithic coccoid bacteria that form thicker, lithified laminae (Reid et al. 2000).

A variety of extremely adaptable cyanobacteria species have been preserved in many marine, freshwater, and later also terrestrial environments, ranging from locations of low light and low temperature to hot rainless deserts and hot springs. This amazing cyanobacterial diversity—embracing extremes of biospheric salinity, acidity, high and low temperatures, desiccation, low oxygen, and even high radiation—is still very much evident in many surviving genera (see chapters 3 and 6). Two other common groups of Archean microorganisms included methanogens and sulfate-reducing bacteria. The oldest specimens of the second group were discovered by Rasmussen (2000), who believes that pyritic filaments, fossil remains of threadlike microorganisms in a 3.235-Ga-old deep-sea volcanogenic sulfide deposit of the Pilbara Craton of Australia, were thermophilic chemotrophic prokaryotes that inhabited sub–sea floor hydrothermal environments.

Indisputable fossil evidence of eukaryotic microalgae goes back to only 1.8 Ga ago, when they appear as simple large, balloon-like cells whose uncertain taxonomy earned them the name acritarchs. But this is not a reliable indication of eukaryotic origins, as nuclei are almost never preserved in fossilized tissues. Rather than being such relatively recent adaptations, eukaryotes may be only a bit younger than prokaryotic species (for more on phylogeny, see chapter 3). Indeed, the presence of steranes derived from cellular and membrane lipids in 2.7-Ga-old shale from Australia's Pilbara Craton makes it clear that some types of simple eukaryotes had existed for nearly a billion years before the time of their first extant fossils (Brocks et al. 1999). Martin and Müller (1998) argue that the first eukaryote evolved from an association between a methanogen and a hydrogen-producing bacterium in an oxygen-free and hydrogen-rich environment (see also chapter 3).

In any case, the current understanding of early eukaryotic evolution is in a state of flux (Roger 1999). Microfossils from the Roper Group in northern Australia show that cytoskeletal and ecological prerequisites for eukaryotic diversification were already in place nearly 1.5 Ga ago (Javaux, Knoll, and Walter 2001). The fossil record shows a slow diversification of algal eukaryotes between 1.8 and 1.1 Ga ago, followed by a rapid growth peaking at about 900 Ma ago, a shift best explained by the advent of sexual reproduction. About 300 Ma later, depending on which evidence to believe, either some marine ecosystems were already supporting the first microscopic animals or the stage was set for the rapid Cambrian eruption of highly organized and highly diversified terrestrial life.

I will describe these eon-making developments, which separated mostly microscopic, slowly evolving prokaryotic generalists from macroscopic, eukaryotic, rapidly developing short-lived specialists, in the next chapter. But before closing this one I will offer at least some comments and clarifications regarding a fundamental question: has this all been an inevitable process or an exceptional development? Is the Earth's biosphere just one of myriads of similar environments sprinkled throughout our galaxy and beyond, or is it a miraculous singularity?

Quantifying the Odds

Perhaps the most revealing way to illuminate the fundamental disparity between the positions of those who advocate life's inevitability and the conclusions of those who see the biosphere as an amazing and utterly unpredictable singularity is to devote the last two sections of this chapter mostly to quantifications of life's probability. Such quantifications can be arrived at in a number of ways. I will review some of the most publicized examples of the past two generations, ranging from probabilities so staggeringly high that they amount to absolute certainty of life as a universal phenomenon to odds so vanishingly low that they do not fundamentally change by moving the calculation boundaries from one star system to a swarm of galaxies.

Both sides have offered plentiful arguments and rebuttals, and the dispute has generated a rich scientific literature, as well as a much larger volume of pseudoscientific and pop-scientific writings. There is also a new field of study, exobiology, whose distinction lies in the fact that, as George Gaylord Simpson (1964) aptly noted decades ago, it has yet to demonstrate that its subject matter exists. Educational background and personal inclinations play undoubtedly major roles in an individual's choice of a camp. There has been no shortage of enthusiasts not just for extrasolar microbes but for extragalactic intelligence eager to intercept our signals, and the singularity argument resonates with many prominent scientists who cannot be accused of any creationist zeal. Given the profound uncertainties my best conclusion is simply to reiterate our ignorance both in regard to the origin of life on the Earth and to its existence elsewhere in the galaxy.

The Inevitability Argument

Christian de Duve is perhaps the most resolute proponent of the inevitability of ubiquitous life. According to de Duve, processes that generated life under the conditions that prevailed on the Earth some 4 Ga ago are chemical, and hence ruled by deterministic laws and bound to arise wherever and whenever similar conditions obtain (de Duve 1995b, p. 437): "All of which leads me to conclude that life is an obligatory manifestation of matter, bound to arise where conditions are appropriate . . . there should be plenty of such sites, perhaps as many as one million per galaxy. . . . Life is a cosmic imperative. The universe is awash with life." Similarly, David Schwartzman (1999, p. 3) believes that

the biosphere has evolved deterministically as a self-organized system given the initial conditions of the sun-Earth system. The origin of life and the overall patterns of biotic evolution were highly probable outcomes of this deterministic process. . . . Thus, the main patterns would be conserved if "the tape were played twice."

Although they are not often so forcefully stated, such beliefs in the deterministic (or near-deterministic) origin of life are not new, and during the past two generations there has been a substantial effort to quantify the odds of extraterrestrial life, or better yet of intelligent civilizations. In 1961 Frank Drake suggested that the probability of communicative civilizations in the galaxy could be calculated as the product of seven variables (Drake 1980). The first three are astronomical: the rate of star formation, the fraction of stars with planetary systems and habitable zones, and the average number of planets per planetary system in habitable zones. The next two are evolutionary: the fraction of the habitable planets where life arises and the fraction of the latter where intelligence evolves. And the last two combine technical and social aspects: the fraction of intelligent species capable of interplanetary communication and the lifetime of such communicating civilizations.

A number of scientists estimated all of these factors shortly after Drake published the equation. Comparisons of their values show a high degree of similarity: prevailing ranges are 50–100% of stars assumed to form planets, 0.3–1 planet per star in habitable orbits, 20–100% of these developing some form of life, 10–100% of life-bearing planets evolving intelligence, 10–50% of intelligent cultures communicating interstellarly, and 10^6–10^8 years being their lifetime (Dick 1996). Not surprisingly, with such estimates the order of magnitude for the number of communicating civilizations in galaxy would be 10^6, a conclusion most effectively advocated since the early 1960s by Carl Sagan.

A set of more skeptical estimates by Rood and Trefil (1981) came up with only 0.003 communicative civilizations per galaxy—but I am not at all persuaded by any of these calculations. I agree entirely with Dyson (1979), who rejected "as worthless all attempts to calculate from theoretical principles the frequency of occurrence of intelligent life forms in universe. Our ignorance of the chemical processes by which life arose on earth makes such calculations meaningless" (p. 209). Estimates of high probabilities of not just any forms of life but of highly developed, interstellarly communicating civilizations beg the question famously asked by Enrico Fermi in 1950: "Where are they?" Half a century later the question can be repeated with even stronger incredulity, because the search for extraterrestrial intelligence, carried out since the 1960s by listening to signals from space and excellently chronicled by Dick (1996) and Wilson (2001), has not come up with any candidates—and it may never find any (Cohen and Stewart 2001). But during the 1990s a new endeavor revived the enthusiasm of believers in many inhabited worlds as the search for extrasolar planets achieved its first successes and then rapidly expanded to find many more possibilities.

Extrasolar planets cannot be photographed, as the scattered light from the central star overwhelms the minuscule share, typically less than 10^{-9}, of the light reflected by one of them. Detection methods used to identify extrasolar planets have included Doppler spectroscopy, which measures the periodic velocity shift of the stellar spectrum caused by an orbiting giant planet; astrometry, which observes the periodic wobble of a parent star induced by a planet; and photometry (Boss 1998). The last method, measuring the periodic dimming of a star caused by a planet passing in front of it, is the only practical way to find Earth-sized planets in the continuously habitable zone (NASA 2000).

The first detection of an extrasolar planet was made by Wolszczan (1994), who reported a body orbiting a pulsar (an extinct star) and causing periodic variation in the star's period. By 2000, astronomers had discovered more than fifty extrasolar planets, most them massive (Jupiter-sized or larger), orbiting either too close to their stars (some with orbits lasting just a few days) or in ellipses elongated too much to support life (Burrows and Angel 1999; Schneider 2000; Doyle et al. 2000). Williams et al. (1997) argued that rocky moons orbiting some of these newly discovered giant planets are also possible candidates for habitable bodies, providing, of course, that they are massive enough (> 0.12 Earth mass) to retain substantial and long-lived atmospheres.

The much-publicized claim that Martian meteorite ALH84001, found in Antarctica, contains microfossils providing strong evidence that Mars had and indeed may still have microbial life (McKay et al. 1996) appeared to be another piece of good news for the advocates of extraterrestrial life. NASA was a heavy promoter of this possibility, but the claim has not fared well under subsequent scientific scrutiny, as it presupposed the existence of cells much smaller (volume less than 1/1000 that of parasitic mycoplasma, the smallest living organism) than anything compatible with cellular organization (Kerr 1997, 1998). But NASA continues its quest: its Kepler mission was designed to search for habitable planets at nearby main-sequence stars.

Nothing has changed for those who advocate a high probability of extraterrestrial life. In one of the most recent restatements of their hopes, Aczel (1998) concluded that even when one assumes that the probability of life on any single planet orbiting within its star's habitable zone is as low as one in one trillion, the compound probability that life exists on at least one other planet is a virtual certainty, because with 300 billion stars per galaxy and 100 billion galaxies in the universe, there are so many places to look.

The Most Improbable Singularity?

Jacques Monod, who shared the 1965 Nobel Prize for physiology for explaining the genetic regulation of protein synthesis in cells, asked in his influential book *Chance and Necessity* what the chances of life's appearance on the Earth were before the event. He concluded that the a priori probability of the Earth's biosphere was virtually zero. This statement, he knew, would be anathema to modern science, which does not deal with unique occurrences (Monod 1971):

Now through the very universality of its structures, starting with the code, the biosphere looks like the product of a unique event. . . . The universe was not pregnant with life, nor the biosphere with man. Our number came up in the Monte Carlo game. Is it any wonder if, like a person who has just made a million at the casino, we feel strange and a little unreal? (p. 145)

Monod argued that the biosphere is a particular phenomenon that is not made of a predictable class of objects or events and that its existence is compatible with first principles (it has a right to exist) but not deducible from these principles (it has no obligation to exist). Science can make the biosphere explicable, but the biosphere itself is unpredictable. Theodosius Dobzhansky, one of the founders of modern genetics, echoed Monod's arguments, concluding that a biologist living 50 Ma ago could not have predicted the evolution of humans and chastised natural scientists who have been unwilling to consider "that there is, after all, something unique about man and the planet he inhabits" (Dobzhansky 1972, p. 173).

Ernst Mayr, one of the leading biologists of the twentieth century, also stressed the incredible improbability of the evolution of intelligent life on Earth (Mayr 1982, pp. 583–584): "A full realization of the near impossibility of an origin of life brings home the point how improbable this event was. This is why so many biologists believe that the origin of life was a unique event . . . no matter how many millions of planets in the universe." And writing nearly two decades after Monod, Gould (1989, p. 14) offered pretty much the same arguments from a paleontologist's perspective, seeing evolution as a

staggeringly improbable series of events, sensible enough in retrospect and subject to rigorous explanation, but utterly unpredictable and quite unrepeatable. . . . Wind back the tape of life . . . let it play again from an identical starting point, and the chance becomes vanishingly small that anything like human intelligence would grace the replay.

Clearly, these conclusions in favor of the Earth's singular, or highly improbable, or at least astonishingly contingent and hence basically unrepeatable life offered by thoughtful scientists of various backgrounds cannot be dismissed as the wishful thinking of creationists. Probability calculations offered to demonstrate either the virtual impossibility of life's random origin or to bolster the theory of life's extraterrestrial provenance are easier to criticize. Perhaps the most famous of these quantifications originated with Fred Hoyle, one of the most accomplished and controversial astronomers of the twentieth century (Hoyle 1980; 1982).

Hoyle (1980) pointed out that the chance of producing a functioning enzyme from the random ordering of the twenty amino acids is, obviously, 1 in 10^{20}, but the chance of getting all of about 2,000 enzymes through a random assembly is only 1 in $(10^{20})^{2000}$, or $10^{-40,000}$, "an outrageously small probability that could not be faced even if the whole universe consisted of organic soup." But this theoretical calculation is questionable. Did the earliest life require the concerted presence of all 2,000 enzymes? Would not their assembly become vastly more probable after an initial pattern was established? And was not evolutionary selection acting as a selective sieve rather than in a manner of random shuffling? Hoyle then considered the essentially unchanged composition of enzymes across nearly 4 Ga from the most primitive single cells to humans. He concluded that the only way to allow for the miracle of life's information to arise is by giving the universe a history much longer than ten billion years (Hoyle 1982).

When Hoyle published his first low-probability estimate, he was not aware of calculations done by Harold Morowitz during the late 1960s. Morowitz (1968) calculated the probability of a living cell's spontaneously forming in an ocean containing all the monomer units necessary for biomass construction at about 1 in 30 billion. Again, this kind of calculation demolishes an entirely fictitious target, presupposing that life would arise by a random construction of a complete cell, a most unlikely course. There are other calculations of this kind, demonstrating vanishingly low probabilities of random generation of life from inorganic constituents, including those by Yockey (1995), but all are just theoretical musings, not replications of real evolutionary histories.

And there is also an in-between position, allowing for possibilities of extraterrestrial life, but still recognizing the likely uniqueness of the Earth's biosphere. Drawing on a variety of astronomical, geological, and biological perspectives, Ward and Brownlee (2000) pled against the continued marginalization of our planet and its place in the universe. They argued that although primitive bacterial life may well exist elsewhere in the universe, the chances of the type of complex life that evolved on our planet existing elsewhere are quite small. Similarly, Crawford (2000), acknowledging the failure to discover any extraterrestrial intelligence and focusing on the long time lag between the first prokaryotic life and the appearance of animals, concludes that the transition to multicelled animals may take place on only a tiny fraction of planets inhabited by single-celled organisms.

I have contrasted the two views on the probability of life in the universe to show how irreconcilable they seem to be. In the absence of any clear-cut evidence, my personal preference is very much in favor of seeing complex life as an incredible singularity, but I find extended arguments for or against this position tedious and unproductive. We should be stressing again and again our ignorance of so many relevant matters, ranging from the conditions and compounds on the early Archean Earth and the actual process that led to the emergence of life, to the frequency of potentially habitable planets and the chances of prokaryotic life evolving into highly encephalized beings. As long as our ignorance of these matters persists, there is room for antagonistic conclusions—but not for any confident opinions. This is fortunately no longer the case with examining life's metabolic paths and species, as scientific advances of the last century have made it possible to appreciate their remarkable complexity and stunning diversity.

3

Life's Diversity and Resilience

Metabolisms, Species, Catastrophes

The diffusion of life is a sign of internal energy—of the chemical work life performs—and is analogous to the diffusion of gas.
Vladimir Ivanovich Vernadsky, *Biosfera*

So then the new-born earth first reared her shrubs and herbage; then in turn in order due did bring to birth the tribes of mortal creatures, risen in many ways by means diverse.
Lucretius, *De rerum natura*

At this point it might be appropriate to have a chapter about life's evolution, a fairly detailed chronological account of the biosphere's complexification that would pause at major milestones and emphasize the emergence of first organisms, development of photosynthesis, appearance of first multicellular bodies, diffusion of land plants, diversification of animals, and rise of mammals and hominids. But the emphasis of this chapter will be not chronological and paleontological, but functional. The first reason for this choice is a surfeit of recent books on evolution. Those published since 1995 cover basic processes (selection, competition, symbiosis) in all-encompassing narratives (Fitch and Ayala 1995; Bell 1997; Fortey 1997; Margulis 1998; Schwartz 1999; Smith and Szathmary 1999). Others offer more detailed examinations of the quality and adequacy of the fossil record (Donovan and Paul 1998; Gee 1999; Schopf 1999) and of the rise and diffusion of major phyla and classes (Chatterjee 1998; Morris 1998; Zimmer 1998). Even more importantly, I do not want to get enmeshed in contrasting and evaluating a great deal of uncertain and often contradictory evidence, some of which may be disproved in a matter of years.

A recent reexamination of the overall quality of the fossil record does not support the logical expectation of diminished quality of preserved species backward in time for the Phanerozoic eon. Benton et al. (2000) based this conclusion on congruencies between stratigraphy and phylogeny: although ancient rocks preserve less information than do the more recent ones, scaling to the stratigraphic level of the stage and the taxonomic level of the family shows a uniformly good documentation of life during the past 540 Ma. Proterozoic evidence is a different matter, as reliable dating of many earlier evolutionary milestones remains elusive, and hence an account stressing particular chronologies, rather than processes, may rapidly become indefensible. This reality is perfectly illustrated by recent claims regarding the origin of animals.

For decades a seemingly sudden (in geological sense) appearance of new life forms has been dated to the early Cambrian, the period whose beginning is usually given as 570 Ma bp. Standard accounts had this Cambrian explosion beginning about 544 Ma ago, with only a few tens of millions of years elapsing before virtually all of the animal lineages known today showed up in the fossil record (McMenamin and McMenamin 1990). Discovery of the Ediacaran fauna (soft-bodied organisms named after the site of their first finds, the Ediacara Hills of South Australia) pushed that Great Divide back just a few million years. New measurements published in 1993 shifted the onset of the explosion to 533 Ma bp and set its duration to just 5–10 Ma (Bowring et al. 1993). Grotzinger et al. (1995) concluded that simple discoid animals might have appeared at least 50 Ma earlier.

An attempt to estimate the divergence time for main animal phyla by comparing genes produced dates up to 1.2 Ga ago (Wray et al. 1996). Ayala et al. (1998) rejected this estimate for methodological reasons and suggested a date of about 670 Ma ago for the divergence of protostomes (arthropods, annelids, and mollusks) and 600 Ma ago for the appearance of chordates. Then the discovery of tiny fossil animal embryos and adult, or embryonic, sponges in Doushantuo phosphorites in southern China placed these animals at 570 Ma bp (Xiao et al. 1998; Li et al. 1998). Doushantuo sponges are tiny, a mere 150–750 μm in maximum dimension, and other animal fossils from the site, showing cells in a cleavage embryo, are no more than 0.5 mm in diameter. The exquisite preservation of such small features is due to calcium phosphate. Later in 1998 came a claim that burrowing wormlike animals moved through shallow sea sediments 1.1 Ga ago (Seilacher et al. 1998). The wiggly grooves found in a central Indian sandstone may be of inorganic origin, but the gap of 400 Ma between the Indian find and other well-identified worm burrowings may be explained by a later extinction of the earliest species. Plausible dates of animal origins thus differ by more than 500 Ma!

Major evolutionary milestones, be they generally accepted or very much contested, will be noted in the context of explaining life's metabolic arrangements, describing the increasingly more sophisticated attempts at reconstructing the tree of life, and assessing the impact of recurrent planetary catastrophes. As explained in chapter 2, biochemical similarities and differences of metabolic pathways define two fundamentally distinct modes of the planet's life—autotrophic and heterotrophic metabolism—and their various modifications, and enormous specific differentiation encompassed by these basic functional modes is the most obvious demonstration of life's variability.

Recent genetic analyses have made it possible to trace the kinships of all organisms at close to their evolutionary beginnings. Explanation of these advances and uncertainties will take up the second part of this chapter. Finally, a modern look at life's diversity must deal with the role of recurrent catastrophes whose global impact has shaped the development, diffusion, and survival of species no less than have the eons of random mutation. Studies of these evolutionary discontinuities have yielded many fascinating, as well as contentious, conclusions, and a better understanding of the frequency, scale, and duration of these events — ranging from relatively abrupt climatic changes to globally devastating encounters with extraterrestrial bodies — is imperative for appreciating their role in the biosphere's past and future evolution.

Metabolic Paths

Metabolic possibilities arise from permutations of key aspects of cellular nutrition, energy sources driving the process, substrates that donate electrons, and compounds that supply carbon. The nomenclature devised for microorganisms at the Cold Spring Harbor Symposium in 1946 has been extended to all life forms, and it provides accurate definitions by combining the prefixes "photo" and "chemo," used to identify energy sources, with the terms "litho" and "organo," which describe electron donors (Lwoff et al. 1946). Organisms energized by solar radiation (able to convert electromagnetic energy into the high-energy phosphate bonds of ATP) are phototrophs; those tapping chemical energy, be it of simple inorganic compounds or organic macromolecules, are chemotrophs. Organisms that derive electrons from elements (H_2, S) or simple inorganic compounds (H_2O, H_2S) are lithotrophs, whereas those using complex organic substrates — be they proteins and lignin in dead biomass being decomposed by bacteria or fungi or carbohydrates in grasses grazed by ungulates or grains eaten by people — are organotrophs. Land plants, algae, phytoplankton, and cyanobacteria are thus photolithotrophs; so are green and purple sulfur bacteria, which use H_2S rather than water as their source of electrons. Photoorganotrophs require solar energy but use organic compounds: purple nonsulfur bacteria can switch to this mode from anaerobic photolithotrophy; when they do so they become heterotrophic, whereas plants and cyanobacteria, deriving their carbon from CO_2, are autotrophic.

Most bacteria and all fungi as well as all animals are chemoorganotrophs, gaining ATP by oxidation-reduction reactions. Higher organisms gain ATP by using complex organic substrates as electron donors (as well sources of carbon) and oxygen as electron acceptor. Such common bacterial genera as *Bacillus* (see fig. 2.18 for one of its common species) and *Pseudomonas* do the same, but many chemoorganotrophic bacteria use nitrates or sulfates as their electron acceptors. And there are also chemolithotrophs, remarkable organisms that thrive in the complete absence of any light and of all organic matter: all they require is the presence of CO_2 and an electron acceptor (most commonly O_2, but some use CO_2 or nitrate [NO_3^-]), together with an oxidizable element (H_2, Fe^{2+}) or inorganic compound (H_2S, NH_3, nitrite [NO_2^-]). Nitrifying bacteria that oxidize ammonia to nitrites and nitrites to nitrates, sulfur bacteria that oxidize sulfur or sulfides, and methanogenic archaea are the biosphere's most important chemolithotrophs. Their metabolism is indispensable for the continuous functioning of biogeochemical cycles of carbon, nitrogen, and sulfur on scales ranging from local to global (see chapter 5).

Photosynthesizers

No other process has shaped the evolution of the biosphere as much as oxygen-producing photosynthesis, whose origins go back to the early Archean eon. Excellent accounts of advances in the far-from-accomplished challenge of unraveling the details of the photosynthetic process can be found in Rabinowitch (1945), Myers (1974), and Somerville (2000). The last author calls the mechanism underlying the photolysis of water "the central enigma of photosynthesis," whose solutions remains a challenge for the twenty-first century. (For more on the water-breaking, oxygen-evolving manganese cluster in chloroplasts, see Yachandra et al. 1993 and Szalai and Brudvig 1998.)

We may be never able to reconstruct reliably any events predating 4 Ga, but as already noted (chapter 2), the first preserved traces of photosynthetic activity are as old as the oldest identified rocks from 3.5 to 3.8 Ga bp (Schopf 1994). If the early Archean atmosphere had no oxygen, then the first phototrophic prokaryotes had to assimilate carbon in ways akin to those of the still-extant anoxygenic bacteria, which were pushed into marginal niches by high levels of atmospheric O_2 resulting from the later evolution of water-splitting photosynthesis. Both obligatory and facultative anoxygenic phototrophs can be found in four bacterial phyla (Blankenship et al. 1995; Peschek 1999).

Chromatiaceae (purple sulfur bacteria) and Chlorobiaceae (green sulfur bacteria)—which thrive particularly in sulfur springs, with the green *Chlorobium* often appearing in mass accumulations—can use H_2 as their hydrogen donor but prefer to use reduced sulfur compounds, mostly H_2S and thiosulfates ($S_2O_3^{2-}$). Sequencing of genes involved in photosynthesis and phylogenetic analyses of the major groups of photosynthetic bacteria indicate that green nonsulfur bacteria (heliobacteria) are the last common ancestors of all photosynthetic lineages (Xiong et al. 2000). Rhodospirillaceae (purple nonsulfur bacteria displaying vivid colors ranging from the red of *Rhodospirillum rubrum* to the brown of *Rhodopseudomonas sulfidophila*) and Chloroflexaceae (green gliding bacteria, the earliest diverging branch of the bacterial tree) rely almost exclusively on hydrogen as their donor of electrons.

All phototrophic bacteria can photoassimilate such simple organic substrates as acetate or butyrate, but CO_2 is their usual source of carbon. Ancient anoxygenic phototrophs would have also needed some protection against relatively high levels of ultraviolet radiation: a few millimeters of translucent mud or dead organic matter would suffice. During the late 1970s came the discovery of bacteria that perform anoxygenic photosynthesis under aerobic conditions. These organisms grow heterotrophically, but they can use their bacteriochlorophyll to harvest light as a source of auxiliary energy when their respiration is suppressed because of insufficient amounts of appropriate substrate (Harashima et al. 1989). Kolber et al. (2000) used a newly developed infrared fast repetition rate fluorometer to search for the evidence of bacterial photosynthetic electron transport and discovered that aerobic bacterial photosynthesis is widespread in tropical surface waters of the eastern Pacific Ocean as well as in temperate coastal waters of the northwestern Atlantic.

But if the Archean atmosphere already contained more than a trace of oxygen (see chapter 2), then the earliest phototrophs might have been more like today's cyanobacteria. Many strains of these prokaryotes can perform anoxygenic, bacterial-type photosynthesis in hypoxic or anoxic sulphide-rich environments (hot springs, ocean sediments) and then shift to water-splitting, O_2-releasing photosynthesis in aerobic niches (Peschek 1999; Whitton and Potts 2000). As noted in chapter 2, oxygen-evolving cyanobacteria, present in the fossil record from about 2.8 Ga ago, were thriving in the Archean ocean, forming matlike growths in shallow marine environments and leaving behind fossil stromatolites. Old doubts about biotic origins of stromatolites were revived by an analysis that showed the sedimentary laminae obeying the same power law over three magnitudes in size (Grotzinger and Rothman 1996). Although abiotic processes may explain the existence of stromatolites, there is still the preponderance of evidence to argue that most of them are of biogenic origin rather than being abiotic fractal frauds resembling living organisms (Walter 1996).

The shift to oxygenic, water-cleaving photosynthesis freed bacteria from their dependence on limited amounts of reduced S, Fe^{2+}, Mn^{2+}, and H_2 and CH_4 and provided photosynthesizers with an unlimited source of electrons and protons. Des Marais (2000) estimated that microbes dependent on hydrothermal energy could sustain an annual fixation of less than 25 Mt C, whereas oxygenic photosynthesizers, able to tap a virtually unlimited supply of hydrogen from water, could eventually fix about 100 Gt C a year. Dismukes et al. (2001) suggested that bicarbonate (formed by dissolution of CO_2) was the thermodynamically preferred reductant before water in the evolution of oxygenic photosynthesis and found that Mn-bicarbonate clusters are highly efficient precursors for the assembly of the tetramanganese oxide core of the water-oxidizing enzyme.

Descendants of the earliest cyanobacteria continue to fill almost every aquatic and terrestrial niche (Rai 1990; Bryant 1994; Whitton and Potts 2000; fig. 3.1). They are abundant as both unicellular and colonial forms, the latter shaped as slender filaments (*Anabaeana, Nostoc*), as strings or clumps whose massive blooms are discernible from space (*Trichodesmium*), as spheres bonded by secretions (*Gloeocapsa*), or as scum layers (*Lyngbya*). They also live symbiotically with numerous species of protists, sponges, worms, and aquatic and land plants (*Anabaena* symbiotic with floating *Azolla* ferns, *Sphagnum* moss, tropical *Gunnera* and *Cycas*). Although clearly prokaryotic, cyanobacteria have an elaborate system of photosynthetic lamellae (analogous to thylakoid membranes of plants) containing pigments that color the cells not just bluish-green but also green (edible *Spirulina*), yellow, red (*Oscillatoria*), and even deep blue.

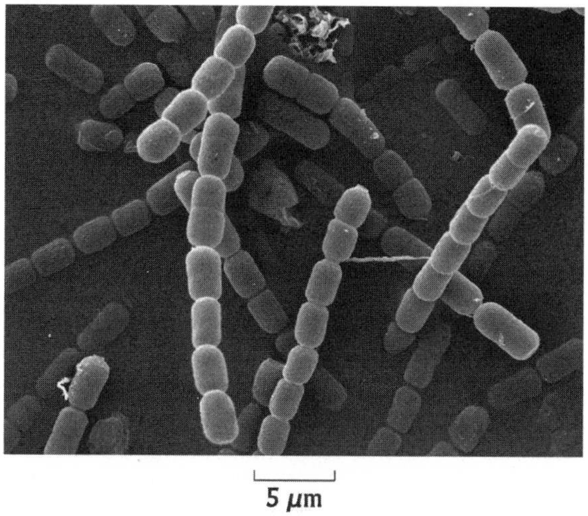

3.1 *Anabaena* is one of the most common filamentous cyanobacteria found in freshwater ponds and lakes. Electron microscope image courtesy of James M. Ehrman, Digital Microscopy Facility, Mount Allison University, Sackville, New Brunswick.

Cyanobacteria are also able to fix atmospheric nitrogen, but as nitrogenase, the enzyme catalyzing this conversion, cannot tolerate O_2, they do so in heterocysts, cells with thickened walls creating anaerobic microenvironments. Nitrogen-fixing ability makes cyanobacteria well adapted for oligotrophic conditions of surface ocean waters. Unicellular *Prochlorococcus,* the smallest and most abundant photosynthesizer in the world's oceans, was discovered only in 1988 (Chisholm et al. 1988), and it contributes 30–80% of the total primary production in oligotrophic waters (fig. 3.2). Colonial *Trichodesmium,* which can produce toxic blooms, is plentiful in tropical and subtropical seas (Ferris and Palenik 1998; Capone et al. 1997); the Red Sea owes its name to its colored tufts or puffs.

Ignorance about particulars of photosynthetic evolution extends through the late Archean to the early Proterozoic. *Grypania,* the first fossil alga to grow in long cylindrical coils, dates to 2.1 Ga ago, which means that

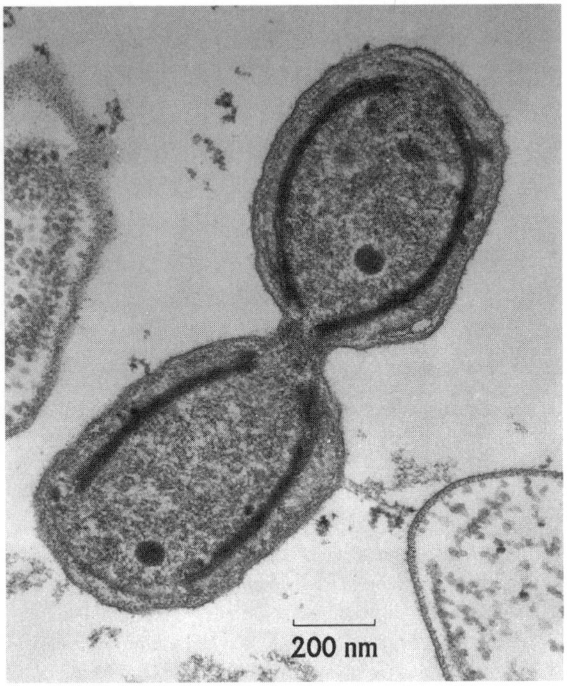

3.2 Electron micrograph of a dividing cell of *Prochlorococcus,* the smallest and most abundant photosynthesizing cyanobacterium, which dominates primary production in oligotrophic oceans. Photo courtesy of Claire Ting, Department of Biology, MIT.

eukaryotic phototrophs had to evolve sometime during the preceding 500 Ma (Runnegar 1994). One solid phototrophic ability of plants is not their original evolutionary achievement — but rather an import: chloroplasts of green algae, red algae, and glaucophytes were derived directly from cyanobacteria through a primary endosymbiosis (for more on this process, see chapter 8).

The fossil record shows microbial mats declining and green and red algae increasing in abundance only about 1–0.9 Ga ago. New molecular evidence suggests that the colonization of continents by eukaryotes was proceeded by a symbiosis between phototrophs and fungi and that land plants may have appeared by 700 Ma, predating the great Cambrian radiation of animals (Heckman et al. 2001).

Both comparative morphology and rRNA sequencing (explained in the next section) strongly support the monophyletic origin of land plants (Bateman et al. 1998). They indicate that both of the two great classes of land plants — bryophytes (nonvascular plants without distinctive water-conducting tissues, including liverworts, hornworts, and mosses) and tracheophytes (vascular plants including lycopods, horsetails, ferns, and seed plants) — evolved from charophytes, freshwater green algae whose fossils are documented from more than 600 Ma ago. Tracheophytes use phloem, a complex single cellular conduit system, to distribute large amounts of sugars and other photosynthates from their places of origin to such sinks as developing leaves, flowers, or root tips, where they are used to build new tissues. Plasmodesmata are cytoplasmic channels between cells that readily transmit small solutes (inorganic ions and sugars) but restrict the movement of large proteins, mRNA, and organelles (Pickard and Beachy 1999).

The earliest land plants may have appeared during the mid-Ordovician, after 450 Ma bp. Comparisons of three mitochondrial introns — pieces of silent DNA within genes — were used to identify liverworts as the earliest land plants (Qiu et al. 1998). Introns are present in all vascular plants, hornworts, and mosses but are entirely absent in liverworts and algae. Fossils of *Cooksonia,* the first accepted vascular plant, and clubmosses go back to the mid-Silurian, after about 430 Ma bp. All early plants were either leafless or had only small spine-like tips, and leaves with a broad lamina evolved only in response to a pronounced decline of atmospheric CO_2 during the Devonian period between 410–363 Ma ago (Beerling, Osborne, and Chaloner 2001). Discovery of fossilized hyphae and spores of glomalean fungi in 460-Ma-old Ordovician chert (Redecker et al. 2000) indicates that fungi were present before the first vascular plants arose. This indicates that fungus-plant symbioses, endophytic and mycorrhizal, were essential in aiding the invasion of plants into nutrient-poor and desiccation-prone environments (Blackwell 2000).

Not long after their emergence, during the early Devonian period from about 410 Ma ago, land plants underwent such a rapid diversification and diffusion that 50 Ma later, by the end of the Devonian period, they were the dominant photosynthesizers on land (Gensel and Andrews 1987; Gray and Shear 1992; Kenrick and Crane 1997; Bateman et al. 1998). This evolutionary spurt was quite remarkable, because by its end land plants had solved all challenges associated with terrestrial life: besides laminate leaves with stomata, they also acquired stems with complex fluid transport, structural tissues that enabled them to reach unprecedented heights, roots for respiratory exchange of gases, and specialized sexual and spore-bearing organs and seeds. Although cellulose was initially just one of many structural materials—including silica and chitin, an amorphous polysaccharide—the evolution of terrestrial plants selected for its universal presence, and this microfibrillar polysaccharide now accounts for about half of all phytomass (Duchesne and Larson 1989; fig. 2.3).

Gymnosperms—plants with seeds exposed on the surface of cone scales, including conifers, cycads, gingkophytes, and gentophytes—were the dominant land plants 200 Ma ago. The first described angiosperm (flowering plant, reproducing by forming ovules in an enclosed cavity) is *Archaefructus* from the Upper Jurassic strata (about 140 Ma old) in China's Lianoning (Sun et al. 1998). Angiosperms have been dominant since the mid-Cretaceous, about 90 Ma bp. Analyses of DNA sequences of plastid and nuclear genes from 560 species produced a well-resolved and well-supported phylogenetic tree and demonstrated that *Amborella trichopoda,* a shrub of the monotypic family from New Caledonia, is the earliest extant angiosperm and that water lilies (Nymphaeales) are the next diverging lineage (Soltis et al. 1999; fig. 3.3). This conclusion was confirmed by another set of mitochondrial, plastid, and nuclear sequences of all gymnosperm and angiosperm lineages (Qiu et al. 1999; Kenrick 1999).

Once the basics of photosynthesis became understood during the nineteenth century, the standard scientific description saw the process as CO_2 fixation and O_2 evolution. As Tolbert (1997) noted, it will not be easy to overcome this 150-year-old dogma. In reality, photosynthesis is a complex process of O_2 and CO_2 exchange energized by the absorption of specific wavelengths of solar radiation (Hall et al. 1993; Lawlor 1993; Baker 1996; Raghavendra 1998; Hall and Rao 1999). Excitation of pigment molecules (mainly chlorophylls a and b, bacteriochlorophyll, and carotenoids) takes place in two different reaction centers located in photosynthetic (thylakoid) membranes inside chloroplasts in specialized leaf (in some species also stem) cells. Both chlorophyll a and b also have two absorption maxima in narrow bands between 420–450 and 630–690 nm: photosynthesis is thus directly energized by blue and red light (fig. 3.4).

Both parts of this carbon metabolism are initiated by dual activities of Rubisco, a large water-soluble enzyme that acts as a carboxylase as well as an oxygenase. In the first role it catalyzes the addition of CO_2 to a five-carbon compound, ribulose 1,5-bisphosphate (RuBP), to form two molecules of a three-carbon compound 3-phosphoglycerate (fig. 3.5). The sequence of this multistep C_3-reductive photosynthetic carbon cycle was unraveled by Melvin Calvin and his colleagues by the early 1950s

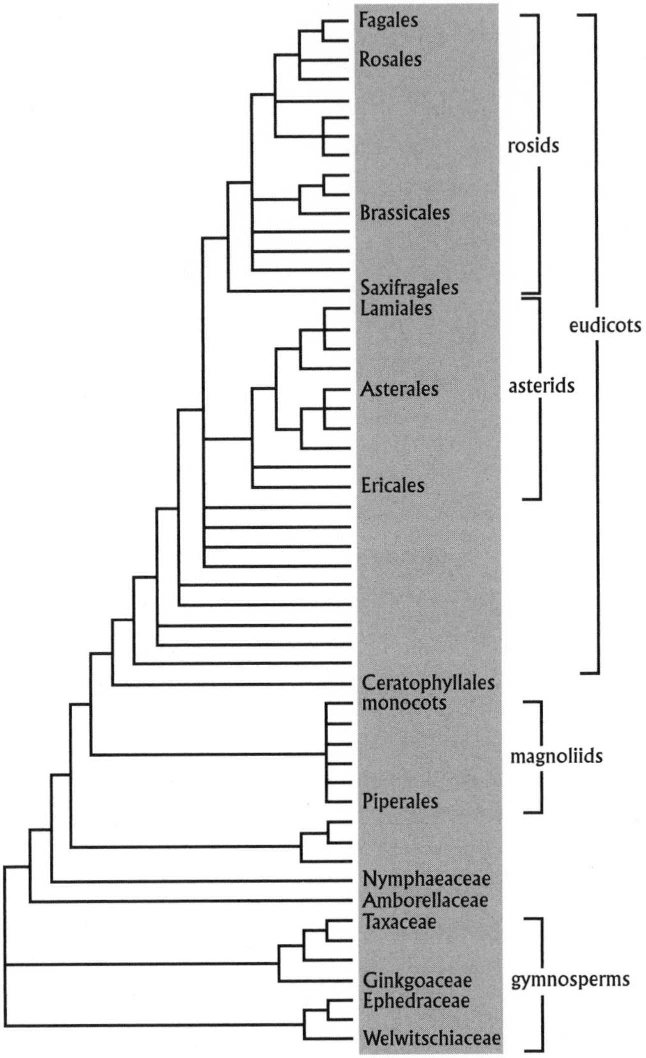

3.3 Phylogenetic relationships for angiosperms determined by Soltis et al. (1999).

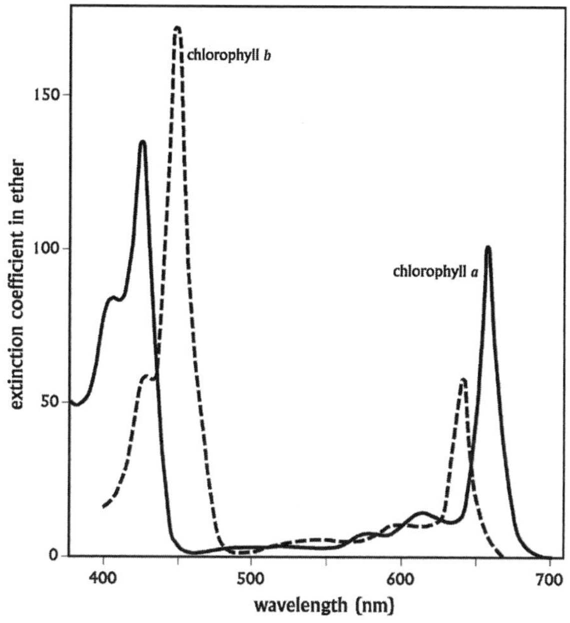

3.4 Absorption spectra of chlorophylls extracted in ether show two distinct peaks in blue and red light, around 450 and 650 nm (Smil 1999d).

(Calvin 1989). In its second role Rubisco catalyzes the binding of O_2 to RuBP to produce 3-phosphoglycerate and a two-carbon compound (2-phosphoglycolate) during the C_2-oxidative photosynthetic cycle.

This oxygenation does not appear to be of any use for the plant, but photorespiration, a complicated set of reactions aimed at recovering the reduced carbon and removing phosphoglycolate, is not a separate process. There is just one CO_2 and one O_2 pool, and the C_3 and C_2 cycles create a necessary balance for the net exchange of the gases during photosynthesis (fig. 3.6). Although the idea is controversial, some experimental results with transgenic plants indicate that photorespiration may protect C_3 plants from photooxidation among species living under high-intensity light (Kozaki and Takeba 1996). Because of the relatively low CO_2 concentrations and high O_2 levels in today's atmosphere, about half of all photosynthetic energy is now used by the C_2 cycle (Tolbert 1997). Only a drastic reduction of atmospheric O_2 (to about 2%) or greatly elevated ambient CO_2 levels would eliminate C_2-cycle losses in C_3 plants.

Photorespiration is an entirely different process from the nighttime evolution of CO_2 by mitochondrial respiration in leaves. In some species photorespiration may amount to as much as half of net photosynthesis (Hall and Rao 1999). Consequently, both the maximum and average efficiencies of converting sunlight into chemical energy by the C_3 cycle are rather low (for details see chapter 4), and primary productivities (the annual increments of phytomass) vary widely with nutritional and climatic factors (see chapter 7). Water availability is the factor most commonly limiting carbon fixation. Photosynthesis entails an extremely lopsided exchange of CO_2 and H_2O. The difference between the water vapor pressure inside and outside plants is two orders of magnitude higher than the difference between external and internal CO_2 levels. As a result, C_3 plants need 900–1,200 moles (and some up to 4,000 moles) of H_2O to fix one mole of CO_2.

Plants with much high water utilization efficiency—using "just" 400–500 moles of H_2O for each mole of CO_2 fixed—do not initially follow the fixation sequence of the C_3 cycle. Instead of reducing CO_2 with Rubisco, they use the enzyme phosphoenol pyruvate carboxylase (PEP) in their mesophyll cells to form oxaloacetate, a four-carbon acid (fig. 3.7). This acid is reduced to malate (another four-carbon acid) and transported into chloroplasts of the bundle sheath cells, where CO_2 is recovered by decarboxylation in the C_3 cycle (Sage and Monson 1999). These C_4

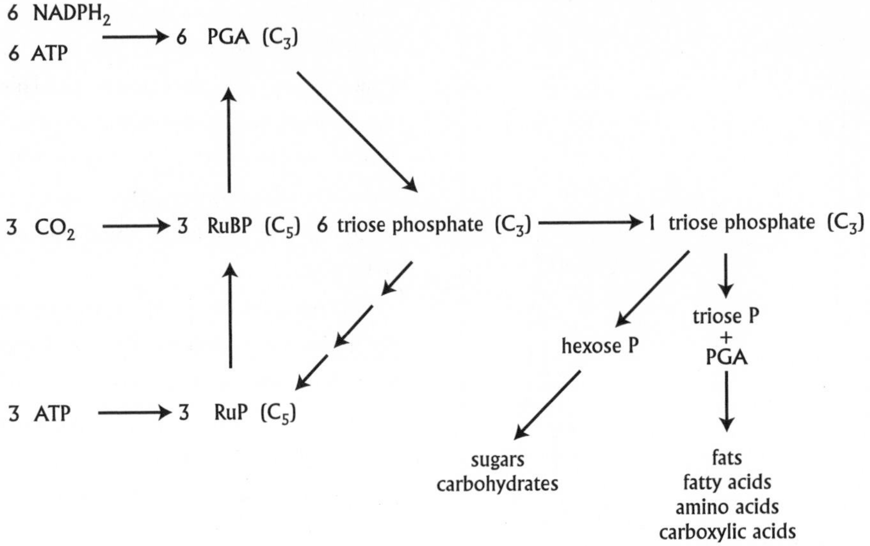

3.5 Basic reactions of the photosynthetic carbon reduction (Calvin) cycle (Smil 1999d).

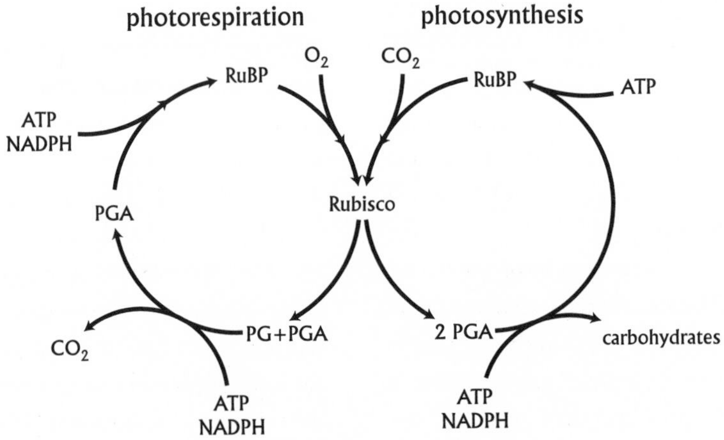

3.6 Rubisco mediates both photosynthesis and photorespiration. The first process takes place when RuBP is carboxylated by the enzyme, the second when RuBP is oxygenated by it. Based on Tolbert (1997).

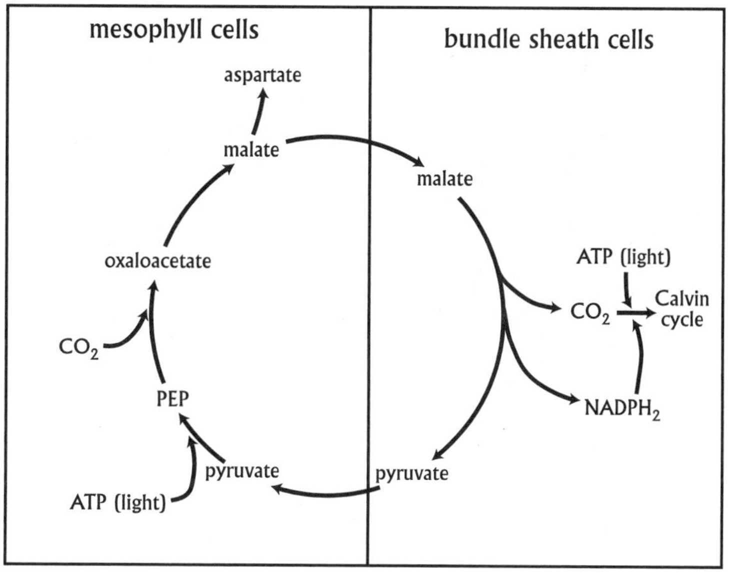

3.7 Carbon fixation in C_4 plants is divided between mesophyll and bundle sheath cells.

plants also differ anatomically from C_3 species; the latter have no significant differentiation in mesophyll and bundle sheath, whereas the vascular conducting tissue of C_4 species is surrounded by a bundle sheath of large, thick-walled cells with chloroplasts (fig. 3.8).

PEP carboxylase has a greater affinity for CO_2 than Rubisco; moreover, O_2 levels in the bundle sheath are low, whereas CO_2 concentrations are near what is required to saturate Rubisco, whose oxygenating action is practically eliminated as it catalyzes the C_3 cycle. All of this results in appreciably higher photosynthetic conversion efficiency than in the C_3 plants. There is also no light saturation in C_4 species, whereas C_3 plants saturate at irradiances around 300 W/m^2, optimum temperature for net photosynthesis is 15–25 °C in C_3 plants but 30–45 °C in C_4 varieties. The C_4 pathway thus appears to be an obvious adaptation to hot climates and aridity. But Reinfelder at al. (2000) discovered that a coastal marine diatom, *Thalassiosira weissflogii,* also uses C_4 photosynthesis to cope with low aquatic concentrations of CO_2, and they believe that unicellular C_4 assimilation may have predated the evolution of higher C_4 plants.

Corn, sugar cane, and sorghum are the most important C_4 crops, and some of the worst weeds, including crab grass (*Digitaria sanguinalis*), also follow this assimilation path. A global comparison of annual phytomass accumulation shows C_4 plants taking eleven of the top twelve places (Hatch 1992). Maximum daily growth rates reported for field crops (all in g/m^2) range from just over 50 for corn and sorghum to around 40 for sugar cane, above

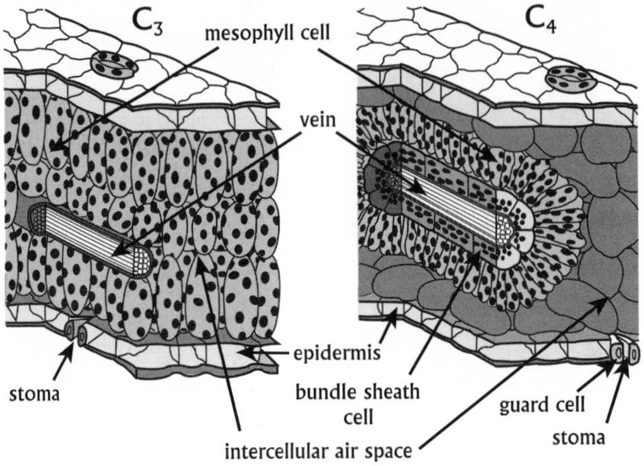

3.8 Leaf sections contrasting the structural differences between C_3 and C_4 plants (Smil 2000a).

35 for rice and potatoes, between 20 and 25 for most legumes, and just short of 20 for wheat. Analyses of carbon isotopes in tooth enamel of large mammals show that their diet was dominated by C_3 plants until about 8 Ma ago. The subsequent increase in the biomass of C_4 plants led to their dominance in equid and other mammalian diets in the Americas, Africa, and parts of Asia by 6 Ma bp (Cerling et al. 1997). This shift is attributed to declining atmospheric CO_2 levels related to increased continental weathering. Consequently, the past 7 Ma of the biospheric evolution have favored the C_4 species, whereas rising levels of anthropogenic CO_2 (see chapter 9 for details) are tilting the balance once again in favor of a C_3 world.

Crassulacean acid metabolism (CAM) is the other important modification evolved to minimize H_2O losses. Succulents—Crassulaceae and cacti, and also some orchids and bromeliads—use CAM by absorbing massive amounts of CO_2 during night and converting it initially into C_4 acids. During the day, with their stomata closed, sunlight energizes decarboxylation of these acids and C refixation into carbohydrates via the reductive pentose phosphate (RPP) cycle. Unlike in C_4 species, these processes are not spatially separated: they take place at different times in the same cells. Many CAM plants can totally suspend any gas exchange for weeks, even months. Pineapple (*Ananas sativus*), *Aloe,* and *Opuntia* are the only notable CAM crops. A high degree of metabolic plasticity among most CAM plants has led Lüttge (1998) to ask whether there is something like obligatory CAM in a given species.

Chemolithotrophs

The ability to use atmospheric CO_2 as the source of carbon for biosynthesis energized by oxidation of simple inorganic compounds is limited to prokaryotes. The first bacterial group using this mode of metabolism was identified by Sergei Nikolaevich Winogradsky (1856–1953) during his studies in Strasbourg. Winogradsky observed the colorless bacterium *Beggiatoa* accumulating globules of elemental sulfur in its filaments and correctly interpreted this storage as the transient phase during the oxidation of hydrogen sulfide to sulfuric acid, but he had not concluded that these bacteria use CO_2 as their only source of carbon (Winogradsky 1887; fig. 3.9).

3.9 Sergei Nikolaevich Winogradsky (1856–1953), the most famous Russian microbiologist, who discovered chemolithoautotrophic metabolism. Photo from the author's collection.

Winogradsky wondered why *Beggiatoa* grown in a medium containing H_2S would fill itself rapidly with droplets of sulfur that would disappear as soon as the filamentous bacteria were taken out of the solution. By concluding that sulfur is being oxidized by *Beggiatoa* to sulfuric acid he discovered a new mode of life.

Soon afterward Winogradsky isolated chemolithotrophic nitrifiers, which oxidize NH_3—and only then did he clearly formulate the concept of bacterial autotrophy independent of solar radiation (Winogradsky 1890). The third class of this unique prokaryotic group includes methylmonads oxidizing methane or methanol; finally, there are hydrogen-oxidizing bacteria, which, naturally, require the presence of H_2 besides O_2 and CO. Both methylmonads and hydrogen oxidizers, however, are facultative autotrophs: they can also grow as heterotrophs. Autotrophs can also be found in every group of methanogens, organisms formerly put into a separate bacterial phylum and now classified as a major group of Archaea (details on prokaryotic classification follow in the next section).

Beggiatoa abounds in sulfureta, sulfur-rich soil layers smelling of H_2S, where it may form intricate filaments arranged in rosettes. Many other sulfur-oxidizing bacteria—*Thiobacillus, Sulfolobus,* and *Thiospira* being among the most common genera—are found in solfataras (vents releasing volcanic gases); in sulfur-rich waters, including hot springs up to 75 °C and with very low pH; in marine muds; and in soils (Kuenen and Bos 1989). Some of these microbes prefer to oxidize just elemental sulfur, the reaction producing sulfate and H^+ ions; others are eclectic in their choice of substrate and use sulfur, sulfides, sulfites, and thiosulfates.

Nitrifying bacteria take care of a critical link in the biosphere's complex nitrogen cycle: they are the only organisms that can convert ammonia to nitrite and nitrite to nitrate, a highly soluble compound that is the principal source of the nitrogen for all plants (for more on nitrates, see chapter 5). Winogradsky isolated the two microbial genera carrying the terrestrial nitrification by cultivating them on an inorganic substrate: *Nitrosomonas* derives its energy by oxidizing ammonia to NO_2^-, and *Nitrobacter* by converting NO_2^- to NO_3^-. Ammonia is always the starting point, as all forms of organic nitrogen, be it complex proteins or simple urea, must be first decomposed to NH_3 before further bacterial conversion.

Nitrosospira, Nitrosococcus, and *Nitrosovibrio* are other ammonia-oxidizing species; *Nitrococcus* is the most abundant nitrite oxidizer in the ocean. Except for some *Nitrobacter* strains, all of these bacteria are obligatory chemolithotrophs. Nitrification can proceed fairly rapidly in both terrestrial and aquatic ecosystems. The process slows down with temperatures below 30 °C (optima are between 30 and 35 °C, and nitrification virtually stops below 5 °C), and on land it is inhibited in acid (pH below 6), poorly aerated, and very wet soils. Fire, a frequent presence in seasonally arid ecosystems, increases nitrification.

Aerobic hydrogen-oxidizing bacteria, including species of such common genera as *Pseudomonas, Alacaligenes,* and *Nocardia,* are often called Knallgas bacteria ("explosive gas" in German, referring to the inflammable mixture of H_2 and O_2) to distinguish them from anaerobic hydrogen-oxidizing microbes (Gottschalk 1986). They can be found in volcanic gases and in hot springs totally or largely devoid of organic substrates. But autotrophy is not their primary mode of metabolism: because they are nutritionally very versatile and use various organic substrates, taxonomists usually place Knallgas bacteria with their heterotrophic relatives.

Most methanogens, be they hetero- or autotrophic, can grow using the reduction of CO_2 (or HCO^{3-}) to CH_4, with H_2 as their sole energy source; some can grow, albeit very slowly, on CO, and all autotrophic species assimilate CO_2 by a unique variation of the acetogenic pathway (Ferry 1993). Some methanogens also have a limited capacity to reverse their normal metabolism and oxidize CH_4 to produce CO_2 and H_2, but most of the molecular hydrogen usually comes from the decomposition of biomass by other anaerobic microbes: only in geothermal fluids is it of inorganic origin. Finally, I must mention a group of unique chemolithotrophs that do not need CO_2 to synthesize their tissues. Some species of carboxydotrophic bacteria oxidize highly poisonous and flammable CO to CO_2 and assimilate a part of the formed gas via the RPP cycle; others just oxidize the gas without assimilating CO_2 (Meyer 1985). These bacteria, including species of *Pseudomonas* and *Alcaligenes,* live in soils and waters and around submarine hydrothermal vents.

Heterotrophs

Heterotrophy requires organic compounds synthesized by autotrophs to support an organism's growth, reproduction, and mobility. Carbon is usually absorbed by heterotrophic cells in relatively simple compounds. This means that enzymes must first be used to destroy the structure of biopolymers by severing glycoside bonds of complex sugars (breaking them down to their constituent monosaccharides, with glucose being the dominant nutrient), ester bonds in lipids (fats are hydrolyzed into glycerol and fatty acids), and amide bonds of proteins, whose amino acids can be used both as energy sources and for protein synthesis in consumers' bodies.

There are three basic strategies for oxidizing organic molecules: aerobic glycolysis, anaerobic fermentation, and dissimilatory anaerobic oxidation (Anderson 1986). Atmospheric oxygen is the most common electron acceptor, and the highly exergonic (–2870 kJ/mol) oxidation of biomass, producing water and CO_2, is used by the vast majority of heterotrophs. Some anaerobic organisms convert oxidized carbon compounds to lactate or ethanol. These fermenters include both bacteria (*Lactobacillus,* with its yogurt-making species) and fungi (most notably *Saccharomyces,* the indispensable bread and wine yeast; fig. 3.10) and they gain no more than 197 kJ/mol by those oxidations. Other anaerobes are able to use nitrates, nitrites, sulfates, and ferric iron present in their niches as electron acceptors.

Pseudomonas and *Clostridium* are common bacteria that reduce nitrate to nitrite. Thiopneute bacteria, including red-colored *Desulfovibrio,* common in muds and in estuarine brines, reduce sulfates to H_2S or S while incompletely oxidizing such simple organic compounds as lactate and acetate. Released H_2S returns to the atmosphere, but in iron-rich waters the gas reacts to form pyrite in a process that was responsible for a part of Archean iron deposits. These dissimilatory reductions—less exergonic than anaerobic glycolysis and more exergonic than fermentations—are critical for continuous functioning of global biogeochemical cycles (see chapter 5). Medically important anaerobic bacteria, include several species of *Clostridium*—*C. botulinum* and *C. perfringens* in contaminated food and *C. tetani* in infected wounds. Many species of *Bacteroides* found in eye, middle ear, lung, abdominal, urinary, and skin infections, are also implicated in circulatory bacteremia and endocarditis (Holland et al. 1987).

Methanogens, belonging to Archaea (see the next section for details), perform the final task of biomass degradation in those environments where oxygen, nitrate, sulfate, and ferric iron have all been depleted. They use CO_2 as the final electron acceptor in producing CH_4 and other simple C_1 compounds (Ferry 1993; Holland et al. 1987). Anaerobic fermentation proceeds in such common natural environments as marine and freshwater sediments, marshes and bogs, flooded soils, gastrointestinal tracts, and geothermal habitats—and it is also widely used to treat organic wastes in sewage plants.

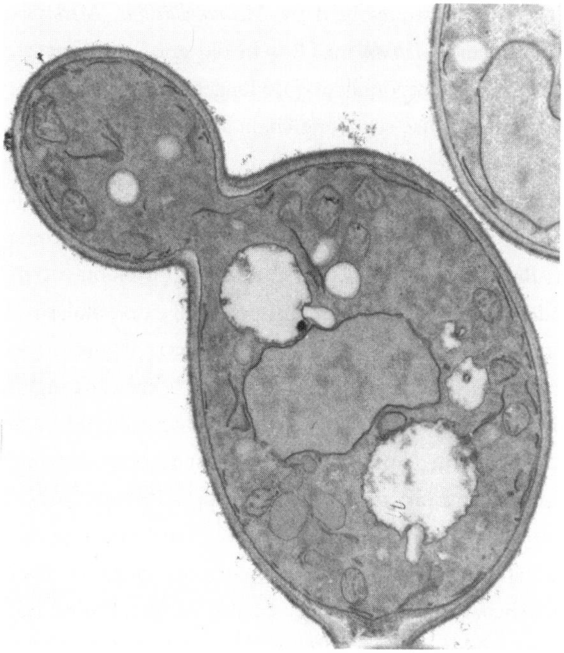

3.10 A budding cell of *Saccharomyces cerevisiae,* the best known species of Ascomycota, which is responsible for fermentations of bread and alcoholic beverages. Transmission electron microscope image courtesy of E. G. Vrieling, Rijksuniversiteit Groningen.

Methanogenic activity was first described by Alessandro Volta in 1776, but identification and studies of it matured only after World War II. Easily degradable organic molecules produced by the decomposition of cellulose, including formate, acetate, butyrate, and ethanol, are the most common substrates used by *Methanobacillus, Methanosarcina,* and *Methanotrix.* But in old anoxic sediments, where such compounds may no longer be available, methanogens can use such long-chain alkanes as hexadecane, $n\text{-}C_{16}H_{34}$ (Zengler et al. 1999).

Heterotrophs rely on a variety of feeding strategies. Microbial decomposers, including such ubiquitous bacteria as *Bacillus, Clostridium,* and *Proteus,* digest biomass outside of their cells by excreting powerful extracellular hydrolytic enzymes. Fungi and many insects also resort to extracellular hydrolysis. In contrast, most metazoa ingest their food and break down polymers internally, although many of them do not rely solely on their digestive enzymes but also on a variety of microbial helpers. Perhaps the best known are the rumen-dwelling anaerobic bacteria, protozoa and fungi aiding the metabolism of ruminants. They get their carbon and energy by degrading cellulose and hemicelluloses in grazed phytomass and produce acetic, butyric, and propionic acids and generate microbial biomass that, with undegraded dietary protein, serves as the protein source for ruminants: a prominent role in herbivory. Parasites absorb their nutrients directly from their host's living cells; so do mutualistic symbionts, but they provide their host with a valuable service: exchange of carbohydrates from plants for fixed nitrogen from rhizobial bacteria is among the most important symbiotic swaps in the biosphere.

The choice of consumed substrates ranges from extreme specialization to indiscriminate omnivory: strains of *Bacillus fastidiosus,* growing only when uric acid is available, represent the first extreme; strains of *Pseudomonas putida,* proliferating on scores of organic compounds, or omnivorous humans illustrate the other. Saprovores, mostly microbial, feed on dead biomass. Herbivores consume only phytomass: some, such as *Schistocerca* migratory locusts are indiscriminate polyphages, others, such as koala (a monophagous folivore eating only leaves of some eucalypts), are highly specialized.

Carnivores include both scavengers and hunters, and some omnivores exploit the entire feeding spectrum from detritus to meat. Whereas the smallest animals must feed almost incessantly, large animals can go without feeding for extended periods of time (weeks or months) during hibernation. But many small eukaryotes and many more prokaryotes can become cryptobiotic, drastically reducing or totally suspending their metabolism (for more on cryptobiosis, see chapter 6). Many bacteria have a marked resistance to starvation and exceedingly long survival capability in the near absence of any nutrients (Fletcher and Floodgate 1985).

All previous claims of bacterial survival were overshadowed by the recent isolation and growth of a 250-million-year-old halotorelant spore-forming bacterium found in a brine inclusion in halite crystals in a New Mexico shaft (Vreeland et al. 2000). Sequencing of its ribosomal RNA showed it to be related to *Bacillus marismortui* and *Virgibacillus panthotenicus.* Not surprisingly, the dating of this find has been questioned (Hazen and Roedder 2001), and as we have learned from past DNA work, such findings must be replicated in different laboratories and in different systems before they can be believed. Only convincing replications would justify the question asked by Parkes (2000, p. 844)—"If bacteria can survive for this length of time, why should they die at all?"—and would have to be considered when thinking about the possibility of panspermia (see chapter 2).

Thermoregulation splits heterotrophs into ectotherms and endotherms: cold- and warm-blooded animals. Portability of a constant thermal environment has conferred enormous survival and competitive advantages on endotherms, which have radiated to virtually every terrestrial niche. But this success comes at a high energy cost. To be above the ambient temperature most of the time, core body temperatures of endotherms must be relatively high (36–40 °C for mammals, up to 42 °C for birds) and maintenance of these temperatures requires high feeding rates and limits the share of energy that can be diverted to reproduction and growth. As a result, zoomass productivity (the assimilated energy fixed as new tissue in growth and reproduction) of endotherms is generally an order of magnitude lower than in ectotherms (Bakker 1975). Endothermy also limits the size of the smallest animals, as the rising specific metabolic rates of tiny organisms require a higher frequency of feeding.

But endotherms are unsurpassed masters of both rapid and long-distance locomotion: their maximum aerobic power is commonly an order of magnitude higher than that of ectotherms. Both massive mammals and tiny birds are capable of astonishingly long seasonal migrations powered by lipid stores (Elphick 1995). Swimming is the most efficient way of locomotion and running the most demanding, and most of the time spent in motion is in search of feed (fig. 3.11). Ectotherms, whose thermoregulation is overwhelmingly behavioral, cannot be active in thermally extreme environments, nor can they be as competitive in optimal temperatures as endotherms. Still, as the hyperdiverse insects show, ectothermy has not been an insurmountable obstacle to the conquest of many terrestrial niches.

Yet another obvious heterotrophic divide is between asexual and sexual reproduction, the latter being, of course, the norm among animals. Evolutionary biology has paid a great deal of attention to sexual reproduction, explaining its dominance as due to advantages of genetic recombination, which introduces desirable variation and

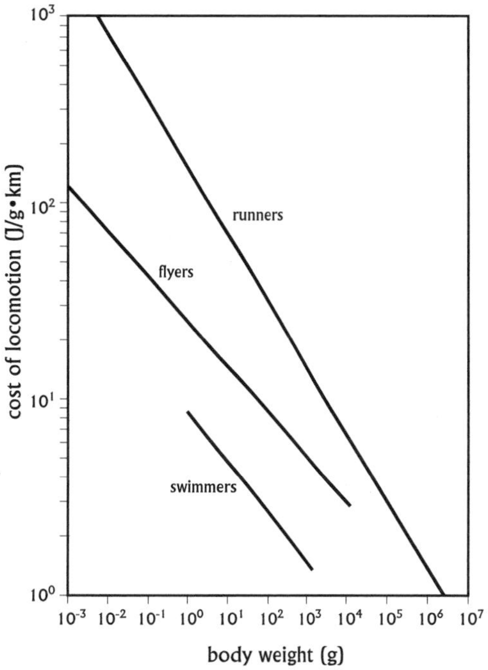

3.11 Energy cost of swimming, flying, and running. Based on Smil (1991).

purges deleterious mutations (Stearns 1987; Barton and Charlesworth 1998). Yet many extremely successful and ancient organisms have remained asexual. For example, Glomales, mycorrhizal fungi whose symbioses with plants are helpful in securing mineral nutrients' ecosystemic services, have remained asexual ever since they colonized the continents some 400 Ma bp (Sanders 1999). And several parasitic protozoa have clonal population structures with independently evolving lineages (Ayala 1998).

As noted in the opening of this chapter, the evolutionary sequence of heterotrophs remains to be clarified for the late Proterozoic eon, but it is relatively noncontroversial after 530 Ma bp. Prokaryotic heterotrophs were among the earliest Archean organisms, and nearly four billion years of their evolution have resulted in an astonishing array of metabolic, feeding, and niche adaptations. As already noted, multicellular heterotrophs evolved between 1,100 and 600 Ma ago, and the fossil record from about 530 Ma ago documents a spectacular emergence of diversified skeletonized fauna in a span of just a few million years (Erwin et al. 1997). Sponges are the most primitive surviving animal phylum; other simple phyla include Ctenophora (comb jellies), Cnidaria (jellyfish and sea anemones), and Platyhelminthes (flatworms).

More complex animals are divided in two basic groups determined by the different fate of the initial opening of the primitive digestive tract in the embryo: Arthropoda, Annelida (earthworms), and Mollusca (snails, clams, squids) belong to protostomes, Echinodermata (star fish, sea urchins) and Vertebrata (fish to mammals) to deuterostomes. There are other essential differences concerning the development of mesoderm and coelom (from a solid block of tissue in protostomes, initially as pouches off the primitive digestive tract in deuterostomes) and the cleavage of the developing embryo (spiral and determinate in protostomes, radial and indeterminate in deuterostomes).

Until recently the oldest undeniably vertebrate remains were those of ostracoderms, small, fishlike creatures from the Ordovician period (about 460 Ma bp) that became extinct about 100 Ma later (Forey and Janvier 1994). But a recent discovery of fishlike animals from the early Cambrian (Shu et al. 1999) puts vertebrates among the organisms emerging during the sudden radiation of new life forms that occurred at that time. A major event in the subsequent history of vertebrates was the transition between fish and amphibians (tetrapods) that took place during the Late Devonian, about 365 Ma bp (Janvier 1996). Some 310 Ma ago, mammal-like reptiles split from birdlike reptiles, and molecular clocks confirm that modern orders of mammals go back to the Cretaceous period, more than 100 Ma ago, and diversified before extinction of dinosaurs (Kumar and Hedges 1998).

The Tree of Life

The enormous variety of life on the Earth begs for a comprehensible classification, for reasons ranging from intellectual curiosity to practical needs. Foundations for this task were laid before the middle of the eighteenth century, and most of the additions took place before 1900. By the mid-twentieth century, phylogeny—the term was coined by Ernst Haeckel (1834–1919) in 1866—had become a marginal part of biology preoccupied with physiology, evolution, and genetics (Hillis 1997). The phylogenetic data that began trickling, and then streaming, from the new molecular biology beginning in the 1960s reversed this trend. New tools of genetic analysis, especially automated DNA sequencing, and deployment of sophisticated statistical tests have made it possible to examine both the unity and the diversity of life in unprecedented and astonishing detail. A regrettable, but powerful, impetus for renewed interest in classification has come from the realization that the biosphere is experiencing a rapid loss of its diversity, driven, unlike all of the previous episodes of catastrophic extinctions, by the actions of the dominant species.

After noting first the milestones of biological classification, I will explain the essentials of RNA sequencing, the preferred method used since the late 1970s to derive new phylogenies that have subverted a number of long-held assumptions about the evolution of life. Not surprisingly, this taxonomic revolution has encountered some pointed criticism, and I will note the strongest arguments offered by both sides of this far-from-settled dispute. In closing, I will assess the recent claims concerning the unknown extent of the Earth's biodiversity, which may be made up of "just" around three million species — or which may be an order of magnitude richer than the currently described total.

From Linné to SSU rRNA

As scientific classifications go, the Linnean system is a venerable one: it arose from the author's youthful fascination with the variety of plants, and it was first outlined in 1735 when Carl Linnaeus (1707–1778), a physician and Sweden's most famous naturalist, published the first edition of his *Systema naturae* in Leiden (Linnaeus 1735; fig. 3.12). After he was ennobled in 1757, Linnaeus changed his name to von Linné (for more on "God's registrar," see Koerner 1999 and Frängsmyr 1983). His diligent predecessor, John Ray (1627–1705), whose *Natural History of Plants* included more than 18,000 species, is now remembered only in specialists' accounts (Ray 1686–1704). Ten editions of *Systema* followed during Linné's lifetime, and although the classification has been superseded in many ways, modified, and substantially expanded, its binomial nomenclature and its hierarchical taxonomy remain the dominant way of sorting out the nearly two million named species.

3.12 Carl Linnaeus (Karl von Linné, 1707–1778), author of the famous *Systema naturae*. Photo from E. F. Smith Collection, Rare Book & Manuscript Library, University of Pennsylvania, Philadelphia.

Appearance, be it of whole organisms or of some of their parts, has been the guiding choice of the Linnean classification, which orders species into nested categories. These categories (their progressively less inclusive sequence is kingdom, phylum, class, order, family, genus, species) designate rank in hierarchy. Taxa designate named groupings of organisms: they may include a single organism in a monospecific genus, such as koala (*Phascolarctos cinereus*), or more than 20,000 species of such a large family as Leguminosae. Names of genera (kinds) must be unique, and duplicate names are avoided in other taxa; species (appearance) names must be binomial, with both names in Latin or in latinized form (for a highly informative review of nomenclatorial challenges, see Trüper 1999).

The shortcomings of Linnean classification are obvious, ranging from exclusion to arbitrary criteria. Living organisms were assigned to just two kingdoms, plants and animals (minerals were the third, inanimate kingdom), and Linné's plant taxonomy, based solely on the appearance of the reproductive system, soon became obsolete. A new kingdom, Protista, was introduced by Haeckel (1894) to classify a multitude of relatively simple organisms that appeared to be neither plants nor animals and that have been commonly called protozoa. Some aquatic animals also use photosynthesizing organisms: sea slugs and giant clams engulf and modify cyanobacteria, and in 1998 *Daphnia obtusa* (the water flea, an arthropod) was found to have chloroplasts in its gut.

Perhaps the most influential reorganization of multiple-kingdom systems that respected this basic divide was the one developed during the late 1950s by Robert H. Whittaker (1959). Whittaker's system had five kingdoms—prokaryotic Monera (bacteria) and eukaryotic Protoctista (algae, protozoans, slime molds, and a variety of other aquatic and parasitic organisms), Fungi (mushrooms, molds, and lichens), Animalia (with or without backbones) and Plantae (mosses, ferns, conifers, and flowering plants). Margulis and Schwartz (1982) became perhaps the most enthusiastic promoters of this classification. Their book came out five years after Carl Woese began subverting the old classification order by comparisons of ribosomal RNA, a fact the two authors barely acknowledged by merely stating that some biologists would give a separate kingdom to methanogenic and halophilic bacteria.

Woese proposed more than that: his new kingdom of Archaebacteria (Woese and Fox 1977) was not intended to be on the same level as the traditionally accepted kingdoms within the eukaryotes (fungi, plants, animals). Later he made the distinction explicit by creating three primary domains (superkingdom categories) of Archaea, Bacteria, and Eucarya, thus advocating the third realm of life (Woese et al. 1990; fig. 3.13). His classification met first with considerable resistance, but by the mid-1990s most biologists had been converted by impeccable evidence that relies on the sequencing of about 1,500 base pairs in a universal ribosomal gene that codes for a part of the cellular machinery assembling proteins (Woese and Fox 1977).

3.13 Carl Woese, the discoverer of Archaea and the originator of a new classification of life. Photo courtesy of the University of Illinois News Bureau.

Woese's tribulations and eventual triumph are recounted in Morell (1997).

Ribosomes, as already noted in chapter 2, are the sites of polypeptide production in cells (for details on ribosomes and the synthesis of proteins, see de Duve 1984, Hill et al. 1990, and Frank 1998). Ribosomal RNA (rRNA) is present largely in two unbroken strands, one of which, labeled in biochemical literature as 16S in prokaryotes and 18S in eukaryotes, was used by Woese and Fox (1977) as the substrate to produce the oligonucleotide sequences released by digestion with T1 ribonuclease. The Svedberg unit (S) is used to measure the rate of ultracentrifugal sedimentation, that is, indirectly, the molecular size: 16S rRNA has about 1,540 nucleotides (its model can be viewed at smi-web.stanford.edu/projects/helix/ribo.html). This study of oligonucleotides was superseded first by reverse-transcriptase sequencing of rRNA and eventually by the polymerase chain reaction cloning of DNAs encoding rRNA.

The polymerase chain reaction, whose 1983 discovery by Kary Mullis was rewarded in 1993 by a Nobel Prize in chemistry, makes it possible to amplify specific sequences of DNA from just a few template molecules (Mullis et al. 1986). But the initial assumptions that replication can extend to DNA going back millions of years proved false: about 100,000 years seems to be the maximum for intact survival (Wayne et al. 1999). There are several reasons why rRNA makes a good molecular clock. As a very ancient molecule present in the progenote, it can be found in all living species conserved both in structure and function. Its slow changes allow the entire evolutionary spectrum to be studied, and there has been no evidence of its lateral transfer between different species. And because even a bacterial cell has 10,000–20,000 ribosomes, it is not too difficult to isolate workable amounts of rRNA.

Oligonucleotide catalogues generated by rRNA sequencing are compared to determine the degree of relatedness among studied species or used as inputs into classification programs. A variety of advanced statistical methods—maximum parsimony, maximum likelihood, cluster analysis—and computer simulations have been used to construct complex phylogenetic trees and to test hypotheses in an evolutionary context (Huelsenbeck and Rannala 1997; Pace 1997; Pagel 1999; fig. 3.14). With the diffusion of automated DNA sequencing it has also become possible to compare complete genomes.

The first complete genomes for each of the three domains of life became available in a quick succession starting in 1995: first for a bacterium, *Haemophilus influenzae* (Fleischmann et al. 1995), then for the first archaeon, *Methanococcus jannaschii* (Bult et al. 1996), and the first eukaryote, the budding yeast *Saccharomyces cerevisiae* (Goffeau et al. 1996). By 2000 there were thirty complete genomes of prokaryotes, including archaeons *Archaeoglobus fulgidus, Methanobacterium thermoautotrophicum,* and *Thermoplasma acidophilum,* a thermoacidophile thriving at pH 2 in an unusual niche of self-heating coal refuse piles; and a ubiquitous bacterium, *Escherichia coli* (Klenk et al. 1997; Doolittle 1998; Fraser et al. 2000; Ruepp et al. 2000; TIGR 2000). And just before the end of 2000 came the first complete genome of a flowering plant, a little brassica *Arabidopsis thaliana* related to broccoli (Arabidopsis Genome Initiative 2000). Another molecular approach used to reconstruct phylogenetic relationships has used a "protein clock" measuring the rate at which proteins change over time (Doolittle et al. 1996).

But rather than removing remaining uncertainties and strengthening the emerging consensus, this flood of new genetic information has posed new questions. Earlier expectations that comparing phylogenies based on genetic analyses of individual species will make it possible to reconstruct an unambiguously branching tree of life have been replaced with a rather confusing image of an unruly, intertwined bush hiding many surprises.

How Many Realms?

Advocates of genetically based classification see traditional phylogenies, whether based on obvious macroscopic features or on microscopic structures, as inferior and maintain that only the deciphering of molecular information should guide the reconstruction of relationships among species. That is why Woese insisted on going against the traditional division of life into prokaryotic and eukaryotic organisms and defined the third domain of life.

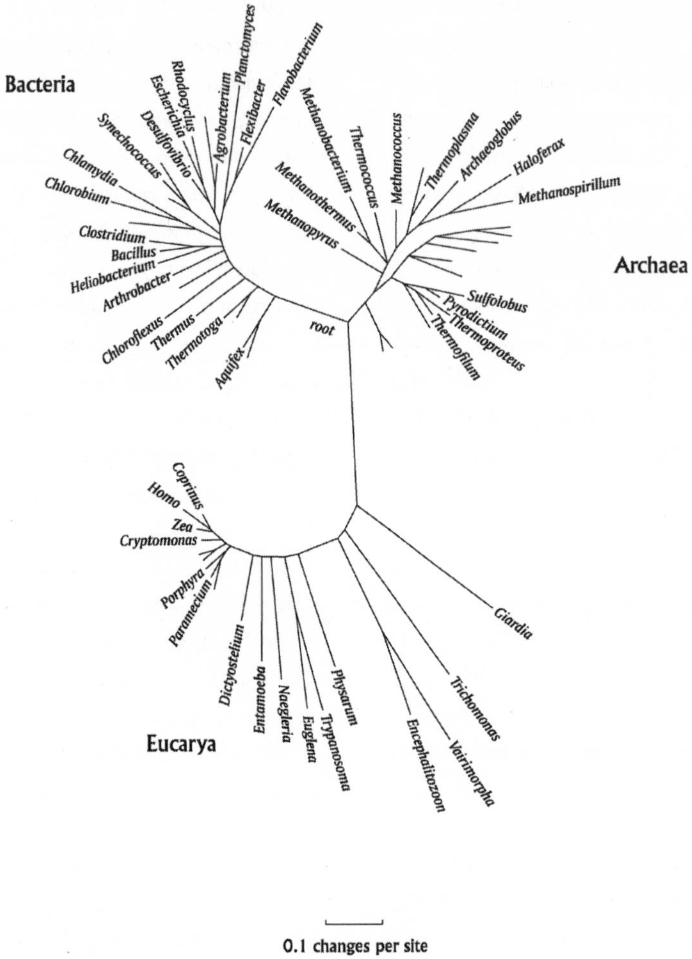

3.14 Universal phylogenetic tree based on SSU rRNA sequences. Based on Pace (1997).

As outlined by Woese et al. (1990), the archaeal domain includes two kingdoms, Euryarchaeota — mostly methanogens, extreme halophiles, and some hyperthermophiles — and Crenarchaeota, defined phylogenetically and including mostly hyperthermophiles and hyperacidophiles (DeLong 1998). In 1996 rRNA sequences of uncultivated organisms from a hot spring in Yellowstone National Park led to the recognition of the third archaeal kingdom, Korarchaeota (Barns et al. 1996).

Supporters of this new division point out that the distinct classification of Archaea now rests on much more than a comparison of rRNA sequences. The availability of the first complete genome sequences revealed that a significant proportion of the archaean genome is composed of unique genes. For example, *Methanococcus jannaschii* shares only 11% of homologies with *Hemophilus influenzae* and 17% with *Mycoplasma genitalium*. The domain's unique position is intriguingly illustrated by its way of transcribing DNA using RNA polymerase. Like bacteria, archaea have just a single polymerase RNA (eukarya have three), but it resembles eukaryotic RNA polymerases in its complexity and sequence homology. In W. F. Doolittle's apt characterization, it is "as if the archaea were preparing to become eukaryotes" (Morell 1996). Unlike bacterial cells, archaea have ether-linked lipids and their walls lack murein, and the two domains have differently composed flagella, common organelles for motility among prokaryotes, and different antibiotic sensitivities. And unlike any organism in the other two domains, only archaea are extremely hyperthermophilic, able to metabolize at temperatures exceeding 100 °C (Stetter 1996; Jarrell et al. 1999; Rothschild and Mancinelli 2001; see also chapter 6).

Criticism and outright rejection of this genetically based sorting have come from two very different sources: from the advocates of a holistic approach and from new genetic analyses going beyond SSU rRNA. The first group resents the setting aside of a separate domain and three new kingdoms for extremophilic microbes (a kingdom each for plants and animals with their enormous outward variety). Margulis (1998), an eloquent adherent of this view, strongly disagrees with the rRNA-based phylogenetic approach and argues that any classification of organisms should be based on the entirety of life forms rather than on any particular characteristic.

Taking this perspective, archaea are obviously very similar to bacteria, and both are much less like eukaryotes. This similarity now extends to their niches. Archaea, unculturable but detected by the amplification and cloning of rRNA and comparing it with kingdom- or family-specific 16S rRNA sequences, appear to be common also in such nonextreme environments favored by bacteria as agricultural soils (Bintrim et al. 1997) and ocean water (DeLong et al. 1994). Samples collected during the late austral winter in Antarctic coastal waters have shown that 21–34% of all prokaryotes and 18–30% of all picoplankton are composed of archaeal biomass, and similarly high shares (15–40%) of archaea in picoplankton have been found in the Puget Sound in the Northwest Pacific. Archaea appear to be no less abundant in greater depths: Karner et al. (2001) discovered that offshore Hawai'i crenarchaeota accounted for more than one-third of all DNA-containing picoplankton at depths below 1 km. And archaea, either as single-species aggregates or in consortia with metabolically interdependent bacteria, are also abundant in anoxic ocean sediments, where they consume more than 80% of all CH_4 produced deep below the ocean floor (Orphan et al. 2001).

Margulis (1998) concluded that the creation of the two prokaryotic groups obscures the diversity of the four other kingdoms in her preferred classification. Perhaps the most dismissive critic of the three realms is Ernst Mayr (1998), who faults Woese's cladistic approach on many points, but two arguments carry the greatest weight. Mayr believes that the eukaryotic acquisition of a nucleus (and all the accompanying features) was "perhaps the most important evolutionary event in the whole history of life" (p. 9, 722), a reality militating against subdividing prokaryotes. And he stresses that even combined bacterial and archaeal species diversity is only a small fraction of eukaryotic richness, and hence the principle of balance clearly favors combining the two prokaryotic groups. Unfortunately, his critique also included such irrelevant barbs as "Here it must be remembered that Woese was not trained as a biologist and quite naturally does not have an extensive familiarity with the principles of classification" (p. 9, 721).

In contrast, Drapeau (1999) noted that the grand "diversity" of the eukaryotes may be simply a human bias. Woese (1998b) countered Mayr's arguments by stressing that

> *the disagreement . . . is not actually about classification. It concerns the nature of Biology itself. Dr. Mayr's biology reflects the last billion years of evolution; mine, the first three billion. . . . His is the biology of visual experience, of direct observation. Mine cannot be directly seen or touched; it is the biology of molecules, of genes and their inferred histories. . . . For me, evolution is primarily the evolutionary* process, *not its outcomes. (p. 11, 046)*

The second source of questioning of the three realms has come from our expanding genetic understanding. New phylogenies have been searching for the cenancestor, the most recent common ancestor of all existing organisms, and mapping the subsequent branchings of the evolutionary tree (Fitch and Upper 1987; Doolittle and Brown 1995). Initially, rRNA-based analyses supported a triple-branched, treelike structure, with bacteria and archaea forming the two lower branches and eukarya—diverging first to diplomonads, microsporidia, and slime molds and eventually to animals and plants—at the top (Pennisi 1998). But by the late 1990s it became clear that molecular classifications do not converge on any clearly identifiable tree with sequential branchings.

Whole genomes present a rather untidy picture, as many archaea contain a substantial store of bacterial genes, and nuclear genes in eukaryotes derive both from bacteria and archaea. This means that the cellular evolution has not been solely a vertical process but has been repeatedly affected by lateral (horizontal) transfers of individual genes and perhaps even of entire genomes (Pennisi 1998; Katz 1999; Nelson 1999; Doolittle 2000). As a result, prokaryotes have received a significant share of their genetic diversity through the acquisition of sequences from distantly related organisms (Ochman et al. 2000), and eukaryotes are chimeras, organisms containing genetic material from a variety of ancestral lineages.

The latest evidence has thus both uprooted the traditional Darwinian tree of life (there is no single cell at the root) and radically changed its appearance. Treelike branches are seen only at the top levels of eukaryotic diversity, but below them is a tangle caused by rampant lateral gene transfers and arising not from a single cenancestor but from a community of primitive cells.

Assessing Biodiversity

An entirely different debate concerning life's diversity began at about the same time Woese started to challenge the traditional classification of organisms. Although the answers to its two key questions—how diverse is life in terms of individual species, and at what rate is this biodiversity declining because of human actions—rest largely on simple counting, rather than on sophisticated molecular analyses, consensus has been elusive. Answers to the first question range from less than two to more than thirty million, and those to the second from an unprecedented rate of extinction to a regrettable but far from catastrophic diminution of life's diversity. I will deal with this loss in some detail in chapter 9, and in chapter 8 I will review the relationships between biodiversity and function, productivity and resilience of ecosystems; here I will concentrate on the extent, attributes, and preconditions of biodiversity.

The first edition of *Systema naturae* (Linnaeus 1735) listed about 4,000 species; the tenth (Linnaeus 1758) contained about 9,000 binomial plant and animal names. The half-million mark was surpassed before 1880, a century later the count was close to 1.4 million, and a major international review sponsored by the UNEP concluded that about 1.75 million species had been scientifically described by the mid-1990s (Heywood and Watson 1995). Metazoa, with about 1.2 million, or 68%, of species, dominate; plants (including algae) add 310,000 species (nearly 18%), fungi 72,000 (4%), and prokaryotes just 4,000 (a mere 0.2%). The vertebrate count stands at about 45,000, including just over 4,600 mammals. The easiest way to review the Earth's biodiversity is to visit *The Tree of Life* Web site, the United Nations University's *Biodiversity and Biological Collections Web Server,* or the *Deep Green* project's site (see appendix H for URLs).

There are obviously many more species than the number described so far (May 1990). Most of these unknown organisms are concentrated among arthropods and prokaryotes, particularly in habitats that have been difficult to access whether deep crustal rocks (see chapter 6 for more on the subterranean biosphere), tall rain forest canopies, or deep ocean sediments. As for this last habitat, less than 1% of species living there are known, and the difference between the described and estimated total for benthic bacteria may be more than three orders of magnitude (Snelgrove 1999). Heywood and Watson (1995) opted for an ultimate total of 13.6 million living species. This count would require a number of insects an order of magnitude higher (8 million vs. the described 950,000), a twenty-fold increase in the fungal count, and a 250-fold enlargement of the bacterial and archaeal total to one million species.

Erwin's (1982) studies of tropical forest canopy insects in Panama led him to estimate that there are 30 million insect species alone, and some published guesses of the total species count have been as high as 80 million. Erwin's extrapolation was based on samples obtained by insecticidal fogging of a few high canopies in Panama. But Gaston (1991) pointed out that estimates of 30–50 million insect species seem incompatible with the best available taxonomic and ecological information and suggested the most feasible total of about 5 million. Regrettably, our ignorance is such that we cannot offer reasonably constrained estimates with a high degree of confidence. Whatever the actual species total might be, the existing global workforce of about 7,000 systematists, especially with the decline of systematic studies at many universities, is entirely inadequate to describe and classify the declining tropical biodiversity (Blackmore 1996).

Several major conclusions will not change with any further additions to the current list of species. Increases in the number of individuals encompassed by a count and in the size of the sampled area result in higher diversity (Humphries et al. 1995). Large contiguous areas are more diverse than small, and especially remote, islands. Diversity has a pronounced latitudinal gradient: tropical rain forests and tropical and subtropical coral reefs are the biosphere's richest ecosystems. The richest tropical rain forests have more than 200 tree species per hectare, and the records from the Ecuadorian and Peruvian Amazon are, respectively, 307 and 289 species (Valencia et al. 1994; Gentry 1988; Wright et al. 1997). In contrast, there are about 700 native trees in all of North America. Consequently, the countries harboring the greatest biodiversity—Brazil, Colombia, Peru, Bolivia, Mexico, Congo, Madagascar, Indonesia, and Australia—are situated entirely or partially between the two tropics. So are, naturally, most of the biodiversity hot spots, areas harboring extraordinary concentrations of endemic species that have exceptionally high rates of habitat loss (Myers et al. 2000; for more on these hot spots, see chapter 9).

The diversity of terrestrial species, including freshwater organisms, is greater than that of marine life, but the reverse is true for animal phyla: all but one of them are present in the ocean, but only half are found on land or in lakes and rivers. And diversity is highly unevenly distributed at all taxonomic levels. Key factors promoting hyperdiversity are small size (allowing for highly specialized niche subdivision), metamorphism (different life stages make it possible to occupy several niches), herbivory (providing abundant resource base), specialized parasitism (offering unique niches), and relatively rapid mobility (facilitating dispersal and colonization of empty niches) (May 1988).

Not surprisingly, insects, possessing all of these attributes, are the most hyperdiverse organisms. With some 950,000 species they account for nearly 90% of all arthropods, the most hyperdiverse animal phylum, and for more than half of all described organisms. In turn, beetles (Coleoptera), with at least 350,000 species, are the most diverse insect order (Gaston 1991). Coevolution between beetles and flowering plants has been proposed as the best explanation of this extraordinary diversity, as numerous opportunities for specialized herbivory were added to older modes of detritivory and fungivory (Farrell 1998). But Labandeira and Sepkoski (1993) concluded that the great radiation of modern insects that began 245 Ma ago was not accelerated by the expansion of flowering plants during the Cretaceous period, as the basic trophic modes of insects were in place 100 Ma before any angiosperms appear in the fossil record. Other notable examples of hyperdiversity are rodents (the richest mammalian order) and orchids (the most diverse monocotyledonous plant family).

Biodiversity is a fundamentally multidimensional concept, and hence no single variable can measure it objectively (Humphries et al. 1995; Purvis and Hector 2000; Wilson and Perlman 2000). Whittaker (1972) divided the total diversity within a large area (γ-diversity) into local (α) diversity (the number of species in a specified area) and the turnover of species between habitats or localities (β-diversity). Shannon's diversity index measures how individuals are apportioned within a particular area (Magurran 1988). Purvis and Hector (2000) offer another useful threefold division of commonly used biodiversity measures into numbers (species or population richness), evenness (the extent to which individuals are spread evenly among species), and difference (disparity and character diversity of phenotypes among the species) indicators.

As defined by the Convention of Biological Diversity, the concept should encompass diversity within species, between species, and of ecosystems. Genetic differences between the individuals of a species can be analyzed along a continuum ranging from nucleotide sequences to obvious macroscopic or functional traits, and variability can be assessed both within and between populations. Although the total species count has attracted most of the research attention, there is no comprehensive listing for all of the described species, and the knowledge of their niches and life histories ranges from excellent to rudimentary. Appraisals of ecosystemic diversity can encompass categories ranging from populations to continental-scale biomes. What we need above all, given the rate at which the species have been disappearing, is to know how many species or functional groups are needed for good ecosystem functioning (Purvis and Hector 2000). I will have more to say on this in chapters 8 and 9.

Surviving Recurrent Catastrophes

Existing biodiversity has always been just a transient reality to be changed, either gradually or abruptly, by extensive extinctions of not just individual species, but whole families, and by the evolution of new organisms. Although we now have in many ways quite remarkably detailed accounts of such changes during the Phanerozoic eon, we have much less confidence in explaining the causes of those developments. I will note briefly the mechanisms and effects of both kinds of these catastrophes, ranging from relatively slow but often long-lasting climatic changes to encounters with extraterrestrial bodies (Budyko 1999; Ager 1993; Sharpton and Ward 1991; Berggren and Van Couvering 1984).

I will not look at these changes against the background of the new catastrophism, which is, with its emphasis on discontinuities and sudden qualitative changes evident in the geological record, so much in conflict with the classic Darwinian belief in uniformitarianism, the process of gradual, incremental, evolutionary shifts (for its history in geology, see Albritton 1975 and Berggren and Van Couvering 1984). I call attention to recurrent catastrophes in the biosphere's evolution in order to emphasize life's admirable resilience. Surface life has come very close to extinction on a number of occasions, and yet the biosphere eventually emerged from these global close calls, and from other less momentous catastrophes, richer. Indeed, Courtillot and Gaudemer (1996) concluded that larger mass extinctions that came after a long period of stability seem to have been followed by larger equilibrium levels of biodiversity.

Climate Changes

Climate changes range from gradual and moderate warming or cooling to truly catastrophic excursions, and some of these extreme events can take place within surprisingly short (on geological time scales) periods of time and can be followed by equally extreme reversals (Budyko 1999). The earliest episode of snowball Earth, some 2.3 Ga ago, was associated with the rise of atmospheric O_2 levels, which precipitated a global cooling by conversion of previously high concentrations of CH_4. By far the most pronounced repeat of catastrophic cooling happened during the Neoproterozoic, between 750 and 580 Ma ago. At that time, the combination of small continents scattered near the Equator and increasing rainfall accelerated erosion and the removal of atmospheric CO_2 and triggered a deviation-amplifying process of global cooling.

The most extreme hypothesis for explaining the Neoproterozoic event posits a conversion to snowball Earth within a millennium as an average global temperature of –50 °C put even oceans under more than one kilometer of ice (Hoffman et al. 1998). Only the sturdiest prokaryotes—perhaps akin to today's oscillatorian cyanobacteria perennially frozen into Antarctic ice and revived for just days or weeks in late summer when meltwaters form on or in the ice—could have survived that event (Vincent and Howard-Williams 2000). Eventual accumulation of volcanic CO_2 triggered a bout of warming, with average temperatures rising to 50 °C. Other simulations of the event see a less extreme outcome, with an equatorial belt of open water that would have provided a refuge for multicellular organisms (Hyde et al. 2000).

None of the subsequent ice ages has come close to the Neoproterozoic freeze, but life has been imperiled at other times for other reasons. Most notably, the late Permian mass extinction of 250 Ma ago brought the rapid demise of more than half of the families of marine genera and of more than 90% of all marine species as the immobile Paleozoic sea-bottom communities dominated by brachiopods, bryozoans, and crinoids were replaced with mollusk-dominated ecosystems (Erwin 1993). The terrestrial effect was similarly quick and widespread, with 70% of vertebrate genera disappearing and a massive loss of rooted plants (Ward, Montgomery, and Smith 2000). Massive volcanic activity that ejected close to two million km^3 of lava and created the vast basaltic Siberian Traps appears to be the most obvious cause of this mass extinction. Giant volcanic eruptions have been linked with a number of other major extinctions during the past 500 Ma (Renne and Basu 1991; Courtillot 1999).

But because dates for both basalt floods and extinctions cannot be narrowed to less than a few million years, other explanations must at least be considered. They include injection of CO_2 into surficial environments by the overturn of anoxic deep oceans, a consequence of reduced greenhouse gas effect and the resulting growth of ice sheets and sea ice (Knoll et al. 1996; Wignall and Twitchett 1996), and global warming so potent that it killed most of the plant life (McElwain et al. 1999). Explanation of more recent ice ages has relied heavily on astronomical theory linking the eccentricity of the Earth's orbit to the changes in insolation and climate (Milanković 1941; Hays et al. 1976; Berger 1991). This link, never as close as originally claimed, now appears to be much more complex.

Henderson and Slowey's (2000) dating for the end of the penultimate ice age confirms orbitally driven deglaciation either in the Southern Hemisphere or in the tropics, but not in the Northern Hemisphere. Recent findings also indicate a surprisingly high frequency of abrupt climate changes, some taking place in less than a decade, during the last 100,000 years (Dansgaard et al. 1993; Broecker 1995; Alley 2000). And yet counterintuitively, the link between climate and evolution has not been all that strong. The best available evidence shows that the rate at which new species have been accumulating throughout the Phanerozoic eon has been virtually constant, at about 400 per million years (Rosenzweig 1997). Other research confirms that the rate may have been roughly constant (Alroy 2000; Newman 2001). This conclusion is in contrast to a long-held belief that biodiversity has increased substantially during the past 250 Ma. At the same time, there is abundant proof that climatically induced changes including the Quaternary ice ages, have had major genetic consequences through colonization and mixing of species (Hewitt 2000).

And a more detailed look at the fossil evidence of mammalian evolution of the past 80 Ma shows that since the diversity of mammals rose sharply during the 10 Ma following the Cretaceous low, it has settled into oscillations around a more or less stable plateau, with almost no correlations with numerous climatic shifts of the past 50 Ma (Alroy et al. 2000). Recent millennia have been climatically benign—unusually stable and thankfully devoid of any massive basalt flows or sustained series of volcanic eruptions—but one study of an ecosystem's destruction by Krakatau's 1883 eruption and its subsequent recovery demonstrates the capacity for life's impressive comeback on a local scale (Thornton 1996).

Encounters in Space

The Earth is constantly showered with myriads of microscopic dust particles and with larger bits of universal de-

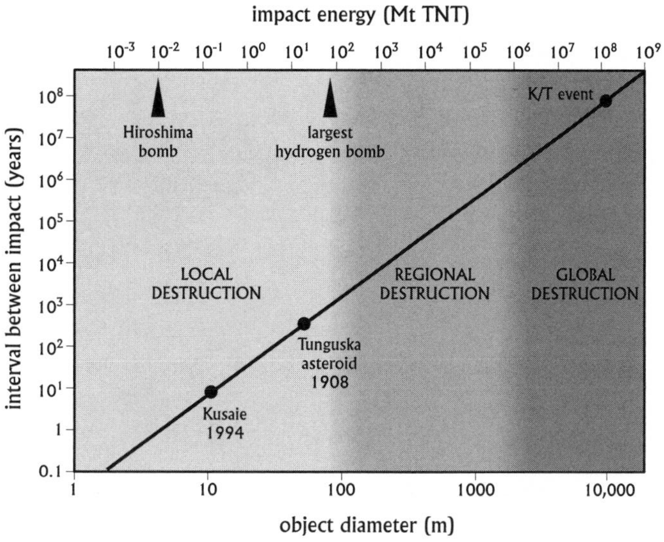

3.15 Intervals between impacts of extraterrestrial bodies of various diameters. Tunguska-like asteroids, whose energies are comparable to those of large hydrogen bombs, come once every few centuries. Based on Jewitt (2000).

bris. Although the latest reevaluation shows that there are only about half as many near-Earth asteroids with diameters larger than 1 km as previously estimated (Rabinowitz et al. 2000), the chances of destructive encounters remain high. Once every thousand years comes an object with a diameter of 100 m, leaving behind such impressive craters as Arizona's famous Barringer. Once every million years arrives a 2-km object whose impact, equivalent to 1 Mt of trinitrotoluene (TNT), is felt globally; and once every 100 Ma is the Earth is jolted by an asteroid, or a comet, 10 km across (fig. 3.15). Such impacts, equivalent to more than 10^8 Mt TNT, leave behind multiring craters with diameters in excess of 150 km—and a biosphere transformed by immense tsunamis, huge volumes of debris floating in the atmosphere, acid rain, reduced insolation and temperatures, and extensive fires.

More than 60 large (diameter in excess of 10 km) impact craters have been discovered so far, most being younger than 200 Ma (Pilkington and Grieve 1992; Dressler et al. 1994; Hodge 1994). Erosion, sedimentation, metamorphosis of rocks, and plate tectonics have combined to eliminate the record of numerous impacts that took place during the Archean eon. One of the three known craters with diameters in excess of 150 km, Vredefort in South Africa (1,970 Ma bp), is deeply eroded; another one, Sudbury in Ontario (1,850 Ma bp), is tectonically deformed. The structure of the Chicxulub crater, buried under 300–1100 m of Tertiary carbonate rocks in the northern Yucatan

Peninsula, was reconstructed by means of gravimetric and magnetometric investigations, and its identification bolstered the most famous theory about the consequences of large impacts on the biosphere, an explanation of a mass extinction that occurred 65 Ma ago.

Long a matter of speculation, this extinction, exemplified by the demise of dinosaurs on land and ammonites in the sea, was attributed by Alvarez et al. (1980) to a collision with a large asteroid. A thin iridium layer deposited at the Cretaceous-Tertiary (K/T) boundary, discovered by Walter Alvarez near the Italian town of Gubbio, was the main evidence of the impact. A subsequent search uncovered many similar layers, as well as quartz lamellae considered clear signs of impact, worldwide: iridium is extremely rare in the Earth's crust, but it is a constituent of space objects. The postulated mode of extinction was a global shroud of impact-generated dust that led to drastically reduced photosynthesis. Identification of the impact site a decade after the theory was published provided a seemingly incontrovertible confirmation.

The crater was first noticed because of a concentric pattern in gravity and magnetic field data, and its transient cavity was eventually determined to be about 100 km across, with the outermost ring about 195 km in diameter (Morgan et al. 1997; fig. 3.16). If an asteroid created this crater, it had a diameter of about 12 km; if a comet did it, it was 10–14 km across, depending on its impact velocity. The transient cavity was 35–40 km deep, the maximum depth of excavation 12 km, and the maximum uplift of the transient crater rim was about 8 km. But the Chicxulub discovery did not close the debate on K/T extinction.

The controversy generated by the Alvarez theory has led to a large number of publications ranging from emphatic endorsements to caustic dismissals (Albritton 1989; Glen 1990; Archibald 1996; MacLeod 1996; Officer and Page 1996; Powell 1998). Regrettably, Alvarez (1997) offers neither an adequate review of the counterarguments nor a refutal of them. The initially proposed killing mechanism (blocking out sunlight and drastically reducing photosynthesis) is considered too simple. Acid rain from the mass of sulfur injected into the atmosphere or from nitrogen compounds generated by shock heating of the atmosphere, loss of a part of the atmosphere, immense wildfires, and CO_2 release by the impact are among the extinction mechanisms now suggested.

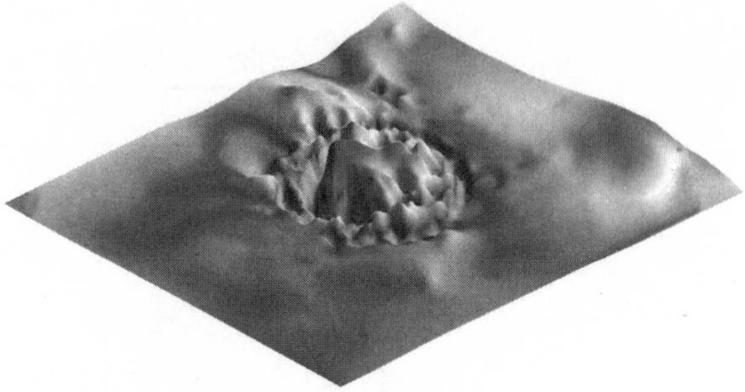

3.16 Chicxulub crater, buried under more than 1 km of Yucatan limestone laid down since the impact. The map shows the 200- to 300-km-wide crater centered near Merida and a 3-dimensional view of the multiring crater derived from aeromagnetic measurements of gravity anomaly. Based on images available at <http://icdp.gfz-potsdam.de/html/chicxulub/ICDP-Chix/>

Critics see iridium anomalies and extinctions as two fundamentally different phenomena and point out a stairstep pattern of extinction before the end of the Cretaceous: sudden mortality is thus an illusion, and the impact hypothesis is not necessary to account for the extinctions (Ager 1993; Smith and Jeffrey 1998). I particularly identify with Ager's (1993) sentiment concerning dinosaurs:

"Always it comes back to the extinction of the dinosaurs. I must admit to being a little tired of those stupid great beasts. . . . Their importance, in my view, is grossly exaggerated" (p. 186). Like all large animals, dinosaurs were exceptionally vulnerable to extinction, however caused, and they were very much in decline before the end of the Cretaceous, with only a handful of the known total of some 350 genera alive by the period's end.

Massive volcanic eruptions are a more likely cause of the extinction, as they could inject enough sulfur and CO_2 into the atmosphere to generate extremely acid rains and cause global warming, and a long period of intensive volcanic activity can also explain the iridium enrichment. The event also had a curiously limited effect on birds and mammals: combined molecular and paleontological evidence shows a mass survival of birds across the K/T boundary, and incremental changes, rather than any explosive post-impact radiation, for both birds and mammals (Cooper and Penny 1997). In any case, the biosphere has had to cope with at least a score of such events, and none of them was crippling enough to terminate it. Neither were high-energy bursts from nearby supernovas (Wheeler 2000). The solar system is within 100 parsecs of a supernova explosion every 2 Ma and within 10 parsecs only once every 2 Ga. Every explosion releases about 2.5×10^{28} Mt of TNT equivalent (Burrows 2000). During the past 500 Ma there should have been as many as ten events exposing the biosphere to 500 roentgens, lethal to most vertebrates, but there is no obvious record of such supernova-related disruptions.

4

Energizing the Biosphere

Solar Radiation and the Earth's Heat

Activated by radiation, the matter of the biosphere collects and redistributes solar energy, and converts it ultimately into free energy capable of doing work on Earth . . .
Vladimir Ivanovich Vernadsky, *Biosfera*

The Sun appears to be poured down, and in all directions indeed it is diffused, yet it is not effused. For this diffusion is extension.
Marcus Aurelius, *Meditations*

The biosphere is energized by two very different sources: by the Sun's radiation, and by the Earth's heat. When compared in terms of average global power density, the first flux is three orders of magnitude larger: insolation (the flux of solar radiation reaching the Earth's surface) averages nearly 170 W/m^2, whereas the mean density of heat lost through the Earth's crust is merely 85 mW/m^2. Solar radiation powers photosynthesis, the biosphere's most important energy conversion, as well the global atmospheric circulation and the cycling of water; in turn, these transformed solar flows are the major determinants of the Earth's climate and the principal sculptors of the Earth's surface. But the Earth's relatively small heat flux has had an enormous impact on the evolution and composition of the biosphere, as it energizes the planet's plate tectonics, the incessant process of formation and destruction of the ocean floor that also reconfigures continents and generates volcanic eruptions and earthquakes. These processes—ranging from the slow formation of new oceanic crust and the uplift of massive mountain ranges to cataclysmic discharges of pent-up tectonic energies—explain many particulars of biospheric history and organization.

Adequate understanding of all of these energetic phenomena is fairly recent. The old puzzle of the Sun's energy origins was solved only during the late 1930s by Hans Bethe, Charles Critchfield, and Carl Friedrich von Weizsäcker (Bethe 1996). Global studies of the Earth's radiation balance could become highly accurate only with measurements from various Earth observation satellites.

Plate tectonics prevailed as the key geological paradigm only during the 1960s (Kahle 1974; Motz 1979; Oreskes 1999). A large number of heat flow measurements from boreholes was amassed only by the early 1970s (Pollack and Chapman 1977). And the first unambiguous detections of solar variability were obtained by radiometers carried by Nimbus 7 satellite launched in 1978 (Fröhlich 1987). Still, many fundamental questions concerning both our star and the interior of our planet remain unanswered. How does the Sun generate its enormous magnetic fields, which engender most of its storms? Why does it concentrate this magnetism into huge dark sunspots, and why does their frequency vary so regularly, in roughly eleven-year cycles? How exactly is the process of ocean floor spreading driven from deep in the Earth's mantle?

Sun and Solar Radiation

Basic astronomic facts are easy to state (Sturrock 1986; Phillips 1991). The Sun is located in one of the spiral arms of our galaxy, about 33,000 light years (ly) from its center and near the galactic plane. Besides the faintly reddish binary α Centauri (4.3 ly away), the other nearby stars are βEridani (10.8 ly), 61 Cygni A and B (11.1 ly), and ε Indi (11.3 ly). The Sun's radius is 696.97 Mm (compared to the Earth's roughly 6.5 Mm), and its mass (2×10^{33} g) is five orders of magnitude larger than that of the Earth. The planet's mean orbital radius is 149.6 Gm (the distance known as one astronomical unit, or AU), with perihelion at 147.1 Gm and aphelion at 152.1 Gm.

The first measurements of solar irradiance go back to the late 1830s, and several decades later, the first electric pyrheliometers gave fairly accurate results of the total solar flux. Studies of the solar spectrum had already made the first real breakthrough in the early nineteenth century. In 1814 Johann von Fraunhofer improved on Newton's famous prismatic dispersal of colors by introducing the light through a fine slit and focusing it with a lens after the prismatic refraction. When the light was viewed in this way, every rainbow color was interrupted by a multitude of fine black lines, and Fraunhofer denoted the strongest ones with letters. The explanation for the lines was supplied only in 1902 by Gustav Kirchhoff and Robert Bunsen: the dark lines, now exceeding 25,000, are spectral signatures of the elements present in stars (Friedman 1986).

In contrast, the correct understanding of stellar energetics came only during the late 1930s, and the era of continuous monitoring of the Sun arrived only in February 1996, when the Solar and Heliospheric Observatory (SOHO) satellite, launched on December 2, 1995, reached its permanent position—the inner Lagrangian point where it is balanced between the Earth's and the Sun's gravity—and its twelve instruments began unprecedented continuous examination of the star. A variety of images acquired by these instruments is available both in real time and for long-term comparisons as well (Fleck et al. 1995; fig. 4.1).

On the Main Sequence

The Sun sits almost perfectly in the middle of the main sequence of the Hertzsprung-Russell diagram, which plots the distribution of absolute visual magnitude against the spectral class for stars of known distance (fig. 4.2). The main sequence contains so-called dwarf stars: besides the binary α Centauri, the Sun's closest neighbor, Sirius and Vega are other well-known stars in this, the fifth (V) class. Hydrogen makes up about 91% of the Sun's huge mass, with helium accounting for all but about 0.1% of the rest; carbon, oxygen, and nitrogen are the most abundant elements among the minor constituents. Its spectral type (G2) means that the Sun belongs to stars of yellow color, with characteristic lines of ionized calcium and metals and

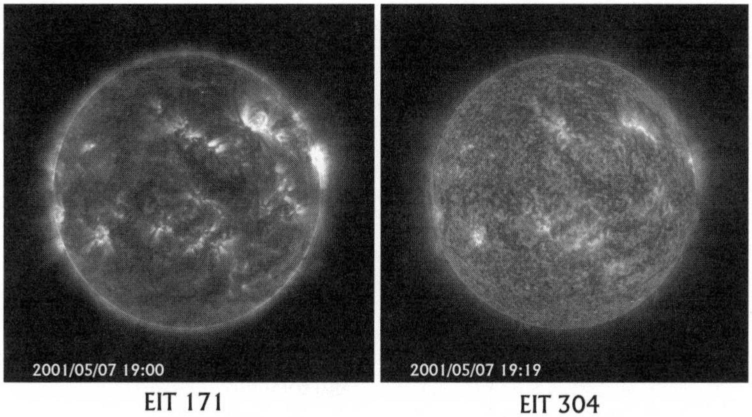

4.1 These SOHO images show the full solar disk captured at 171 Å (Fe IX/X) and 304 Å (He II) by the Extreme Ultraviolet Imaging Telescope (EIT) on May 7, 2001. Color images, archival and in real time, are available at <http://sohowww.nascom.nasa.gov/>

with effective temperature within the range of 4,900–6,000 K. Millions of Sun-like stars among the roughly 100 billion radiant bodies of our galaxy belong to the same (G2 V) category.

Proton-proton (p-p) reactions provide nearly four-fifths of the Sun's energy; the reaction chains proceeds as follows:

$$^{1}H + {}^{1}H \rightarrow {}^{2}D + \text{positron} + \text{neutrino}$$
$$^{1}H + {}^{2}D \rightarrow {}^{3}He + \gamma\ \text{rays}$$
$$^{3}He + {}^{3}He \rightarrow {}^{4}He + {}^{1}H + {}^{1}H$$

The first two reactions must take place twice for every third reaction to proceed; the net result is the fusion of four protons to form a helium nucleus (Phillips 1992).

The probability of the first step in the chain, the collision of two protons to form a deuteron, is exceedingly low, and random collisions that happen at a speed sufficient to overcome the electrical barrier are made possible only by the enormous quantity of protons in the Sun's core. The second step, absorption of another proton by a deuteron, is accomplished almost instantaneously: light helium (^{3}He) formed by this reaction is converted to normal helium (^{4}He) only after several millions of years, resulting in the return of two free protons available for another p-p cycle. This cycle provides the bulk of thermonuclear energy in all stars with core temperatures of up to about 17 MK; above that level the much less important carbon-nitrogen cycle becomes dominant (Cox et al. 1991; Kaler 1992; Phillips 1992).

With about 9.2×10^{37} reactions taking place every second, and with every p-p chain liberating 4.2×10^{-27} J, fusion in the Sun's core consumes 4.4 Mt of matter per second and, according to Einstein's mass-energy relationship, releases 3.85×10^{26} J. This immense energy flux is

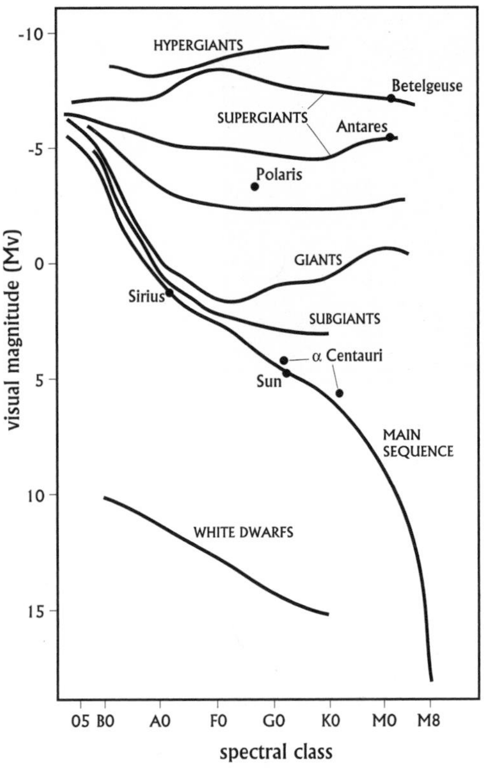

4.2 A modern version of the Hertzsprung-Russell diagram showing the branches of star sizes from hypergiants to white dwarfs as well as a few well-known stars. Based on Kaler (1992).

transported outward through the radiation zone (taking up about 45% of the star's radius) and then through the outermost convection zone (about 30% of the Sun's diameter) to the photosphere. The Sun's temperature declines from 10^7 K in the core to 10^6 K in the radiative zone and 10^4 K in the outer layer of the convection zone.

The finely granulated and opaque photosphere is a relatively thin (about 200 km), sharply defined surface layer that radiates the bulk of both the visible and infrared wavelengths emitted by the Sun. Isotropic radiation would produce an average flux of 63.2 MW/m^2 of the Sun's photosphere, corresponding to an effective photospheric temperature of 5778 K. According to the Stephan-Boltzman law, total flux is proportional to the fourth power of temperature (with σ equal to 5.67×10^{-8} W/m^2):

$$F = \sigma T^4.$$

Total luminosity will be, obviously, the product of surface and total flux:

$$L = 4\pi r^2 \sigma T^4.$$

The photosphere is repeatedly marked by sunspots—regions of reduced temperature that differ widely in duration (hours to several weeks) and size (many are much larger than the Earth)—surrounded by bright, small faculae that expose the hotter underlying gas. Sunspot frequency follows a well-known eleven-year cycle as the locations of the spots in each hemisphere converge toward the solar equator (Radick 1991). A relatively narrow layer of the chromosphere separates the photosphere from the solar corona, an extremely hot (about 2 MK) zone made up mostly of completely ionized hydrogen gas, which is the source of a small part of total solar radiation but sends out X-rays and extreme (10–100 nm) ultraviolet radiation. Local temperature within coronal flares, intense electromagnetic explosions whose frequency also follows the eleven-year sunspot cycle, can reach 29 MK, more than in the Sun's core. These flares are the source of the great terrestrial magnetic disturbances that are so spectacularly signaled by auroral displays in high latitudes.

The Sun's rather modest brightness (visual magnitude +4.83) is still enough to make the star visible with naked

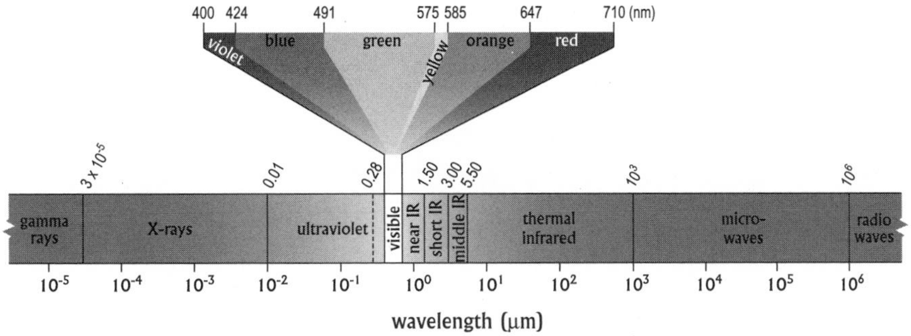

4.3 The electromagnetic spectrum, ranging from gamma rays to radio waves, and its visible part between blue and red light (Smil 1999d).

eye from a distance of 20 parsecs, that is, an equivalent of about 620 Pm, or roughly 4.1 million AU. But the Sun is dim indeed compared to such bright stars as Betelgeuse (α Orioni), whose light reaches us from 320 parsecs away, or Deneb (α Cygni), which is 500 parsecs distant. The heliosphere, the space permeated by the solar wind—a continuous but often-gusting flux of charged particles—extends far beyond the orbit of the solar system's outermost planets, slowing rather abruptly when the outflowing plasma cannot prevail over the small inward pressure of the interstellar gas (Jokipii and McDonald 1995).

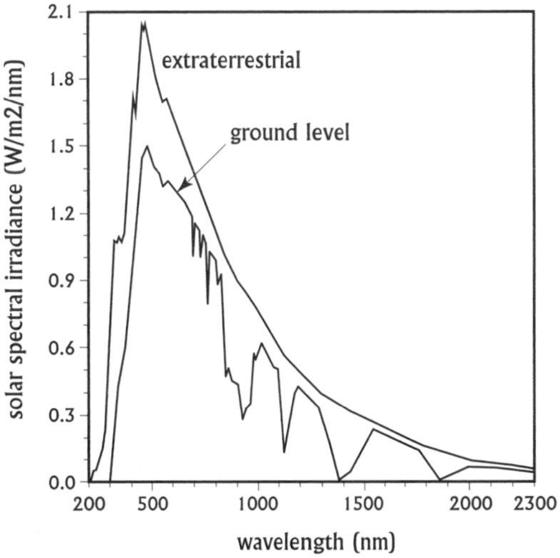

4.4 Comparison of the solar spectrum in space and the radiation received at the ground after its passage through the Earth's atmosphere (Smil 1999d).

The Solar Spectrum

The Sun's spectrum extends from wavelengths shorter than 0.1 nm to those in excess of 1 m, that is from gamma rays all the way to radio waves (fig. 4.3). The distribution of its wavelengths closely resembles that of a black body (an object in thermal equilibrium) radiating at near 6,000 K (fig. 4.4). Two fundamental physical laws explain the property of radiation emitted by black bodies, the term used for materials that have a continuous radiation spectrum in all wavelengths and absorb all radiant energy that reaches them. With actual photospheric temperature at about 5800 K, the wavelength of the maximum emission would be (according to Wien's displacement law, which relates wave frequency to the temperature of a radiating body) almost exactly 500 nm, near the lower limit of green light (491–575 nm). The formula for Wien's law (with the maximum wavelength in μm) is

$$\lambda_{max} = 2989/T.$$

According to Planck's law, expressing the distribution of energy emitted from a black body as a function of its wavelength and temperature, the energy flux density emitted at this maximum wavelength will be about 2 $W/m^2 \cdot nm$, and the spectral irradiance of other wavelengths will diminish rapidly in both directions. The formula for Planck's radiation law (with the total energy radiated from the body measured per cm^2 per unit wavelength at the wavelength λ) is

$$R_1 = 37.418/\lambda^5(e^{14,388/\lambda T - 1}).$$

Integrating the areas below the irradiance curve yields about 38% of all solar energy carried by visible wavelengths, less than 9% by ultraviolet radiation (<400 nm), and 53% by the infrared frequencies (>700 nm). The standard division of the solar spectrum into ultraviolet (UV), visible, and infrared (IR) segments is based on the effects of solar radiation on biota.

Exposure to progressively shorter UV wavelengths is harmful to all but a few admirably radiation-resistant living organisms; the highest frequencies are lethal to most organisms (although specific thresholds vary enormously), and the lower ones are germicidal and skin-burning. The latter effect peaks at 297 nm, within the UVB band (290–320 nm), whereas longer wavelengths between 320–400 nm (UVA) are much less harmful. Marine phytoplankton is particularly sensitive to higher UV levels (Demers, de Mora, and Vernet 2000). Elimination of both UVA and UVB increases the productivity of various phytoplankton two to four times, whereas an increase in UVB substantially reduces photosynthesis. Measurements in Antarctic waters confirmed primary productivity declines of 6–12% during springtime ozone depletion (R. C. Smith et al. 1992). Many higher plants are similarly vulnerable to UV radiation, and, as a result, their crops could be reduced.

In humans and animals, excessive exposure to UV radiation brings increased risks of skin cancers and various eye complications (Tevini 1993; Altmeyer et al. 1997). Every 1% increase in UVB flux causes a 2–3% rise in the incidence of basal and squamous cell carcinomas, two of the most common and, fortunately, usually nonlethal skin cancers. The relationship of higher UVB exposures to rates of malignant melanoma is harder to quantify, but the overall incidence of melanoma also increases with increased UVB exposure. Even small UVB increases may cause higher incidence of cataracts, conjunctivitis, photokeratitis of the cornea, and blepharospasm. Yet another potentially worrisome effect is a weakening of human and animal immune systems through damage to suppressor lymphocytes, which protect the organism against skin cancer and against bacterial and parasitic invasions through the skin.

Absorption of UVB radiation by stratospheric O_3 has been the only effective way of protecting the biosphere from an excessive UVB flux. Consequently, the UVB effect on biota has been relatively unimportant ever since the appearance of oxygenated atmosphere resulted in the formation of a stratospheric ozone layer that prevented all but a tiny share of that radiation to reach the biosphere (see chapter 3). (Threats to the integrity of this protective layer, which until recently seemed to be well beyond the human reach, will be described in chapter 9.) Visible wavelengths, extending from the 400 nm of the deepest violet to the 700 nm of the darkest red, drive photosynthesis and reveal the world to all light-sensitive heterotrophs. Plant pigments, dominated by chlorophylls, absorb the radiation mostly between 420–490 nm and 630–690 nm (Hall and Rao 1999). Photosynthesis is thus driven by blue and red light, but IR radiation participates in the photosynthetic process indirectly, by warming plant tissues and transpiring water. The peak sensitivity of human eyes is for green (491–575 nm) and yellow (576–585 nm) light; the maximum visibility is at 556 nm (3.58×10^{-19} J/photon).

A Variable Star

As attested by the traditional term used for describing the solar flux incident on Earth, the "solar constant," the Sun's radiative flux used to be seen as invariant. Because there is hardly any attenuation of the Sun's radiation as it travels through interplanetary space, the irradiance at the top of the Earth's atmosphere is easily calculated by dividing the Sun's total energy output (3.85×10^{26} W) by the area of the sphere with radius equal to the Earth's mean orbital distance from the star (149.6 Gm). Using the above values gives about 1,370 W/m^2. Terrestrial measurements of the solar constant have been always compromised by the intervening atmosphere, and conventional stellar photometry could generally detect only changes in excess of 1%, although several observatories achieved accuracy of 0.5%. During the 1960s and early 1970s NASA used a solar constant of 1,353 W/m^2 as the design value for its space vehicles, although unsystematic pre-1980 measurements from high-altitude balloons, planes, rockets, and spacecrafts indicated an increase of 0.029% per year after 1967 (Fröhlich 1987).

The first systematic extraterrestrial studies of solar irradiance became possible only through the Earth Radiation Budget (ERB) radiometer on NIMBUS 7, launched in 1978, and by the Active Cavity Radiometer Irradiance Monitor (ACRIM I) on the Solar Maximum Mission (SMM) satellite deployed in 1980. The Earth Radiation Budget Satellite (ERBS) followed in 1984, and ACRIM II, on the Upper Atmosphere Research Satellite, in 1991 (Foukal 1990; Willson 1997). Although we now have more than two decades of overlapping measurements with increasingly more reliable radiometers, the spread of the results is nearly 0.5%, consistent with an uncertainty of about 0.4% associated with each device (fig. 4.5; Quinn and Fröhlich 1999). During 1980, the first year of overlapping measurements, ACRIM I showed a mean of about 1367 W/m^2, now generally accepted as solar "constant," and the less accurate ERB indicated 1371 W/m^2 (Foukal 1987).

Satellite monitoring has revealed a complex pattern of tiny short-term variations that were impossible to record from Earth-bound observations. Spectral irradiance variation from the maximum to the minimum of the eleven-year solar activity cycle is very large, at least 100%, for wavelengths shorter than 100 nm, but the star's total irradiance declines by only about 0.1% between the peak and the trough of the cycle (NRC 1994). Short-term (days to weeks) variations caused by the passage of dark sunspots and bright faculae across the rotating Sun are as much as 0.2% percent around the mean (Foukal 1987). Studies of Sun-like stars also indicate that similar (0.2–0.3%) longer-term changes are also possible. Intercycle comparisons also show that the total irradiance had increased by

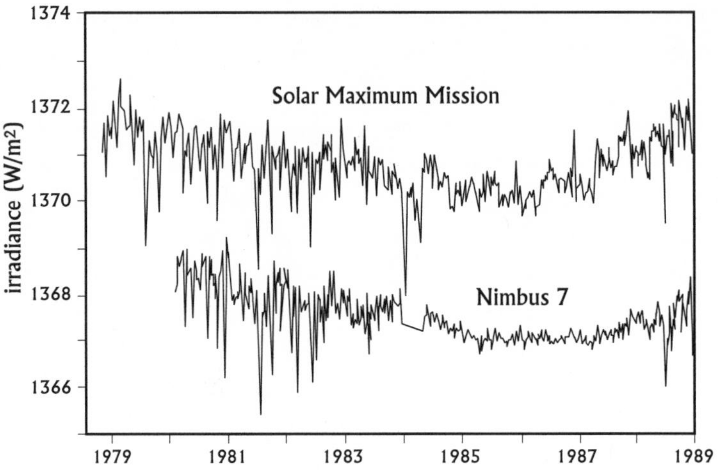

4.5 Comparison of the first decade of measurements by radiometers on two satellites shows the short-term fluctuations of solar output amount mostly to less than 0.15% of the total flux. Based on Willson (1997).

0.036% between the minima of solar cycles 21 (1986) and 22 (1996) (Willson 1997). The origin of these cyclical variations remains uncertain: convection in the solar interior is the most likely explanation, but obviously not one open to observation. Satellite observations show that the Sun's shape and temperature vary with latitude in an unexpectedly complex way, but solar oblateness does not appear to vary with the solar cycle (Kuhn et al. 1998).

Possible effects of the total solar irradiance variability on the Earth's climate will be discussed later in this chapter, together with the implications resulting from a significantly lower energy output of the young Sun. But first I will outline in some detail the radiation balance of the Earth, including complex transformations of solar energy in the biosphere.

The Radiation Balance of the Earth

Although the disc with the Earth's diameter (12.74 Mm) intercepts a mere 4.5×10^{-10} of the Sun's total irradiance, the rate of this intercept, 174.26 PW, and its annual aggregate, 5.4×10^{24} J, are vastly larger than any other natural or anthropogenic energy flow on the Earth. Terrestrial heat flow amounts to about 44 PW, or 1.29×10^{21} J/year. And in 2000 the global consumption of all fossil fuels and primary (hydro- and nuclear) electricity reached about 11.25 TW (355 EJ/year), or a mere 0.0065% of the solar flux. Comparisons made in terms of power densities (W/m^2) and taking into account only that share of radiation that is actually absorbed by the atmosphere and the Earth's surfaces shave only an order of magnitude from these enormous differences. Incoming radiation is distributed over a rotating sphere (geoid, to be more precise), which means that the radiation received at the top of the Earth's atmosphere is only a quarter of the intercept value for the disc of the same radius as the planet, or about 342 W/m^2. But a significant share of this radiation is bounced back into space. This reflected share (albedo) is commonly in excess of 80% for thick clouds or for fresh snow, whereas most vegetated areas have albedos below 20% and seas between 7% and 14%.

Albedo extremes in individual categories range from 2–3% for thin, wispy cirrus clouds to 90% for thick cumulonimbi; less than 10% for dark soils to 25–30% for sandy ground; less than 20% for dry grasslands to 5% for some coniferous forests; and 4% for water at Sun angle 90° to 60% at Sun angle 3°. Planetary albedo is thus seasonally and hemispherically variable. Presatellite calculations of the Earth's average albedo ranged between 28% and 35% (Kessler 1985), and a long-term average derived from satellite observations is 31%, or 102 W/m^2 (Kiehl and Trenberth 1997). The amount of incoming radiation actually absorbed by the atmosphere and by the Earth's surface is then no greater than 235 W/m^2.

This flux is the actual energizer of the biosphere: it puts the atmosphere and oceans in motion, it powers the water cycle and geomorphic processes, and it drives the conversion of simple inorganic compounds to organic matter through photosynthesis. Following its fate is a rather complicated enterprise that has to account for scatterings, absorptions, and reradiations. There must be an eventual balance between the incoming and outgoing radiation, but changing concentrations of greenhouse gases have been altering the Earth's effective radiative temperature for billions of years, and these natural shifts have been recently potentiated by anthropogenic emissions.

Solar Energy Transformations

After its long journey from the Sun, solar radiation arrives at the top of the Earth's atmosphere virtually unchanged, but its transformation begins as soon as it encounters the atmospheric gases. In the thermosphere, at altitudes above 80 km, temperatures vary by hundreds of degrees (from about 600 K during solar cycle minima to 1300 K during the maxima) because of the changing absorption (the transfer of radiant energy to matter) of highly variable X-ray, extreme UV, and UV radiation. The bulk of incoming shortest wavelengths, below 120 nm, is absorbed by atomic and molecular nitrogen. Further penetration of most of the radiation between 120–300 nm is blocked at altitudes below 130 km by molecular oxygen and particularly by ozone, whose concentrations peak at about 30 km.

Visible wavelengths pass largely unhindered through the troposphere, but water vapor and CO_2 (and also CH_4, N_2O, and tropospheric O_3) absorb a great deal of the infrared flux, with absorption peaks between 1.5–2, 2.5–3.5, and 4–8 μm (fig. 4.6). Consequently, the generally depleted and jagged solar spectrum recorded at the Earth's surface departs noticeably from the extraterrestrial curve (fig. 4.4). The global mean of the total insolation—the solar radiation absorbed by surfaces—amounts to about half (168 W/m^2) of the incoming total and is the source of two key biospheric inputs: it heats all rocks, soils, waters, plants, and living organisms, and some of its wavelengths energize photosynthesis. Seasonal and local variations in total insolation correlate highly with overall cloudiness: the winter hemisphere shows a fairly regular poleward decline; the highest values, in excess of 250 W/m^2, are in the cloudless high-pressure belts of subtropical deserts (fig. 4.7).

Only a small part of absorbed shortwave radiation (24 W/m^2) is returned to the atmosphere as sensible heat;

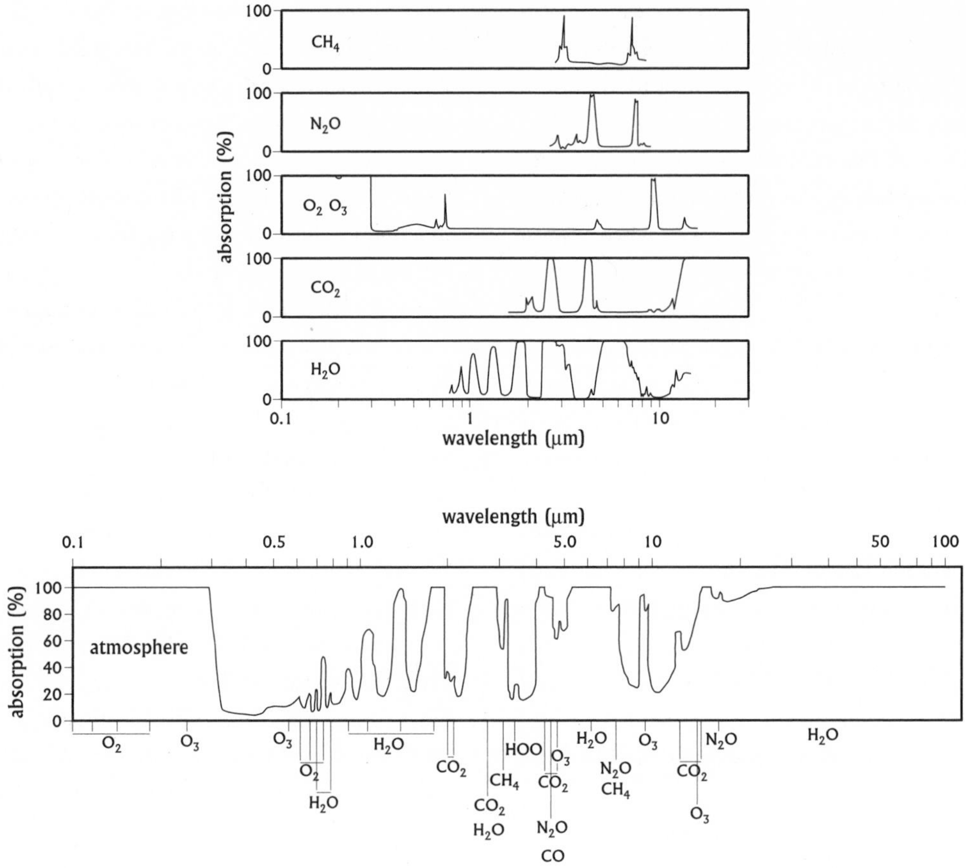

4.6 The top part of the image shows the major absorption bands of six greenhouse gases, and the bottom portion displays their aggregate effect on the UV, visible, and IR wavelengths.

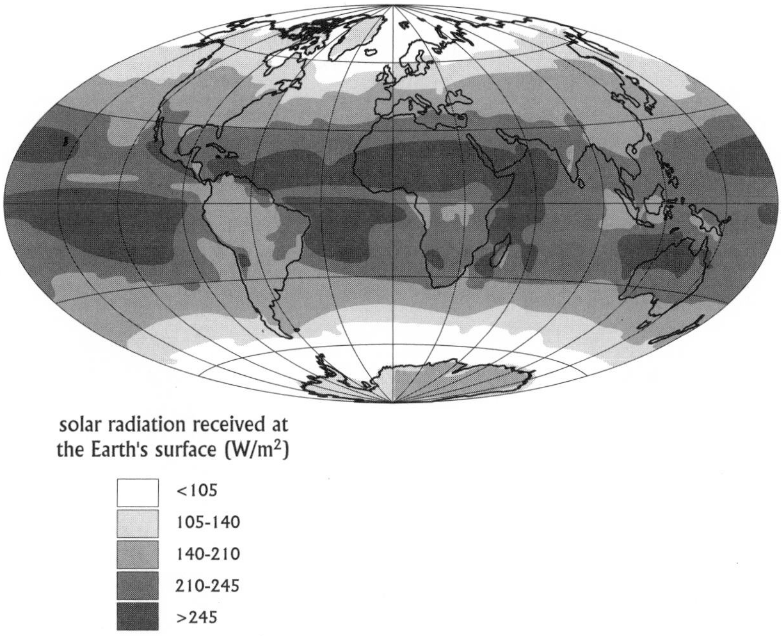

4.7 Solar radiation received at the Earth's surface (downward shortwave flux) is highest in the subtropical latitudes and lowest in polar regions. This image shows annual means based on 96 months of satellite measurements (July 1983–June 1991). Based on the Langley 8-Year Surface Radiation Budget Dataset, available at <http://srb-swlw.larc.nasa.gov>

more than three times as much of it (78 W/m^2) is spent on evaporating and transpiring water, and the rest is reradiated in thermal IR wavelengths. Because of the presence of IR-absorbing gases in the atmosphere, only a small share of this IR flux escapes directly back to space, and most of it is absorbed by atmospheric gases and reradiated both up and down (fig. 4.8). The downward flux is emitted mostly by water vapor (150–300 W/m^2) and by CO_2 (70–75 W/m^2; for more details see the next section). The annual global mean of the flux is 324 W/m^2, with midlatitude continents receiving around 300 W/m^2. This flux is merely a temporary delay of the outward flow of IR radiation, but it is the largest supplier of energy to nearly every ecosystem. After it has warmed the biosphere, the IR radiation must be reradiated to space: its peak flux is at 10 μm.

More than two dozen average global surface-atmosphere energy budgets have been published since the first estimates in 1917, but more accurate balances became

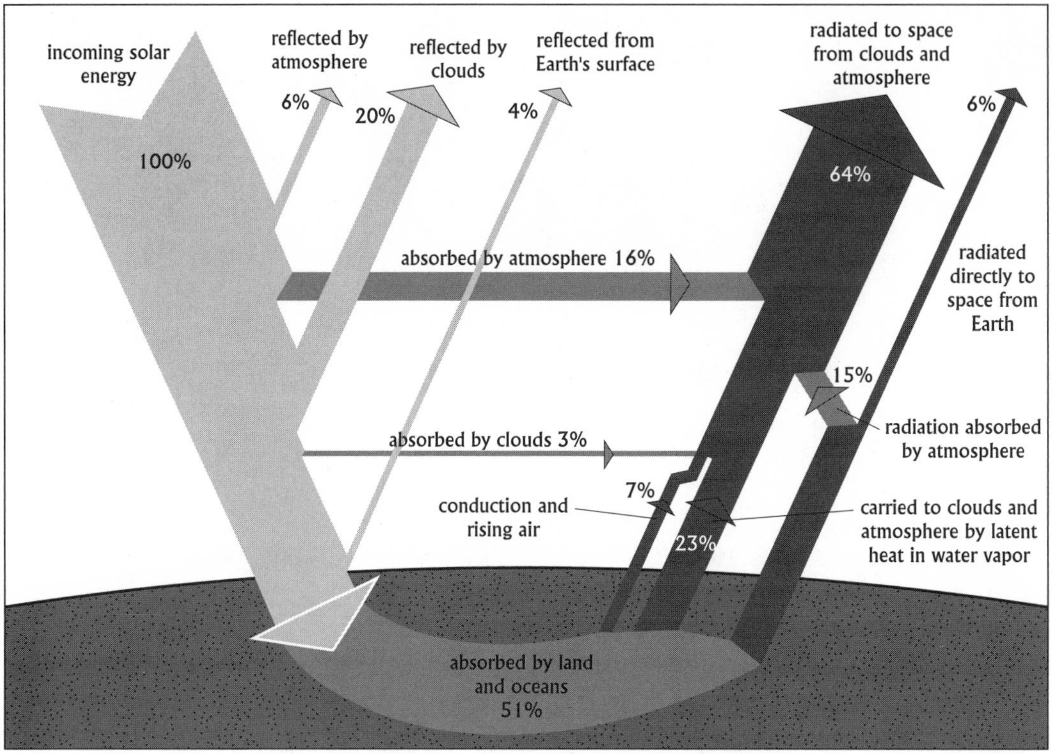

4.8 The Earth's radiation budget, derived from Earth Radiation Budget Experiment satellite monitoring. Based on the graph available at <http://asd-www.larc.nasa.gov/erbe/components2.gif>

possible only with satellite measurements (Dines 1917; Kessler 1985; Ramanathan et al. 1989; Kiehl and Trenberth 1997). The most accurate global information has come from satellite monitoring, above all from the specifically dedicated Earth Radiation Budget Experiment (ERBE), whose instruments were carried by the Earth Radiation Budget Satellite, deployed from a Space Shuttle flight in 1984, and by NOAA-9 and NOAA-10 satellites launched, respectively, in 1985 and 1986. ERBE instruments measure the total solar radiation received by the Earth, the energy emitted by the planet into space, and the Earth's albedo (details on ERBE missions can be found at the ERBE Web site listed in appendix H). The Clouds and the Earth's Radiant Energy System (CERES), whose first instrument was launched in 1997, extends ERBE's measurements to include the top of the atmosphere and global surface radiation.

In spite of this monitoring, not every individual flux of the Earth's complex energy budget is known with a high degree of certainty. Absorption of incoming radiation by the atmosphere is a perfect illustration of an area of fairly significant uncertainty. Rates published during the 1970s were between 61 and 89 W/m^2, a number of recent global circulation models used values between 56 and 68 W/m^2, and numbers derived from satellite-based estimates of surface flux range between 65 and 83 W/m^2, whereas observations at about a thousand Global Energy Balance Archive (GEBA) sites, extended using empirical equations, give the mean value as 98 W/m^2, higher than any of the previously used quantities (Kessler 1985; Arking 1996; Kiehl and Trenberth 1997).

There is also a long history of unexplained anomalous absorption of shortwave radiation by clouds, a phenomenon that cannot be explained by the intermittent and patchy presence of tropospheric aerosols. These anomalies vary surprisingly little with season and location and can amount to as much as 25 W/m^2 higher levels of absorption by the cloudy atmosphere than predicted by theoretical models (Cess et al. 1995; Cess and Zhang 1996). Uncertainties on the order of 20 W/m^2 also prevail for both net shortwave and longwave fluxes at the surface, and extremes of sensible and latent heat rates differ by 10 W/m^2. (The latest rates for all major flows are shown in fig. 4.8.).

Greenhouse Effect

The effective radiative temperature of a planet is simply a function of albedo and orbital distance: the Earth should radiate at 255 K (–18 °C), compared to 216 K for Mars and 229 K for Venus. It can be calculated as follows (Kaula 1968):

$$T_P = 278[(1 - A)^{0.25}/a^{0.5}],$$

where A is albedo and a is mean orbital distance in AU. Venus has an albedo of 0.705, Mars a mere 0.171. The Venusian atmosphere is 96.5% CO_2 and 3.5% N_2, the planet's surface pressure is about 100 times that on the Earth, and its temperature is 457 K; the Martian atmosphere is 95.3% CO_2 and 2.7% N_2, but the planet's surface pressure is a mere 1% of the terrestrial value, and its surface temperature is –53 °C.

But a planet ceases to be a perfect black-body radiator as soon as one or many gases in its atmosphere selectively absorb a part of the outgoing radiation and reradiate it both downward and upward. This absorption of IR radiation is generally known as the greenhouse effect. Svante Arrhenius, one of the first Nobel prize winners in chemistry, was the first scientist to publish a detailed elucidation of this effect (Arrhenius 1896; fig. 4.9). In the absence of water vapor on the Earth's two neighbors, it is the presence of CO_2 that generates the greenhouse effect, a very strong one on Venus, resulting in an average surface temperature of 750 K (477 °C), and a very weak one on Mars, whose surface has a temperature of just 220 K.

The Earth's actual average surface temperature of 288 K (15 °C) is 33 K higher than its black-body temperature because of IR absorption by a number of atmospheric gases. Water vapor is by far the most effective greenhouse gas, accounting for almost two-thirds of the 33 K difference between the Earth's black-body and actual temperature; it has five major absorption bands between 0.8 and 10 μm, the broadest ones centered at 5–8 μm and beyond 19 μm (fig. 4.6). Other major greenhouse gases (listed in descending order of importance) are CO_2, accounting for nearly a quarter of the temperature difference, with isolated absorption peaks centered at about 2.6 and 4.5 μm and a broad band between 12 and 18 μm; O_3, at 4.7 and 9.6 μm; N_2O, absorbing within narrow ranges centered at 4.5, 7.8 and 8.6 μm; and CH_4, absorbing at about 3.5 and 7.6 μm (Kondratyev 1988; Ramanathan 1998). Additional but comparatively minor natural contributions come mostly from NH_3, NO_2, HNO_3, and SO_2.

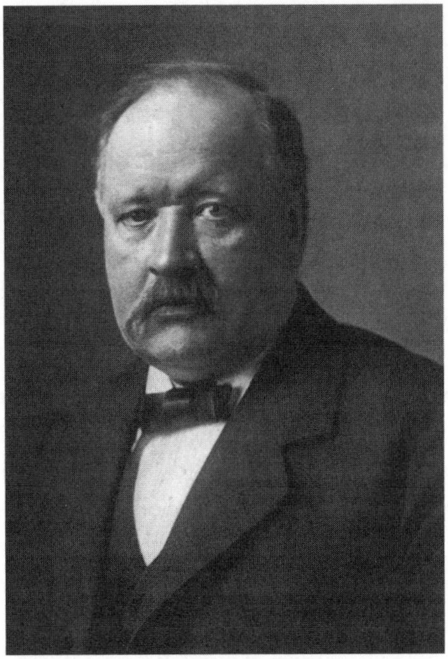

4.9 Svante Arrhenius (1859–1927), one of the first Nobel Prize laureates in chemistry and the first scientist to calculate fairly accurately the eventual tropospheric temperature increase resulting from the doubling of atmospheric CO_2 concentration. Photo from E. F. Smith Collection, Rare Book & Manuscript Library, University of Pennsylvania, Philadelphia.

As I will show later in this chapter, a greenhouse effect amounting to at least 25–30 K must have been operating on the Earth ever since the Archean age. Water vapor, the most important greenhouse gas, could not have been responsible for maintaining relatively stable temperatures on the Earth during the billions of years of life's evolution, because its changing atmospheric concentrations amplify, rather than counteract, the changes of the planet's surface temperatures: water evaporation declines with cooling and increases with warming. Only long-term feedbacks between fluctuating CO_2 levels, surface temperatures, and the weathering of silicate minerals explain the surprisingly limited variability of the Earth's mean tropospheric temperature: lower tropospheric temperatures and decreased rates of silicate weathering will result in gradual accumulation of CO_2—and in subsequent warming (Gregor et al. 1988; Berner 1998).

Approximate reconstruction of CO_2 levels for the past 300 Ma (since the formation of the supercontinent Pangea, whose eventual breakup led to the current distribution of oceans and land masses) indicates first a pronounced rise (to about five times the current level during the Triassic period) followed by a steep decline (Berner 1998; fig. 4.10). Pearson and Palmer (2000) determined the boron isotope ratios of planktonic foraminifer shells first to estimate the pH of surface ocean waters, and then to use these data to reconstruct likely atmospheric CO_2 levels during the past 60 Ma. Their findings point to CO_2 levels above 2,000 ppm 60–50 Ma ago (with peaks above 4,000 ppm), followed by an erratic decline to less than 1,000 ppm by 40 Ma ago, with relatively stable and low (below 500 ppm) concentrations ever since the early Miocene 24 Ma ago (fig. 4.10).

Reliable figures on levels of atmospheric CO_2 are available for only the past 420,000 years, and those thanks to the analyses of air bubbles from ice cores retrieved in Antarctica and in Greenland: they show that preindustrial CO_2 levels never dipped below 180 ppm and never rose above 300 ppm (Raynaud et al. 1993; Petit et al. 1999; fig. 4.11).[1] During the time between the rise of the first high civiliza-

1. Dating of the bubbles must take into account the lag between snowfall and ice formation. In places where snow accumulation can be very slow, as in central Antarctica, the lag time between the age of the ice and the air averages 3,000 years (Delmas 1992).

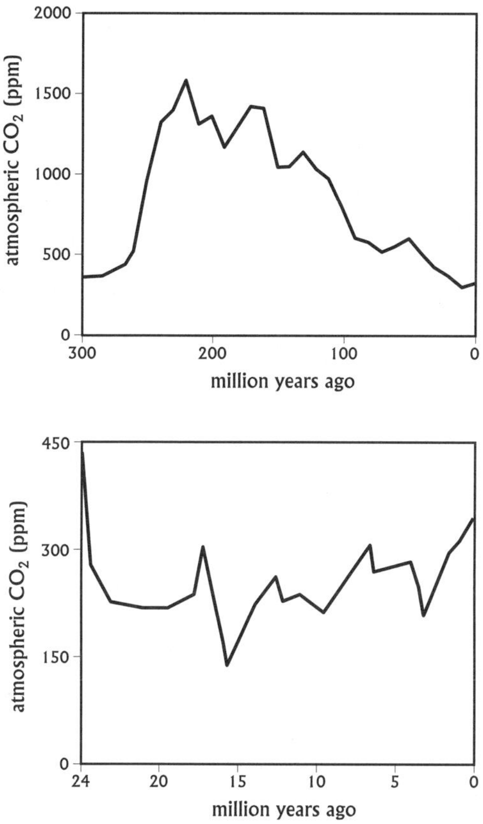

4.10 Atmospheric CO_2 concentrations during the past 300 and 24 million years. Based on Berner (1998) and Pearson and Palmer (2000).

tions (5,000–6,000 years ago) and the beginning of fossil-fueled era, atmospheric CO_2 levels fluctuated within an even narrower range of 250–290 ppm. Subsequent anthropogenic emissions pushed atmospheric concentrations of CO_2 to 370 ppm by the year 2000. Uncertainties concerning the Earth's future radiation balance that arise from increasing concentrations of anthropogenic greenhouse gases and from their complex feedbacks in the biosphere will be discussed in the last chapter.

The Atmosphere and the Ocean

Atmospheric composition is critical for maintaining the biosphere's average temperature well above the freezing point, but life on the Earth is also fundamentally affected by the constant motion of the medium that distributes heat and moisture, carries microbes, spores, pollens, and seeds, and helps erode soils and rocks. Because of the low density of the medium, no more than 2% of the planetary heating is needed to power atmospheric motion in order to offset energy dissipated in turbulence (aloft) and friction at the surface (Lorenz 1976).

The Earth's energy balance would be inimical to life in the absence of water not only because the compound is the principal greenhouse gas, but also because it is a ubiquitous carrier of latent heat and the planet's largest reservoir and regulator of heat. The most remarkable fact regarding water's indispensable role in the Earth's energy balance is that this feat can be accomplished with such small temperature changes. The unique properties of water explain why. Its specific heat (4.185 J/g·°C) is 2.5–3.3 times higher than that of soils and rocks, and its heat capacity (specific heat per volume) of 4.185 J/cm^3·°C is approximately twice as high as that of clays and about six times that of many dry soils (for more on water's attributes, see the next chapter). Consequently, water changes its temperature much more slowly than solids and retains much more heat per unit volume.

Water's absorption in IR wavelengths and its low viscosity make it an excellent heat carrier in ocean currents, and its exceptionally high heat of vaporization (nearly 2.5 kJ/g) makes it an ideal long-distance distributor of latent heat in precipitation. Inevitably, oceans dominate the planetary energy balance, as they intercept about four-fifths of all incoming radiation. And because their average albedo of 6% is lower than that of solid surfaces, oceans receive nearly four times as much incoming solar energy as

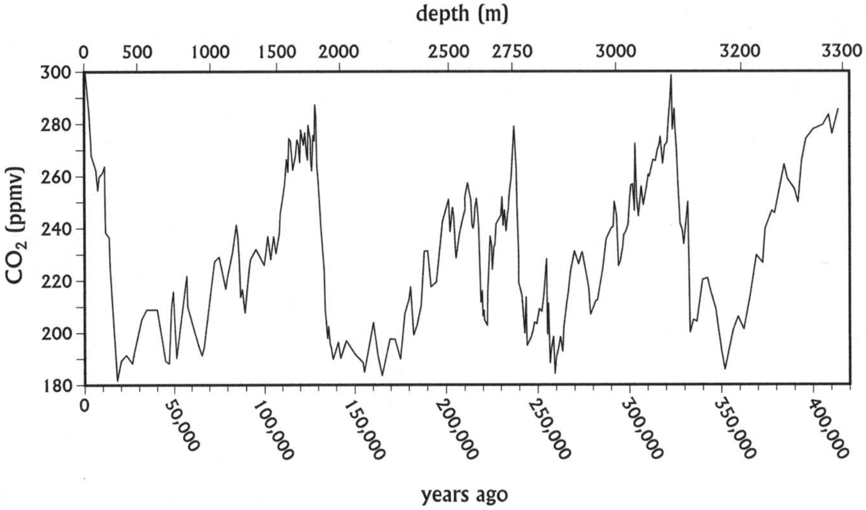

4.11 Atmospheric CO_2 concentrations during the past 420,000 years derived from air bubbles in Antarctica's Vostok ice core. Based on Petit et al. (1999).

do continents and roughly twice as much as is absorbed by the entire atmosphere; they also absorb about two-thirds of all downward longwave radiation. Water's high specific heat restricts the range of temperature amplitudes in the ocean's surface layer. Because of the ocean's pronounced thermal stratification, downwelling is the only effective means of distributing the absorbed energy to greater depths and upwelling is the only way to bring cold deep water to the surface.

Because of water's extraordinarily high heat of vaporization, evaporation of 1 mm/day requires nearly 30 W/m^2. This much power is available as the annual mean even in high-latitude seas (up to 70° N in the Atlantic), and the global mean of the latent heat flux is nearly 80 W/m^2: energizing the Earth's water cycle thus requires about 45 PW. Evapotranspiration, an intensified form of latent heat transfer, is obviously critical for the maintenance of all terrestrial ecosystems. Forests can boost this flux by an order of magnitude: it may be just 15 W/m^2 in late autumnal deciduous woods, but in full foliage it can be as much as 150 W/m^2. Sensible heat flux is important only in regions of high aridity, where its annual peaks are around 80 W/m^2. Plants take advantage of the flux during hot, dry spells to defend themselves against overheating by translocating heat from the soil to leaves, as well as from exposed leaves to the shaded ones. About three-fifths of the global sensible heat flux originates on land.

Solar Radiation and the Biosphere

The entire biosphere, with the notable exception of chemotrophic prokaryotes living beyond the reach of sunlight, is the product of ancient and intricate photosynthetic conversions of solar radiation into complex organic compounds. The Sun's longevity has allowed for billions of years of life's evolution, and its huge mass still has enough hydrogen to maintain the current rate of radiation flux for billions of years to come. But during the early phases of the Earth's evolution the Sun's irradiance was so much lower (at 4 Ga about 30% less than it is now) that without terrestrial compensations the mean temperature would have been well below the freezing point, precluding the existence of life. We may never know exactly how this compensation was accomplished.

In contrast, the cause of the inexorable demise of the Earth's life—the Sun's eventual transformation into a red giant whose expansion will first vaporize all water and then utterly obliterate the planet itself—is fairly well understood from studies of life cycles of other G stars. Understanding the impacts of the Sun's cyclical variations on the Earth's climate is obviously of a much greater practical importance: unequivocal conclusions regarding these effects have been elusive, not because of any shortages of intriguing data, but because of their complexities. It is logical at this point in the discussion to review the essential requirements of the biosphere's most important energy conversion: the transformation of the electromagnetic energy of light into the chemical energy of complex organic compounds.

From the Faint Young Sun to a Red Giant

Much is unknown about the infant Sun. It may have exhibited wildly irregular activity resulting in substantial short-term brightness variations (2–15% in the visible range), and one theory suggests that it lost up to half of its initial mass during the first 100 Ma of its main-sequence life (Guzik et al. 1987; Radick 1991). We are on more solid ground in concluding that the conversion of hydrogen to helium increases the Sun's density and hence the star's core temperature and its rate of thermonuclear reactions. The luminosity of the young Sun has thus been calculated to have been about 30% lower than the present rate (Crowley 1983). Assuming a constant planetary albedo, the Earth's effective radiating temperature for an irradiance of about 950 W/m^2 would be 235 K, compared to the current 255 K. With greenhouse gases at the preindustrial level, the lower luminosity of the faint young Sun would not have allowed liquid water for the first two or perhaps even three billion years of the Earth's evolution (Kasting and Grinspoon 1991). But as noted in chapter 3, the oldest sedimentary rocks can be reliably dated to as far as back as 3.8 Ga ago, and mud cracks and ripple marks are present in rock strata older than 3 Ga. An enhanced greenhouse gas effect is the best solution for this paradox.

Sagan and Mullen (1972) proposed that this enhanced greenhouse gas effect was achieved by about 10 ppm of atmospheric ammonia, but this assumption was questioned because of the short lifetime of NH_3 caused by photochemical decomposition of the compound. Sagan and Chyba (1997) eventually responded to this criticism by suggesting that UV absorption by steady-state amounts of high-altitude organic solids produced from CH_4 photolysis may have shielded ammonia to such an extent that the compound's resupply rates were able to maintain surface temperatures above freezing. Higher atmospheric CO_2 levels provide a much more likely explanation (Kasting and Grinspoon 1991).

Kasting and Grinspoon's argument is built on negative feedback of the carbonate-silicate cycle. Higher levels of evaporation caused by rising temperatures from the brightening Sun would dissolve more tropospheric CO_2 and increase the formation of carbonic acid (H_2CO_3), the chemical agent primarily responsible for the weathering of calcium silicate rocks; Ca^{2+} and HCO_3^- produced by this weathering would be transported by streams to the ocean, where they would form carbonate sediments (for principal reactions, see Appendix D). The gradual decline of atmospheric CO_2 levels would reduce absorption of outgoing radiation, and the planet would cool slightly. Rising CO_2 levels would accelerate the rate of weathering, but they could not change the return of the gas from the grand geotectonic cycle. Eventual subduction of these sediments by moving tectonic plates (carbonate metamorphism) and their subsequent reactions with silica would re-create silicate rocks, which would be re-exposed by tectonic uplift, and CO_2 would be degassed by volcanoes and along the spreading oceanic ridges: its accumulation would enhance the greenhouse effect. The Earth's equilibrium surface temperature would have had to remain above freezing at all times: otherwise silicate weathering would have basically ceased and CO_2 would have accumulated in the atmosphere, leading to extraordinarily high levels in geologically short periods of time.

Not only could these feedbacks have compensated for the comparative dimness of the Sun, they could have also prevented any possibility of runaway greenhouse or icehouse effects. Biota amplified these feedbacks in three ways. In the early ocean, which was either lifeless or contained only prokaryotes, carbonates could be formed only after the two constituent ions reached critical concentrations in the seawater. The evolution of marine biomineralizers, whose death leaves behind accumulations of calcium carbonate ($CaCO_3$), greatly accelerated the carbonate sedimentation. More importantly, oceanic photosynthesis provided an additional means of sequestering carbon in marine sediments.

By far the most persuasive evidence of the impact of biota on long-term carbon cycling comes from the bipartition of sedimented carbon between organic and inorganic compounds as inferred from their isotopic record for the last 3.5 Ga, and more problematically, for up to 3.8 Ga (Berner 1998; Schwartzman 1999). The ratio of these two compounds has been close to one to five during the period, and although the interpretation of this remarkably constant bipartitioning is complicated by such uncertainties as possible variations in the isotopic composition of volcanically derived carbon released to the atmosphere, the sedimentary record leaves no doubt as to the importance of photosynthesis for the carbon cycle almost since the very beginning of the Archean biosphere. The best explanation of this geochemical conservatism is the steering function of phosphorus as the ultimate limiting nutrient, whose availability has been setting the upper limit for primary productivity ever since water-splitting (that is, initially cyanobacterial) photosynthesis became the principal carbon-fixing process (Schidlowski 1991; Schidlowski and Aharon 1992).

Finally, ever since the emergence of vascular plants about 450 Ma ago, terrestrial biota have been accelerating the weathering of silicate minerals through a number of mechanisms and reactions, above all through soil stabilization, through increasing CO_2 levels in soils (from root respiration and litter decomposition carried out largely by microorganisms), through providing organic matter for the formation of humic acids, and through mycorrhizal and fungal digestion of silicates (Schwartzman 1999). Other organisms, ranging from lichens to insects and invertebrates, also participate in this biotic enhancement of weathering, whose cummulative effect is to accelerate natural rates, commonly by one or two orders of magnitude. The pronounced decline in levels of CO_2 during the mid-Paleozoic (particularly during the Devonian, 400–360 Ma ago) is clearly attributable to the terrestrial diffusion of vascular plants, whose deep roots enhanced the rate of weathering (Berner 1998).

So far the Sun has consumed a mere 0.03% of its mass, but the original hydrogen in its core has been depleted from 75% to 35% of the core's mass. A 6-Ga-old Sun will grow about three times larger and some 15% brighter than today, a change that will eliminate any ice from the Earth. As it nears 10 Ga of its age, the Sun's diameter will be about 40% larger than it is now, and the red photosphere will be twice as luminous. Its hydrogen exhausted, the core will contract, and thermonuclear reactions involving protons from outside the core will expand the new red subgiant to ten times its present diameter. But long before a red giant, with 100 times its present diameter, melts the Earth before transforming itself into a brilliant, Earth-sized white dwarf, terrestrial surfaces will be heated to the boiling point, and the oceans will be vaporized (more details in the Epilogue).

Climatic Influences

A generation ago Pittock (1978) concluded his review of solar effects on climate by stating that despite a massive literature on the subject there was still no clear-cut evidence of a statistically significant correlation between sunspot cycles and climate; five years later he could not offer any fundamental revision of that position (Pittock 1983). A decade later came a clear shift: a detailed review of solar influences on global change answered yes to its primary question, "Do solar variations directly force global surface temperature?" (NRC 1994). The first intriguing contribution to this new wave of Sun-climate studies came when Friis-Christensen and Lassen (1991) found a strong correlation between surface temperature and the length of the sunspot cycle for the 130-year period between 1860 and 1990; a few years later, they extended the period for the match to the past 500 years (Lassen and Friis-Christensen 1995).

Other studies have found strong matches between the sunspot cycles and sea surface and subsurface temperature and the abundance of dust particles in cores recovered in the Andes (Kerr 1996). The correlation appears to be particularly strong for the period before 1800, when as much as three-fourths of the warming after the Little Ice Age could be ascribed to the brightening sun (Lean and Rind 1999). A seventy-year period of cooler temperatures related to a virtual absence of sunspot activity between 1645 and 1715 is now known as the Maunder minimum, named after E. Walter Maunder, who reported on the phenomenon in 1890 (Eddy 1977; fig. 4.12). As for the 0.55°C global temperature increase since the mid-nineteenth century, perhaps as much as half of it has been caused by the brightening of the Sun, the rest by higher concentrations of greenhouse gases. Other studies have detected a smaller temperature effect attributable to changing irradiance, but the existence of a link between the two is no longer in doubt.

Explaining the mechanism by which the link operates is challenging, because the irradiance shifts are so small in absolute terms. An irradiance change of 0.1% between the Sun's maximum and minimum activity over the eleven-year sunspot cycle corresponds to 1 W/m^2 at the top of the atmosphere or 0.2 W/m^2 averaged over the Earth's surface (after dividing by four and subtracting the reflected radiation). But because this change is much more pronounced in the UV range, solar variability can affect the atmosphere by increasing temperature in the stratosphere, which, in turn, can cause tropospheric change, including stronger easterly winds and poleward shifts of subtropical jet streams and storm tracks (Haigh 1996). At the same time, we still do not know to what extent this solar high-altitude

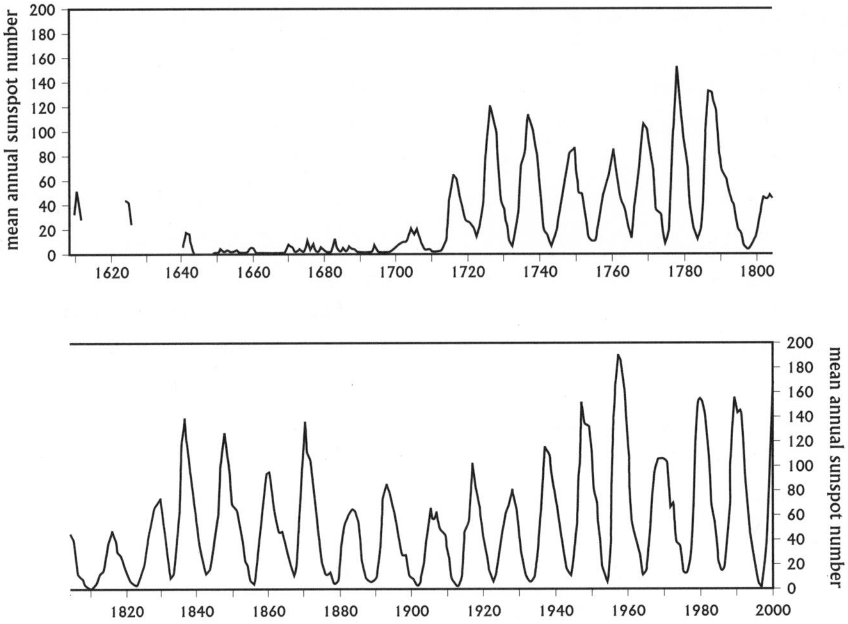

4.12 Observed sunspot numbers show a fairly regular periodicity related to the Sun's magnetic cycle. The paucity of sunspots before 1715 is known as the Maunder minimum. Based on Eddy (1977) and Glanz (1997).

change works its way through the atmosphere to affect the biosphere (NRC 1994).

Undeniably, changes in irradiance can either potentiate or negate the impact of accumulating greenhouse gases. A decrease in the Sun's output between 1980 and 1986 might have canceled the warming produced by anthropogenic greenhouse gases added to the atmosphere during that period (NRC 1994). If sustained throughout the twenty-first century, brightening at the rate calculated for the ten years between 1986 and 1996—0.036% per decade—could produce a warming of about 0.4 °C (Willson 1997). Depending on future emissions of greenhouse gases, this may account for as little as one-sixth but as much as one-third of the total temperature rise during the century (for more on future anthropogenic perturbances of the global carbon cycle and on possible global warming, see chapters 5 and 9). Yet another recent intriguing finding is that not only the number of sunspots, but also the Sun's general magnetic field, has doubled since 1900 (Lockwood et al. 1999), a shift indicating an unknown commonality between the two processes.

Photosynthesis

Photosynthesis is energized by the absorption of light by pigments in thylakoid membranes of bacterial and plant chloroplasts, and the energy efficiency of the entire sequence of converting simple inorganic inputs into new phytomass is surprisingly low. To begin with, reacting compounds contribute 519 kJ, and the difference between the broken bonds in H_2O and CO_2 and newly formed bonds in the photosynthate is about 465 kJ/mole. Eight or nine photons are the minimum theoretical amount of solar energy needed for assimilating a molecule of CO_2; synthesis of ATP per molecule of assimilated CO_2 requires 1,736 MJ, and hence the maximum possible efficiency of photosynthesis would be 465 kJ/1,736 kJ, or almost 27% (Good and Bell 1980).

The actual process is a great deal less efficient: under typical natural conditions at least fifteen to twenty photons are needed to fix a molecule of CO_2. Additional photons are required in order to make up for losses incurred in plants before and after their carbon uptake. Preassimilation losses, amounting to at least 20–25% of incident radiation, are due mainly to inevitable inefficiencies accompanying rapid rates of light absorption and subsequent photosynthetic reactions. Excitation energy in the pigment antenna of the photosystem that is in excess of a plant's capacity for CO_2 fixation is dissipated as heat (Li et al. 2000). Postassimilation losses occur during respiration when newly synthesized carbohydrates are used to produce more complex organic compounds. Autotrophic respiration ranges from less than 20% in immature fast-growing species to virtually 100% in old trees.

Using twenty photons to convert one CO_2 molecule into new phytomass implies an efficiency of roughly 10% of the total insolation. But only photosynthetically active radiation (PAR) can excite the chlorophylls and other pigments, and the average share of these wavelengths is only 45% of incoming sunlight. The fraction of absorbed PAR (FAPAR) has a very high correlation ($r^2 = 0.919$) with the intensity of photosynthetic production measured by the normalized difference vegetation index (NDVI; for details on this index, see chapter 7): FAPAR = 1.16 NDVI – 0.143 (Myneni and Williams 1994). And because a part of the PAR is either reflected by leaves or is transmitted through plant canopies, the peak photosynthetic efficiency would be around 4%. Indeed, such rates of photosynthesis were observed for short periods of time in fast-growing plants under optimal conditions, but they are unattainable over longer periods of time for most ecosystems: the usual production efficiency of new phytomass is overwhelmingly less than a third of this performance. Inadequate water and nutrient supplies are the two most common limiting factors, and cold temperatures will shut down photosynthesis for prolonged periods of time.

Transeau (1926) published the first photosynthetic efficiency calculation, concluding that during 100 days of its growing period, corn (*Zea mays*) on a good Illinois farm converts just 1.6% of the solar radiation it receives. We now know that optimally fertilized and irrigated crops can reach seasonal values above 2%. At the same time, ecosystemic studies have shown typical net conversion efficiencies ranging from less than 0.1% in Alpine and Arctic grasses to 1–1.5% for the most productive temperate and tropical forests (for more on primary productivity, see chapter 7). Plants growing in stressed ecosystems, terrestrial or marine, will have an overall conversion efficiency commonly an order of magnitude lower than similar or identical species growing in more benign environments.

With annual global terrestrial net primary production averaging no more than 18 MJ/m^2 of ice-free land (see chapter 7 for derivation of this rate) and with mean radiation reaching the Earth's surface at 5.3 GJ/m^2, the efficiency of continental photosynthesis would be about 0.33%. The global mean of fixation efficiency for phytoplanktonic net productivity averaging close to 6 MJ/m^2 would be at best 0.1% for the ice-free ocean. The average efficiency of converting solar radiation to new phytomass

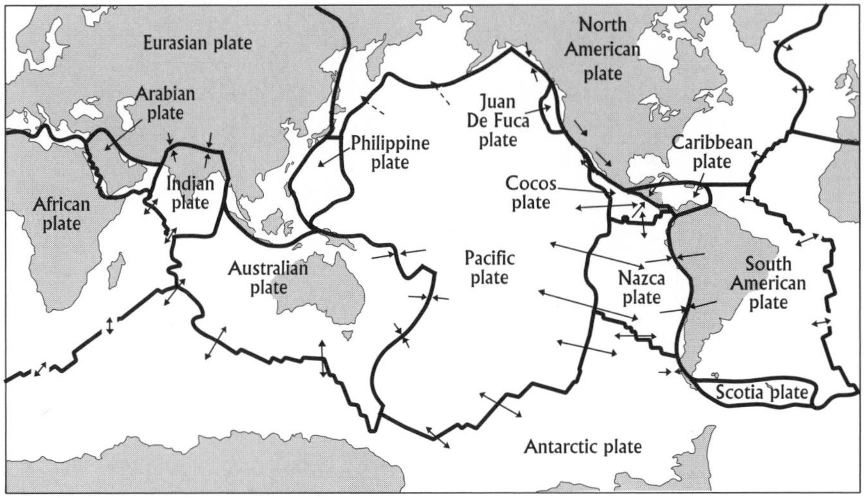

4.13 The Earth's tectonic plates and the directions of their motion. Boundaries according to the United States Geological Survey map available at <http://pubs.usgs.gov/publications/text/slabs.html>

would be then less than 0.2% for the entire Earth: electromagnetic energy of only one out of every 500 photons ends up converted to chemical energy of new photosynthate.

The Earth's Heat

The biosphere's pyramid of life rests overwhelmingly on the conversion of solar radiation to phytomass through photosynthesis powered by solar radiation. But not exclusively: as we have already seen (chapter 3), solar radiation does not energize the metabolism of chemotrophic archaea and bacteria, whether they live deep below the land surface or in marine sediments or at the bottom of deep ocean, in and near hot vents where they oxidize H_2S and support unique ecosystems (for details on these extremophilic microbes, see chapter 6). More notably, whereas solar radiation dominates the total energy input at the Earth's surface, it does not energize the process responsible for the formation of the ocean floor and for the reshaping and redistributing of the continents and the creation of mountain ranges.

The Earth's surface is made up of rigid oceanic and continental plates (fig. 4.13). The former are thinner (mostly between 5 and 7 km) and, when measured on a geological time scale, short-lived and highly mobile (some moving more than 10 cm/year). The latter are much thicker (commonly 35–40 km) and long-lived, and some of them are now virtually stationary. The plates ride on the planet's nearly 3,000-km-thick solid but flowing mantle, which is the source of hot magma whose effusions create new sea floor along about 72,000 km of the Earth-encircling ocean ridge system. This suboceanic mountain chain, rising about 2 km above the surrounding sea floor, is the Earth's single largest surface feature. Old sea floor is eventually recycled back into the mantle in deep ocean trenches (subduction zones). The current shapes and distribution of oceans and continents are thus transitory features, products of incessant geotectonic motions energized by the Earth's heat.

Heat Flows and Plate Tectonics

Three distinct sources account for heating the Earth's surface from the inside of the planet: heat conducted through the lithosphere from the underlying hot mantle (its temperatures, at the core-mantle boundary, are as high as 4,000 K); radiogenic decay of heat-producing crustal elements; and heat transported convectively by magmas and fluids during orogenic events (Vitorello and Pollack 1980). We know that the most important heat-producing radioactive isotopes are long-lived ^{40}K, ^{232}Th, and ^{235}U, and ^{238}U, but we cannot estimate their crustal concentrations with a high degree of certainty (Verhoogen 1980). Models of continental crust published since the early 1980s put potassium concentration (as K_2O) between 1.1% and 2.4% (Rudnick et al. 1998).

Consequently, we cannot accurately apportion the shares of the geothermal flux originating in heat conduction and in radioactive decay (the estimated contribution of the latter process ranges from about half to more than 90%). But after decades of accumulating measurements, we are on a safer ground when quantifying the global total of this flow and its continental and oceanic segments (Davies 1980; Sclater et al. 1980; Rudnick et al. 1998). Not surprisingly, average flow rates are highest along mid-ocean ridges, where new oceanic lithosphere is continually created by hot magma ascending from the mantle: about 3–3.5 km^2 of new ocean floor are added by this process every year. At this rate of crustal spreading today's entire ocean floor could have been added in about 120 Ma. In reality, no part of the sea floor is older than about 210 Ma (Mesozoic era), and most of the oceanic crust is younger than 140 Ma.

The associated heat flux—composed of the latent heat of crystallization of newly formed and cooling basaltic ocean crust, and of the cooling from magmatic temperatures (around 1,200 °C) to hydrothermal temperatures (around 350 °C)—is between 2 and 4 TW (Wolery and Sleep 1988; Elderfield and Schultz 1996). Divergent spreading of oceanic plates eventually ends at subduction zones, some of them marked dramatically by the world's deepest ocean trenches, where the crust and the uppermost part of the mantle are recycled deep into the mantle to reappear through igneous process along the ridges. Movements within the mantle have been a matter of intensive speculation and model building ever since plate tectonics became a universally accepted paradigm of geotectonics during the 1960s (Jackson 1998; Davies 1999). The principal contest has been between two-layer models and whole-mantle flows.

In the first type of model, two sets of convection cells, separated by seismic discontinuity at a depth of about 660 km, are largely isolated through time, and heterogeneities introduced by plate subduction are rapidly mixed into the upper mantle. In the second type of model, subducting slabs penetrate deep into the mantle, and relatively large heterogeneities including the subducted material can persist for long periods of time (Jacobs 1992; Kellogg 1993; Condie 1998). The best explanation of subduction postulates a peculiar way of whole-mantle circulation, with an irregular undulating boundary between the two layers and with the mantle's hypothesized dense bottom layer thinning below colder sinking slabs, in some place to the very

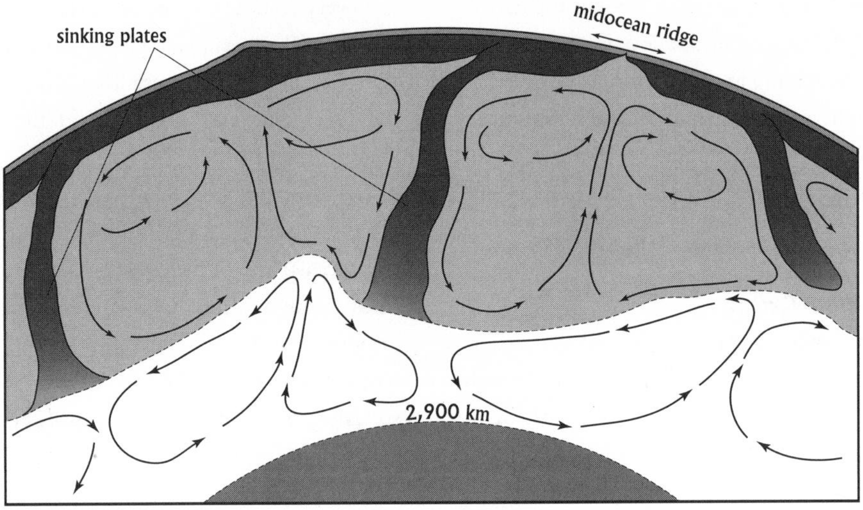

4.14 A model of circulation in the Earth's mantle suggests that the flows of molten rocks are separated into two cells. Based on Kellogg et al. (1999).

core-mantle boundary, and swelling upward in other regions (Kellogg et al. 1999; fig. 4.14).

The total hydrothermal flux of 7–11 (mean of nearly 9) TW accounts for about a third of the global oceanic heat flow of 32 TW (Stein and Stein 1994). In turn the oceanic flux amounts to roughly 70% of the global heat flow of 44 TW. Almost a third of the total oceanic heat loss takes place in the South Pacific, where the crustal spreading rates are the fastest (fig. 4.15). Predictably, heat flows decline with the age of the ocean floor, from as much as 250 mW/m^2 in the youngest crust to less than 50 mW/m^2 through the sea floor older than 100 Ma. A planetary heat flow of some 44 TW prorates to about 85 mW/m^2 of the Earth's surface, with means of almost 100 mW/m^2 for the oceans and only about half as much for the continents. Nonorogenic surface flows through the continental crust range from 41 mW/m^2 in Archean rocks to 49–55 mW/m^2 in Phanerozoic formations (Rudnick et al. 1998).

As already noted, the average global heat flux of about 85 mW/m^2 is a tiny fraction—a mere 0.05%—of the solar radiation absorbed by surfaces (168 W/m^2). But this small power flux, acting over huge areas and across long time spans, has been responsible not only for the opening up of new oceans, the reshaping of continents, and the building of enormous mountain chains, but also for earthquakes and volcanic eruptions. Consequently, the biosphere's evolution has been fundamentally influenced not just by continuous, tectonically driven reconfigurations of the Earth's surface, but also by massive emissions of volcanic gases and particulates repeatedly injected into the atmosphere.

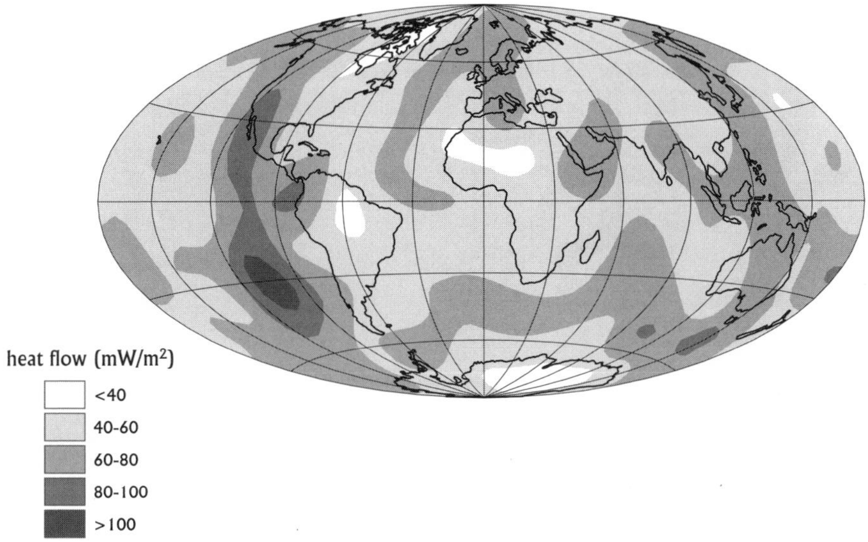

4.15 The Earth's heat flow is highest along the ocean ridges, particularly in the Pacific Ocean offshore Chile. Based on Pollack and Chapman (1977).

Geotectonics and the Biosphere

The production and recycling of oceanic crust is not a geologically new phenomenon. Higher levels of radioactivity and more intensive heating resulted in relatively rapid recycling of oceanic crust on the Archean Earth, where there may have been more than 100 separate but short-lived plates. Later the process assumed a more episodic character. Globally significant events were triggered by catastrophic changes when descending plates, accumulating at the 670-km seismic discontinuity, suddenly sank into the lower mantle and forced a rapid rise of magmatic plumes to the base of the lithosphere and in a new bout of crustal growth (Breuer and Spohn 1995).

Such superevents have been dated to 2.7, 1.9, and 1.2 Ga ago. Prior to the first superevent, juvenile crust was rapidly recycled into the upper mantle, and only after the cooling mantle made the continental crust less easy to subduct did the collisions between continental blocks form the first supercontinent. Smaller slab avalanches with no global effects occurred many times before or after the three superevents. The breakup of a supercontinent typically takes some 200 Ma, and the resulting slab avalanches last for about 100 Ma. They are followed by 100–500 Ma of rising mantle plumes that produce juvenile crust trapped in the growing supercontinent. Eventually, after 200–440 Ma, the new supercontinent is broken up by mantle upwelling beneath it (Condie 1998; fig. 4.16).

During the last 1 Ga the Earth has experienced the formation and breakup of three supercontinents: Rodinia (formed 1.32–1.0 Ga ago; broken up between 700–530

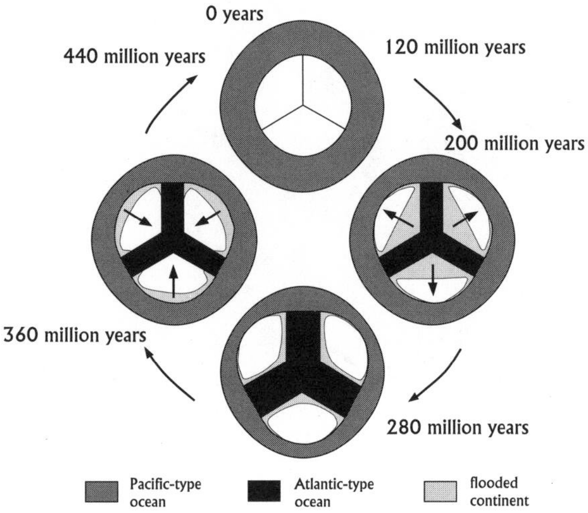

4.16 Simplified diagram of the Earth's recurrent tectonic cycle, proceeding from a breakup of a supercontinent to eventual reassembly of moving plates. Based on Nance et al. (1988).

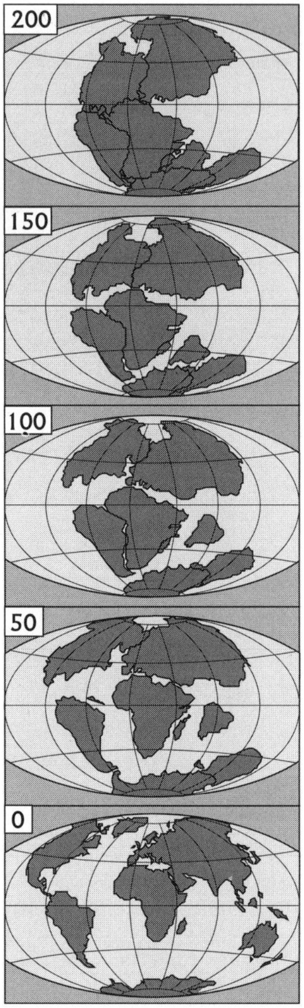

4.17 Intervals of fifty million years show the continuing breakup of Pangea, a supercontinent formed more than 200 Ma ago (Smil 1999d).

Ma ago), Gondwana (formed between 650–550 Ma ago) and Pangea (formed 450–250 Ma ago; began to break up about 160 Ma ago). Pangea spanned the planet latitudinally, from today's high Arctic to Antarctica; its eastern flank was notched by a V-shaped Tethys Sea centered approximately on the equator (fig. 4.17). As the signs of rifting in the Red Sea and East Africa indicate, the process of breaking up is still very much under way (Omar and Steckler 1995; Pollitz 1999). Eventual reassembly of a new supercontinent is a foregone conclusion. Obviously, these grand geotectonic events have had an enormous impact on the evolution of life. Most fundamentally, plate tectonics has made it possible to keep a significant share of the Earth's surface above the sea, making possible the evolution of land plants and heterotrophs and humans (Taylor and McLennan 1995). The changing locations and distributions of continents and oceans have produced very different patterns of global oceanic and atmospheric circulation, two key determinants of climate and hence of life's evolution (Ruddiman 1997). Volcanic eruptions, originating overwhelmingly along the plate boundaries, have been the most important natural source of CO_2, as well as an enormous source of aerosols, whose episodically high atmospheric concentrations have had recurrent global climatic impacts. Formation of the Santorini caldera in the Aegean Sea, about 1500 BC, involved the ejection of about 70 km^3 of fragmentary material. In 1815 the Tambora eruption produced at least 30 km^3, and Krakatau in 1883 released some 18 km^3 (Smil 1991; Thornton 1996). (For the biospheric impacts of prehistoric eruptions see chapter 3.)

The emergence and diversification of land plants and terrestrial fauna during the past 500 Ma years have been much influenced by the changing locations and sizes of the continents. The physical environment of today's biosphere is to a large extent a product of the still-continuing breakup of Pangea (fig. 4.17). For example, both the Himalayas, the world's tallest mountain chain, and the high Tibetan Plateau, the two features that determine the climate for nearly half of humanity, are products of this ongoing breakup. So is the northward transfer of warm Atlantic Ocean waters by the Gulf Stream that has made high-latitude Western Europe so habitable. Only Pangea's breakup can explain many striking identities or similarities of the biodiversity pattern of today's continents. Before the general acceptance of plate tectonics, the only way to explain those very similar biodiversity patterns of regions separated by oceans was to postulate extensive land bridges between continents. The most bizarre of these assumed connections was the Africa-India isthmus via Madagascar, the Seychelles, and the Maldives (Oreskes 1999).

Examination of the divergence among avian and mammalian orders (based on a comprehensive set of genes that exhibit a constant rate of substitution) showed that its timing coincided with the Mesozoic breakup of continents (Hedges et al. 1996). And plate tectonics also plays an essential role in cyclical processes I will examine in the next chapter: the recycling of elements required for the perpetuation of the biosphere.

5

WATER AND MATERIAL FLOWS

Biospheric Cycles

In its life, its death, and its decomposition an organism circulates its atoms through the biosphere over and over again.
Vladimir Ivanovich Vernadsky, *Biosfera*

All process is reprocessing.
John Updike, "Ode to Rot"

The need for the incessant cycling of water and the elements that make up biomass is obvious. Photosynthesis would have to be confined to oceans and lakes if water, the largest constituent of biomass, were not constantly being redistributed around the planet through evaporation, transpiration, and precipitation. And without the cycling of carbon and macro- and micronutrients the biosphere could not have functioned for billions of years, as even the largest reservoirs of these elements would become depleted in relatively short periods of time. Even the atmosphere's enormous reservoir of N_2, amounting to 3.8 Pt, would be exhausted in less than 20 Ma without denitrifying bacteria's freeing the element bound in nitrates.

In spite of water's critical importance in the biosphere, many aspects of its cycling remain poorly understood (Trenberth and Guillmot 1995). The ocean dominates the water cycle in every way, and living organisms have only a negligible role in storing water. But biota have a much greater effect on water flows, as evapotranspiration supplies about 10% of all water entering the atmosphere. Carbon, nitrogen, and sulfur are the only three doubly mobile elements, transported both in water as ionic solution or in suspended matter as well as through the atmosphere as more or less stable trace gases. Theirs are the three true biospheric cycles dominated by microbial and plant metabolism. In contrast, mineral elements move through the biosphere primarily by piggybacking on the grand sedimentary-tectonic cycle. Weathering liberates these elements from parental materials, and they travel in ionic solutions or as suspended matter to be deposited eventually in the ocean. They return to the biosphere only after many millions of years when reprocessed rocks reemerge from the mantle in ocean ridges or hot spots or when the sediments become part of new mountain ranges. On a civilizational timescale, mineral cycles thus appear only as one-way oceanward flows, with human activities enhancing some of these fluxes, most notably the oceanward flow of phosphorus used as crop fertilizer.

The Water Cycle

Water is indispensable for life because it makes up most of the living biomass and because it is an essential metabolic input. Very few organisms are less than 60% water, and many fresh tissues are more than 90% water, with 99% in phytoplankton. Water is also the dominant donor of hydrogen in photosynthesis and the irreplaceable carrier of nutrients and metabolic products. Consequently, it is fortunate that the water molecule is too heavy to escape the Earth's gravity, and hence the only instances when significant volumes of water have ever left the planet were in the aftermath of major extraterrestrial impacts that took place more than 3.8 Ga ago. Today only a very small mass of water is constantly being destroyed in the upper atmosphere through photodissociation. Water from hot springs and water vapor in volcanic eruptions is made up largely of recycled flows, and juvenile water arriving from deeper layers of the crust totals only about 100 Mt/year, an entirely negligible amount.

The water cycle is the biosphere's most rapid and the most massive circulation, driven overwhelmingly by evaporation and condensation (fig. 5.1). Human activities have changed some local and even regional water balances drastically, and pronounced anthropogenic global warming would accelerate water's global cycle. But the only man-made additions to water's circulating mass are globally negligible withdrawals from ancient aquifers and water formation during some chemical syntheses and combustion of fossil fuels. Biomass combustion, no matter whether it is in natural forest and grassland fires or through deliberate burning during deforestation or for household or industrial uses, is obviously a form of accelerated water recycling.

In contrast, combustion of fossil fuels is a source of new water formed by the oxidation of hydrogen. Hydrogen accounts for less than 5% of the dry mass in coals, 11–14% in crude oils, and 25% in natural gas. Its complete oxidation is now producing annually about 9 Gt of water, a negligible amount (less than 0.002%) of annual precipitation. Moisture evaporated from coals and peats adds less than 500 Mt of water a year. Consequently, both the total volume of the Earth's water and its division among the major reservoirs on the Phanerozoic Earth can be considered constant on a time scale of 10^3 years. On longer time scales the periodic ebb and flow of glaciations, including some extreme ice ages, obviously shift a great deal of water among oceans and glaciers and permanent snow.

The Ocean

Just over 70% of the Earth is covered by the ocean, to an average depth of 3.8 km: hence *Aqua,* rather than *Terra,* would be the planet's more appropriate name. The dominance of the ocean in the global water cycle is overwhelming: it stores 96.5% of the Earth's water, it is the source of about 86% of all evaporation, and it receives 78% of all precipitation (fig. 5.2). But the ocean's enormous importance for the biosphere rests more on the extraordinary properties of water than on its enormous area. As already noted (in chapter 4), water has both very high specific heat and heat capacity. Because of its intermolecular hydrogen bonds, it also has an unusually high boiling point; its very high heat of vaporization makes it an ideal transporter of latent heat; and its relatively low viscosity makes it an out-

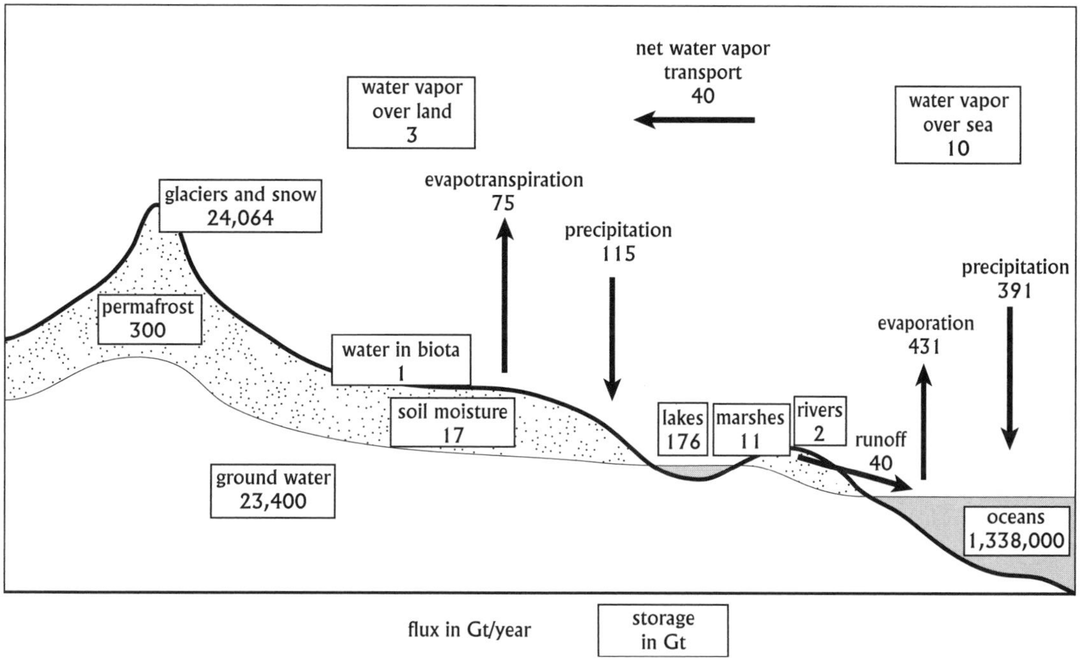

5.1 Annual global flows of the Earth's water cycle; the totals (in Gt/year) are from Oki (1999).

standing carrier of heat in countless eddies and currents (Denny 1993). Its high boiling point also allows for the existence of very hot aquatic ecosystems (see chapter 7). Its high heat of vaporization helps retain plant and soil moisture in hot environments, and its fairly low viscosity makes it an excellent medium for swimming. Water is sufficiently buoyant without being an impediment to a relatively fast movement: yellow-fin tuna (*Thunnus albacares*) can go as fast as 21 m/s in 10- to 20-second bursts (Walters and Fierstine 1964).

Because of the ocean's great average depth, air-sea interactions cannot directly affect the entire water column. Water is densest at about 4 °C, and hence the deep ocean, with a temperature stable near that point, is isolated from the atmosphere by the mixed layer, a relatively shallow column agitated by winds and waves and experiencing both daily and seasonal temperature fluctuations (fig. 5.3). And because of water's extraordinarily high heat of vaporization—about 2.45 kJ/g, although the rate is temperature-dependent, and the exact formula (giving results in J/g) is 2,475 – 2.26T—evaporation transports not just large volumes of water but also huge quanta of thermal energy. Vaporization of 1 mm/day requires inputs of between 28 and 29 W/m^2, and solar radiation averages that much annually even in high-latitudes seas, up to 70° N in the Atlantic and to about 60° S in the Antarctic Ocean. The global mean

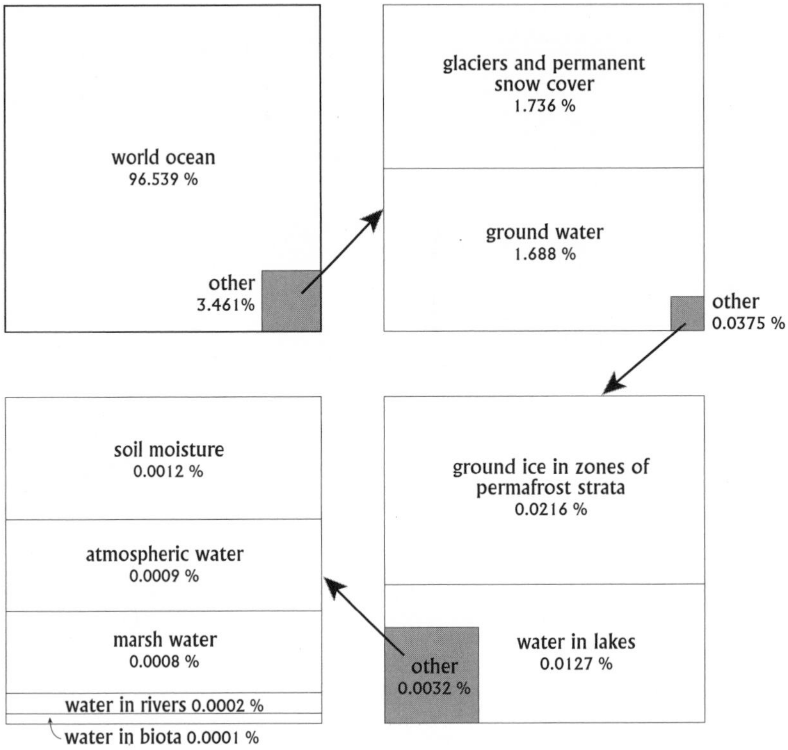

5.2 Nested comparisons of global water reservoirs show the overwhelming dominance of ocean storage and the insignificant, but vital, water content of the biomass. Based on data in Oki (1999).

flux of close to 90 W/m^2 translates into an average daily evaporation of around 3 mm, or an annual total of 1.1 m.

Evaporation of seawater is also affected by the ocean's overturning circulation, a vertical heat exchange whereby cold, dense water sinks near the poles and is replaced by warmer poleward flow from low latitudes (Toggweiler 1994). Macdonald and Wunsch (1996) concluded that there are two nearly independent overturning cells, one connecting the Atlantic Ocean to other basins through the Southern Ocean, and the other connecting the Indian and Pacific basins through the Indonesian archipelago. Complex pathways of Pacific water flowing into the Indian Ocean in the Indonesian seas were traced only recently: the throughflow is dominated by low-salinity water moving through the Makassar and Lombok Straits and more salty water going through the Banda Sea (Gordon and Fine 1996). The World Ocean Circulation Experiment (WOCE), currently under way with the participation of nearly thirty countries, will greatly expand our understanding of the phenomenon. Ganachaud and Wunsch

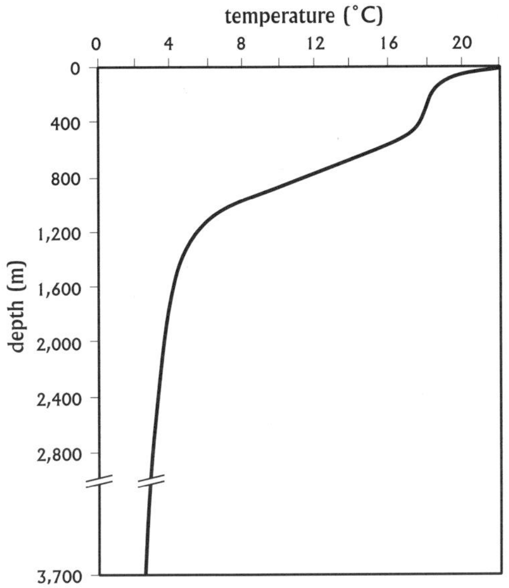

5.3 The normal temperature profile of ocean water shows a rapid decline from surface maxima to a nearly constant value around 3–4 °C below 2,000 m (Smil 1999d).

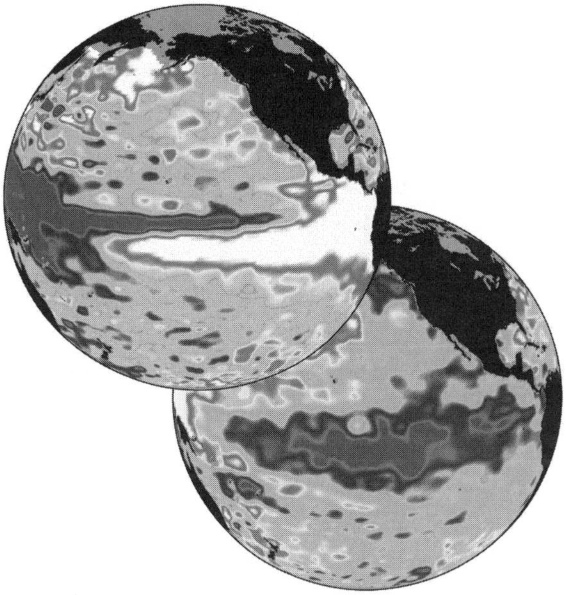

5.4 The dramatic contrast of the cool surface sea temperature in the equatorial Pacific waters during an El Niño event (white wedge on October 28, 1997) and the warm waters of a La Niña episode (dark area on February 27, 1999). The images, from the U.S.-French TOPEX/POSEIDON mission, are available at <http://www.jpl.nasa.gov/elnino>

(2000) used WOCE data to calculate a water and heat flow of, respectively, 505 Gt/year and 1.3 PW between the subtropical (latitude of southern Florida) and North Atlantic and virtually identical rates between the Pacific and Indian Oceans through the Indonesian straits.

Water sinking proceeds in restricted currents, including giant ocean cataracts. The most voluminous of these plunges southward about 3.5 km between Iceland and Greenland, carrying 25 times the flow of the Amazon (Whitehead 1989). In contrast, wind-driven upwelling of deep, cold, nutrient-laden waters goes on along the low-latitude western shores of the Americas, Africa, and India as well as in the Western Pacific equatorial zone, and it is marked by high rates of photosynthesis in what would be otherwise fairly barren surface waters. Even relatively minor changes in ocean surface temperature can have pronounced consequences. This phenomenon is best illustrated by far-flung effects of oscillations between El Niño, a periodic westward expansion of warm surface waters beginning off the coast of South America and by early winter often extending all along the Equator to join warm water off Australasia, and La Niña, a large, wind-driven area of cool water off the South American coast (fig. 5.4).

All but about 10% of water evaporated from the ocean is precipitated back onto sea surfaces. Evaporation exceeds precipitation in the Atlantic and Indian Oceans; the reverse is true in the Arctic Ocean, and the Pacific flows are nearly balanced. Irregular patterns of oceanic evaporation (maxima up to 3 m/year) and precipitation (maxima up to 5 m/yr) require substantial compensating flows from the regions with excess rainfall in order to maintain sea level (Schmitt 1999). The North Pacific, particularly its eastern tropical part, is the largest surplus region (and hence its water is less salty), whereas evaporation dominates the Atlantic waters.

Ocean evaporation produces both constant gentle fluxes of water and latent heat as well as some of the most violent atmospheric flows in seasonal cyclones (American hurricanes, Asian typhoons). The process is also responsible for the Asian monsoon, the planet's most spectacular mechanism for transferring heat absorbed by warm tropical oceans to dry land (Webster 1981). Some of the world's most productive and most diverse ecosystems, as well as nearly half of humanity, are directly affected by this immense seasonal pulse generated by intense heating of equatorial waters.

In the long run oceanic evaporation, and hence the intensity of the global water cycle, is also influenced by the mean sea level: during the last glacial maximum, 18,000 years ago, that level was 85–130 m lower than it is now (CLIMAP 1976). Measurements with an altimeter on the US-French TOPEX/POSEIDON satellite indicate a recent global mean sea level change of almost 4 mm a year, and most of this increase is a short-term variation unrelated to the expected rise caused by global warming and the resulting thermal expansion of water (Nerem 1995).

Other Reservoirs and Global Fluxes

Only two distinctly secondary reservoirs—glaciers and permanent snow, and groundwater—claim more than 1% (roughly 1.7%) each of the Earth's water stores. The tiny remainder of less than 0.04% is stored mostly in ground ice in permafrost strata of soils (0.02%) and in lake waters (0.01%), with the next largest reservoir, soil moisture, being an order of magnitude smaller (0.001%) and with all biomass containing a mere 0.0001% of the total (Oki 1999; fig. 5.2). But such comparisons do not convey the qualitative importance of various water reservoirs, and atmospheric water is perhaps the best illustration of this fact. Water aloft is a mere 0.0009% of the global presence; it makes up only about 0.3% of the atmosphere by mass and 0.5% by volume, and its average residence time is just 10–14 days, yet it suffices to cover some 60% of the planet with clouds at all times, greatly modifying the amount of solar radiation reaching the ground. In addition, as described in chapter 4, water is also by far the most important absorber of outgoing IR radiation, and its high heat capacity and release of latent heat on condensation are critical determinants of the planet's thermal regime.

The net transfer of ocean-derived water vapor to the continents supplies about one-third of terrestrial precipitation. The remainder comes from evaporation from nonvegetated surfaces and from evapotranspiration, the evaporation of water through stomata of leaves. Naturally, dense forests with rich canopies are best at "pumping" water for photosynthesis, and in the process they can increase the latent heat flux by an order of magnitude compared to barren surfaces. Because of its strong absorption of outgoing IR as well as its interaction with aerosols, tropospheric water vapor is a key determinant of the Earth's climate, but global observations of water vapor at higher levels (5–10 km above the surface) have been scarce. Consequently, it is helpful that Price (2000) demonstrated that lightning activity is a good proxy for convection and thus could be used to monitor water vapor concentrations in the upper atmosphere.

Rising air parcels are cooled (dry ones adiabatically, at a rate of 10 °C/km), and rain or snow begin when water drops or ice crystals making up the clouds grow to form

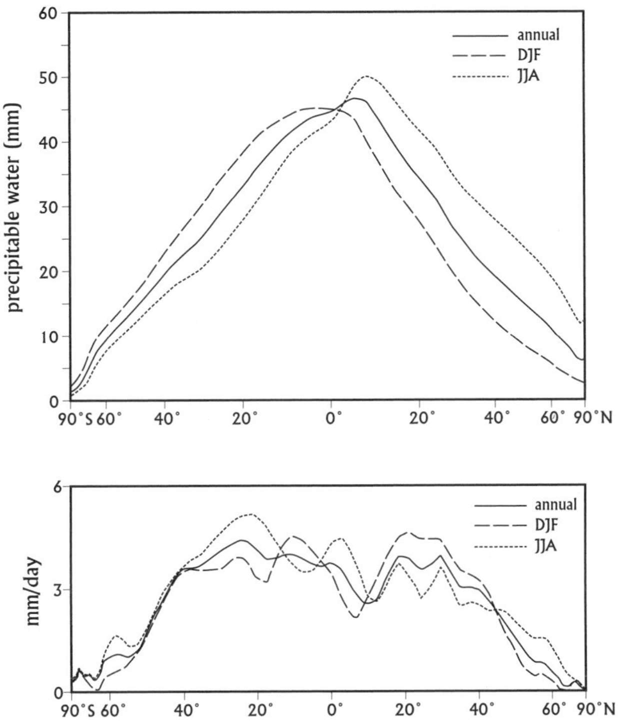

5.5 Meridional profiles of zonal mean precipitable water and evaporation. Based on data in Oki (1999).
Notes: DJF = December, January, February; JJA = June, July, August.

particles of sufficient size to fall as precipitation. These processes take place mostly in the lower two-thirds of the troposphere, but massive thunderclouds penetrate the entire troposphere and can be often seen in space imagery as protruding above the tropopause. Roughly three-fifths of all continental precipitation is evaporated, one-tenth returns to oceans as diffused surface and underground runoff, and just short of one-third is carried by rivers (fig. 5.1). With a mean continental elevation of about 850 m, this implies an annual conversion of some 400 EJ of potential energy to the kinetic energy of flowing water, which is the principal agent of geomorphic denudation shaping the Earth's surfaces.

A global graph of the latitudinal distribution of evapotranspiration shows a weakly bimodal distribution, whereas precipitable water has a clear equatorial peak; the difference between precipitation and evaporation forms a wavy pattern with the maximum surplus at the Equator, secondary maxima almost symmetrically at near 50° N and S, and pronounced minima in the subtropics (fig. 5.5). The equatorial maximum is associated with convective cloud systems, whereas the secondary maxima are the result of extratropical cyclones and midlatitude convection (Jonas 1999). Notable departures from these latitudinal regularities are caused by irregular distribution of continental masses and by major mountain ranges whose orographic uplift results in record rainfalls (extremes in Assam and Hawai'i). Global gridded fields of monthly precipitation patterns can now be constructed based on five different sources of gauge and satellite observations (Xie and Arkin 1996). Because the biosphere's primary productivity is so highly dependent on water supply, these precipitation patterns clearly prefigure the intensity of photosynthesis.

Surface runoff returns the precipitated water to the oceans rather rapidly: average residence times of fresh water range from just two weeks in river channels and weeks to months in soil to years in lakes and swamps. Annual river runoff ranges typically between 200 and 300 mm, but it is as high as 800 mm for Amazonian South America and as low as 25 mm for Australia. Not surprisingly, the Amazon alone carries about 16% of the planet's river water, and the world's five most voluminous streams (the Amazon, Ganges-Brahmaputra, Congo, Orinoco, and Yangzi) carry 27% of all river runoff (Shiklomanov 1999).

In contrast to the rapid surface runoff, water may spend many thousands of years in deep aquifers. Perhaps as much as two-thirds of all fresh water on the Earth is contained in underground reservoirs, which annually cycle an equivalent of some 30% of the total runoff to maintain stable river flows (Ambroggi 1977). Submarine groundwater discharge (SGWD), the direct flow of water into the sea through porous rocks and sediments, appears to be a much larger flux of the global water cycle than previously estimated. Moore (1996) found large enrichment of groundwater-borne ^{226}Ra in brackish waters along the coast of the southeastern United States and concluded that this influx was equivalent to about 40% of the river water flow during the study period, whereas previous estimates of global SGWD ranged from just 0.01% to 10% of runoff (Church 1996). An important consequence of this finding is the need to revise upward estimates of terrestrial fluxes of dissolved materials to coastal waters.

Cycles of Doubly Mobile Elements

The level of high mobility of carbon, nitrogen, and sulfur makes these three elements fairly readily available in spite of their relative biospheric scarcity. Their cycles are complex both in terms of the number of major reservoirs, fluxes, and compounds involved in the cycling processes and in temporal terms as well. Every one of these grand biospheric cycles hides numerous nested subcycles whose fluxes operate on time scales ranging from minutes to millions of years, as the elements may move rapidly among ephemeral reservoirs or be sequestered (assimilated, immobilized) for extended periods of time. Rapid cycling of nutrients in shallow waters is a common example of the first kind of transfers, whereas massive withdrawals of carbon into extensive carbonate deposits illustrate the slow cycling.

And the three cycles are not just essential for the biosphere; they are also fundamentally its creations. Although abiotic transfers and reactions are important in all of them, key fluxes of these cycles would be impossible without living organisms. Rapid cycling of carbon and all essential macro- and micronutrients depends on prompt degradation of dead biomass and the return of elements from complex organic molecules to much simpler inorganic compounds that can be reused in autotrophic production. On land this recycling returns 98–99.8% of all elements incorporated in dead biomass; in the ocean the remineralization puts back 85–90% of carbon and nutrients, with the remainder lost to sediments. Multistep mineralization processes involving archaea, bacteria, protists, and invertebrates cannot be accomplished in any other way—nor could, to give just two examples, nitrogen fixation under ambient conditions or dissimilatory reduction of sulfates.

Unfortunately, the easy mobility of these three elements—the result of many of their volatile and highly water-soluble compounds—also means that human interference in these cycles has become evident on the global level, above all as rising atmospheric concentrations of CO_2, CH_4, and N_2O. Such interference has, as in the case of acid deposition of sulfates and nitrates, major impacts on large regional or continental scales. Environmental problems arising from these changes—potentially rapid global warming, widespread acidification of soils and waters, growing eutrophication of aquatic and terrestrial ecosystems—have been receiving a great deal of research attention in the last decade (Turner et al. 1990; Schlesinger 1991; Butcher et al. 1992; Wollast et al. 1993; Mackenzie and Mackenzie 1995; Agren and Bosatta 1998; Smil 2000a). I will comment on these perils in the closing chapter.

The Carbon Cycle

Water aside, the circulation of carbon is undoubtedly the biosphere's pivotal mass flow, as it has to assure the availability of tens of billions of tons of the element needed annually by plants; hence it is also the best studied of all elemental cycles (Dobrovolsky 1994; Smil 2000a; Wigley and Schimel 2000; Falkowski et al. 2000; fig. 5.6). Atmospheric CO_2 is the only source of CO_2 required for photosynthesis—land plants tap it directly, and aquatic plants use it after it has been dissolved in water—but its reservoir is by far the smallest presence among the element's principal biospheric storages. With about 280 ppm of CO_2 in the preindustrial atmosphere, the airborne carbon amounted to almost exactly 600 Gt, a mass that terrestrial and aquatic photosynthesis would have sequestered in new phytomass in just four or five years!

In reality, carbon assimilated by plants returns—part of it rather rapidly, the other part only after 10^2–10^3 years—to the atmosphere through soil respiration and wildfires. The best proof that past rates of carbon recycling were closely balanced with photosynthesis soil is the total mass of organic carbon stored in the lithosphere: some 15 Pt of carbon are locked mostly in kerogens, transformed remains of buried biomass found mostly in calcareous and oil shales. Assuming that nearly all of this mass was accumulated during the past 550 Ma, since the beginning of the Cambrian, only about 0.03% of newly formed phytomass had to be sequestered every year to arrive at the total. The accumulation rate was particularly high after the appearance of first forests (about 370 Ma during the Devonian) and before the emergence of the first large herbivores (about 250 Ma ago) and colonial herbivorous insects, whose grazing limited the amount of biomass that could be buried. The late emergence of lignin digestion by fungi may have been another important contributing factor (Robinson 1996).

The continuous transfer of a tiny fraction of assimilated carbon from soils to sediments remains the principal terrestrial bridge between the element's rapid and slow cycles, and it has also been responsible for the accumulation of free oxygen in the atmosphere. Only some 3% of all O_2 produced by photosynthesis in excess of respiration remains airborne, however, as the bulk of it is consumed by the oxidation of reduced iron compounds that produced massive banded iron formations. These deposits, the principal source of the world's iron ore, contain about 2×10^{22} g of O_2, or 20 times the present atmospheric level of the gas (see also chapter 2). Only a minuscule part of all sequestered

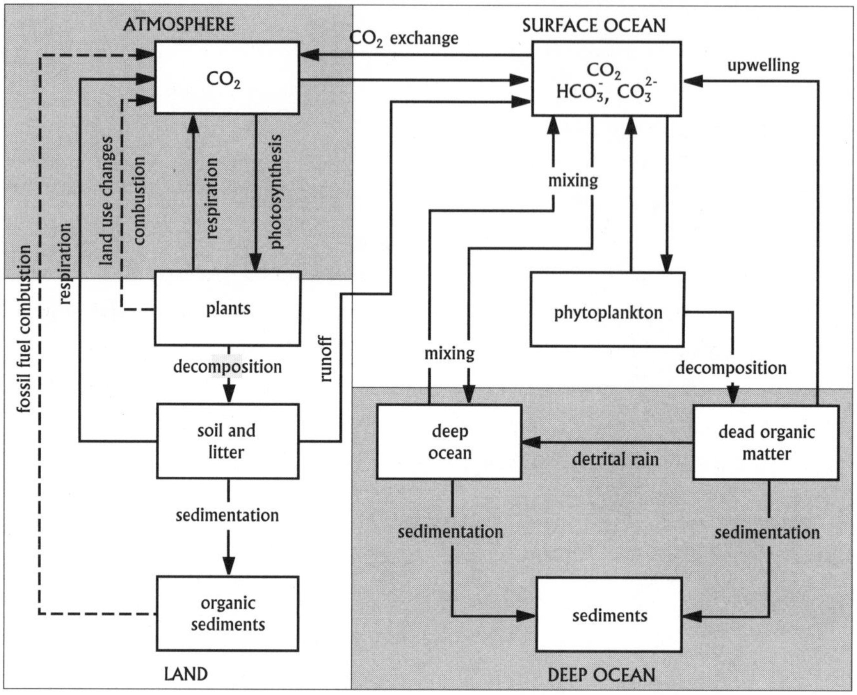

5.6 Principal reservoirs and flows of the biospheric carbon cycle (Smil 2000a).

organic carbon, about 6 Tt, or on the order of 0.04%, is present in traditional fossil fuels (coals and liquid and gaseous hydrocarbons) and only a small part this resource will eventually be recovered. But conversion of more accessible and higher-quality fuels has built the first high-energy civilization in human history and is largely responsible for extensive anthropogenic interference in the biospheric carbon cycle, whose most worrisome outcome may be a relatively rapid rate of global climate change. I will return to this challenge in the closing chapter.

Compared to the atmospheric reservoir of some 600 Gt C, preagricultural ecosystems may have stored nearly twice as much carbon in phytomass (about 1,100 Gt C; see chapter 7), and more than twice as much of it in soils, where it was, and is, overwhelmingly bound in organic matter. Soils now store on the order of 1,500 Gt C, with the storage density going up with higher precipitation and lower temperature and the peak deposit rates in tundra peats (Schlesinger et al. 2000). Soil respiration (oxidation of organic soil matter by microbial and invertebrate decomposers) may result in very rapid biospheric carbon cycling, but the average residence time of carbon in the long-lived humus pool is on the order of 1,200 years. Raich and Potter (1995) estimated annual soil respiration of microbes, fungi, roots, and invertebrates at 77 Gt C/year. Soil respiration is thus higher than our best estimates of net primary productivity (NPP), because it also includes CO_2 evolved by roots and mycorrhizae.

If, as used to be generally assumed, average rates of soil respiration increase with temperature, then global warming should lead to accelerated release of CO_2 from soils. A survey of decomposition in forest mineral soils on five continents, however, found that the rates of CO_2 release were remarkably constant across a global-scale gradient of mean annual temperatures (Giardina and Ryan 2000). Ambient temperature thus appears to be less of a limiting factor for microbial activity than previously believed, and future changes in ambient temperature may have little impact on soil respiration (Grace and Rayment 2000). But these findings cannot exclude the possibility of an eventual positive feedback between temperature and soil respiration resulting in a relatively sudden pulse of CO_2 injected into the atmosphere (Cox et al. 2000).

Standing ocean phytomass is only a tiny fraction, no more than 0.5%, of the terrestrial total. But because of the rapid turnover of phytoplankton, marine photosynthesis assimilates annually almost as much carbon as do land plants, turning parts of the ocean into major carbon sinks during the months of highest productivity. Respiration by zooplankton and other oceanic herbivores returns more than nine-tenths of this assimilated carbon to surface waters to be reused by phytoplankton or to equilibrate with CO_2 in the overlying atmosphere. The rest of the assimilated carbon sinks, with interruptions caused by feeding throughout the water column, to the ocean bottom.

Biomass thus acts as a pump, drawing atmospheric CO_2 into the ocean's depths, and diatoms (about 1,000 species of single-celled algae with silica cell walls and shapes ranging from flat discs to slender rods) make a particularly large contribution to this particle rain from the surface layer (Smetacek 2000). The annual burial rate is on the order of 100 Mt C, or less than 0.3% of all carbon assimilated by phytoplankton. The eastern equatorial Pacific accounts for 20–50% of oceanic net phytoplankton productivity and is also the site of the greatest CO_2 from the ocean to the atmosphere. (Loubere 2000). In the long run the region's high NPP is not governed by aeolian flux of growth-limiting iron or by wind-driven changes in the upwelling rate, but by changes in ocean circulation and the chemical composition of the upwelled water.

Although carbon's slow circulation is overwhelmingly a matter of inorganic reactions, it has important organic components. The process begins with the weathering of silicate rocks (calcium silicate [$CaSiO_3$] is used as a proxy for common Ca and Mg silicates, which are major constituents of the Earth's crust; their principal constituent minerals are plagioclase, biotite, pyroxenes, olivine, and amphiboles) driven by the water cycle and consuming CO_2 from the atmosphere (fig. 5.7; Garrels and Mackenzie 1971; appendix D). In vegetated areas this weathering is accelerated by interactions with the rapid, biota-driven carbon cycling as roots and microbes release CO_2 into soils, where it combines with water and attacks silicates (Schwartzman 1999). The resulting ions and dissolved silica are gradually transported to the ocean, which contains roughly 37 Tt of dissolved inorganic carbon, about 95% of it in highly soluble bicarbonate ions, HCO_3^-, and nearly 98% of it dissolved in dark intermediate and deep waters with temperatures at 2–4 °C.

Inorganic precipitation of carbonates proceeds only after the two constituent ions have reached critical concentrations in the seawater, but the process of carbonate sedimentation is greatly accelerated by marine biomineralizers, which use calcium carbonate ($CaCO_3$) to build their

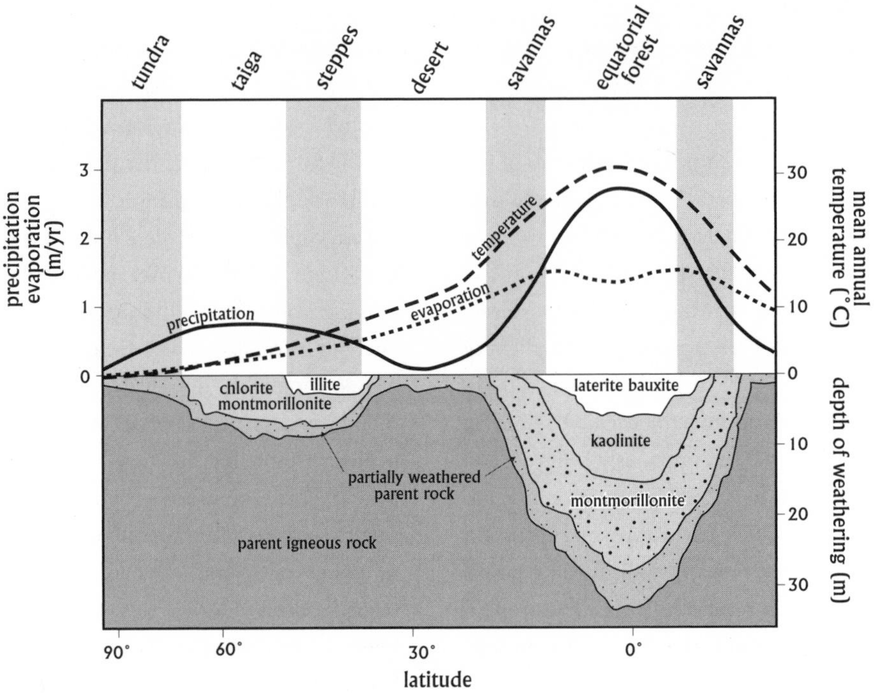

5.7 Weathering of silicate rocks in different environments is affected primarily by precipitation, evaporation, and temperature. Based on Hay (1998).

shells. These and other marine organisms also contribute organic carbon to marine sediments. As noted in chapter 4, the remarkably constant bipartition of sedimented carbon between organic and inorganic compounds—their ratio has been close to one to five for almost four billion years—makes it clear that biota have had a critical role in the element's cycle throughout the biosphere's evolution.

Loose sediments are slowly converted to compact carbonates and hard quartz. $CaCO_3$, as calcite or aragonite, is most abundant, and $CaMg(CO_3)_2$ (dolomite) is the third most common carbonate mineral; $MgCO_3$ (magnesite) and $FeCO_3$ (siderite) are much less common. The eventual fate of these sediments is determined by the large-scale processes of global plate tectonics: they are either pushed up in plate collisions to form mountain ranges or subducted into the mantle as they ride on relatively rapidly moving oceanic plates. The eventual metamorphosis of subducted sediments (by reactions with silica) re-creates silicate rocks that are re-exposed by tectonic uplift, and CO_2 is degassed along the spreading ridges and in volcanic eruptions. In any case, the sediments will reappear after 10^7–10^8 years to be weathered once more.

This cycling of oceanic sediments, and the attendant return of carbon as CO_2, cannot be accelerated by any atmospheric or biospheric influences, but the rates of CO_2 removal from the air may change rather rapidly. Lower tropospheric temperatures and decreased rates of silicate weathering would lead to gradual accumulation of emitted CO_2—and to subsequent warming. Increased temperatures would result in higher rates of evaporation and hence in a faster dissolution of CO_2; the resulting carbonic acid (H_2CO_3), the primary chemical agent of weathering the silicate rocks, would remove more Ca^{2+} and HCO_3^- to be transported by streams to the ocean, where they form carbonate sediments. At the same time, the return of CO_2 from the grand geotectonic cycle would remain unchanged, and the planet would gradually cool.

The loss of atmospheric CO_2 accelerates during the periods of mountain building and greater limestone deposition brought by intensified weathering. Raymo et al. (1988) suggested that the uplift of the Himalayas led to a reduced greenhouse gas effect because of an increased sequestration of CO_2 in what is the planet's most intensive region of denudation, but this proposal has been strongly challenged (Schwartzman 1999). Most notably, without a balancing source of CO_2, all of the gas would have been removed from the atmosphere/ocean pool in a relatively short period of time. Rates of CO_2 release are related above all to sea floor generation. This slow carbon cycle—feedbacks involving CO_2 consumed by weathering and the gas released from metamorphism and magmatism—controls the Earth's temperature on time scales exceeding 1 Ma (Walker et al. 1981; Berner et al. 1983). These feedbacks prevented complete ocean freeze-up even when the early biosphere received 30% less radiation (compared to current levels) from the faint young Sun.

As noted in chapter 4, only approximate reconstructions of CO_2 levels are possible for more distant periods, but air bubbles from Antarctic and Greenland ice cores (the largest one is 3,623 m deep) make it possible to trace CO_2 levels with high accuracy for the past 420,000 years (Petit et al. 1999). During that time CO_2 levels have stayed between 180 and 300 ppm, and during the 5,000 years preceding 1850, they fluctuated only between 250–290 ppm (Petit et al. 1999; fig. 4.10). Their post-1850 rise will be described in chapter 9.

The plant- and microbe-dominated fast cycle also generates pronounced daily and seasonal cycles of atmospheric CO_2 levels. The latter undulation is essentially the biosphere's breath, a globally integrated annual cycle of respiration and photosynthesis. Its inhalations and exhalations are particularly pronounced in the Northern Hemisphere, where most of the terrestrial phytomass is located: CO_2 minima are in August, during maximum carbon storage by forests and grasslands, and maxima in January, when much of the vegetation is either leafless or dormant (fig. 5.8). Both of these cycles have been altered by human actions. Whereas the traditional burning of phytomass fuels (wood, charcoal, crop residues) was essentially just an accelerated form of recycling carbon that was stored in plants for months to decades, combustion of fossil fuels has been emitting increasing amounts of carbon that has been sequestered for 10^6–10^8 years, and an additional net release of carbon comes from the conversion of natural ecosystems (mainly from tropical deforestation) to agricultural, industrial and urban uses (for details, see chapter 9).

Before turning to the nitrogen cycle, I must note that a substantial portion of buried organic carbon is involved in a methane subcycle: it is returned to the atmosphere by anaerobic methanogenic archaea, then oxidized in a stepwise process to CO_2 (Warneck 2000). By far the largest mass of CH_4 generated by microbial decomposition is stored beneath the sea floor in methane hydrates (a combination of ice and the gas). A massive release of this gas,

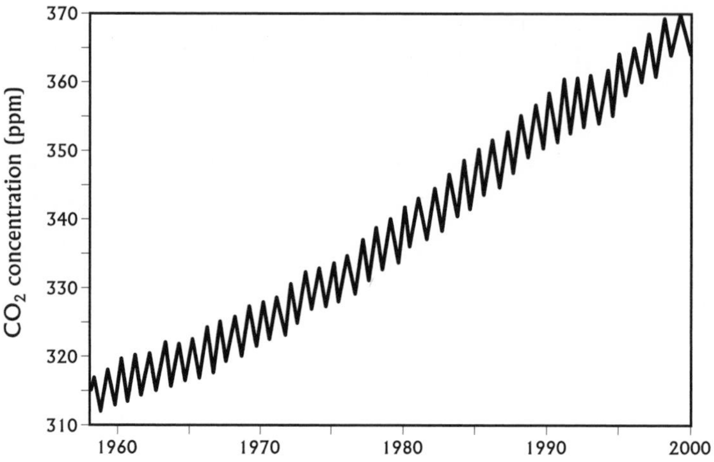

5.8 The biospheric breath is clearly discernible in seasonal fluctuations of CO_2 concentrations monitored at Mauna Loa. Monthly data available at <http://cdiac.esd.ornl.gov/trends/co2/nocm-ml.htm>

followed by its oxidation to CO_2, may have led to the sudden warming experienced 55 Ma ago and associated with the rise of mammals to terrestrial dominance (Katz et al. 1999).

But even noncatastrophic CH_4 oxidation does not perfectly match natural methanogenesis, and the gas released from wetlands, wild ruminants, and wildfires is present in trace amounts in the atmosphere. Air bubbles in polar ice show levels between 300 and 700 ppb during the past 100,000 years, with narrow oscillations around about 700 ppb during the last preindustrial centuries (Etheridge et al. 1998; fig. 5.9). Although low in absolute terms (three orders of magnitude below CO_2 levels), such concentrations are almost infinitely higher than would be their thermodynamic equilibrium value in the Earth's oxygen-rich atmosphere, and they are possible only because methanogenesis is constantly outpacing oxidation. But as with CO_2, current sources of CH_4 exceed its sinks because of human actions ranging from natural gas losses in pipelines to more extensive rice cultivation and growing cattle numbers. At almost 1800 ppb the atmospheric level of CH_4 is now more than twice as high as it was in 1850. And because CH_4 is, over a period of 100 years, an approximately 20 times more potent absorber of outgoing IR than is CO_2, it has become the second most important anthropogenic contributor to the tropospheric warming (Watson et al. 1996).

The Nitrogen Cycle

Although the biosphere is bathed in nitrogen—the gas makes up almost 80% of the atmosphere by volume—the element is not readily available in forms assimilable by organisms. And although biota need the element in only

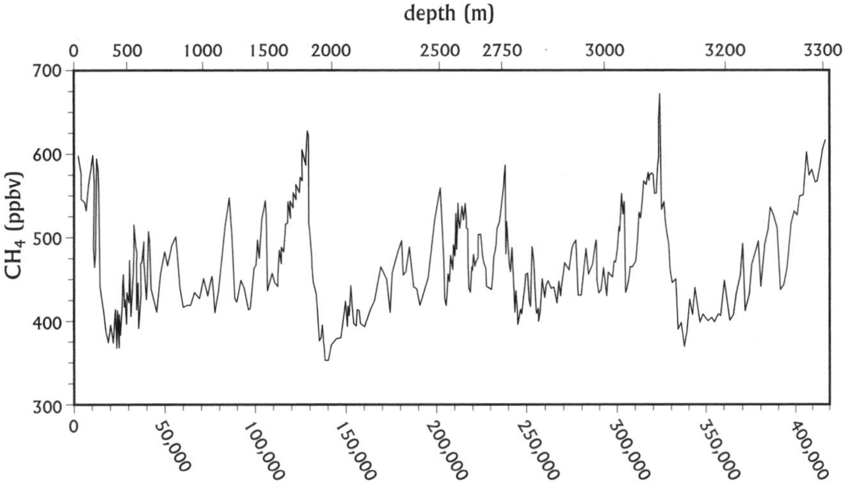

5.9 Methane concentrations derived from ice cores and from recent tropospheric measurements. Based on Petit et al. (1999) and U.S. Environmental Protection Agency (2001).

small quantities, its scarcity is often the most important factor limiting photosynthesis and compromising heterotrophic growth. These paradoxes are explained by nitrogen's atmospheric presence as a nonreactive N_2 molecule held by one of the strongest known triple bonds (–945 kJ/mol) and by a shortage of natural processes that can split it and incorporate the atomic N into reactive compounds. But the relative inertness of N_2 is not entirely unwelcome: a more reactive molecule combining with oxygen might lead to constant nitrate rains and an acid ocean.

Not surprisingly, none of the three principal polymeric constituents of phytomass—cellulose, hemicellulose, and lignin—contains any nitrogen, but life cannot exist without it. The element must be present in every living cell in the nucleotides of nucleic acids (DNA and RNA), which store and process all genetic information; in amino acids, which make up all proteins; and in enzymes, which control life's metabolism; and nitrogen is also a part of chlorophyll, whose excitation by light energizes all photosynthesis.

Nitrogen's biospheric cycle also stands apart from the circulations of the other two doubly mobile elements because its three key fluxes—fixation, ammonification, and nitrification and denitrification—are nearly completely dominated by bacteria (fig. 5.10). Nitrogen fixation, the conversion of unreactive N_2 to reactive compounds, can be abiotic, but it requires the high-energy discharge in lightning to sever the N_2 molecule, and the element then forms nitrogen oxides (NO and NO_2), which are eventually converted to nitrates. In contrast, biofixation, moving N_2 to NH_3, takes place at atmospheric temperatures and pressures and fixes two orders of magnitude more N than lightning, but it is performed only by about 100 bacterial genera

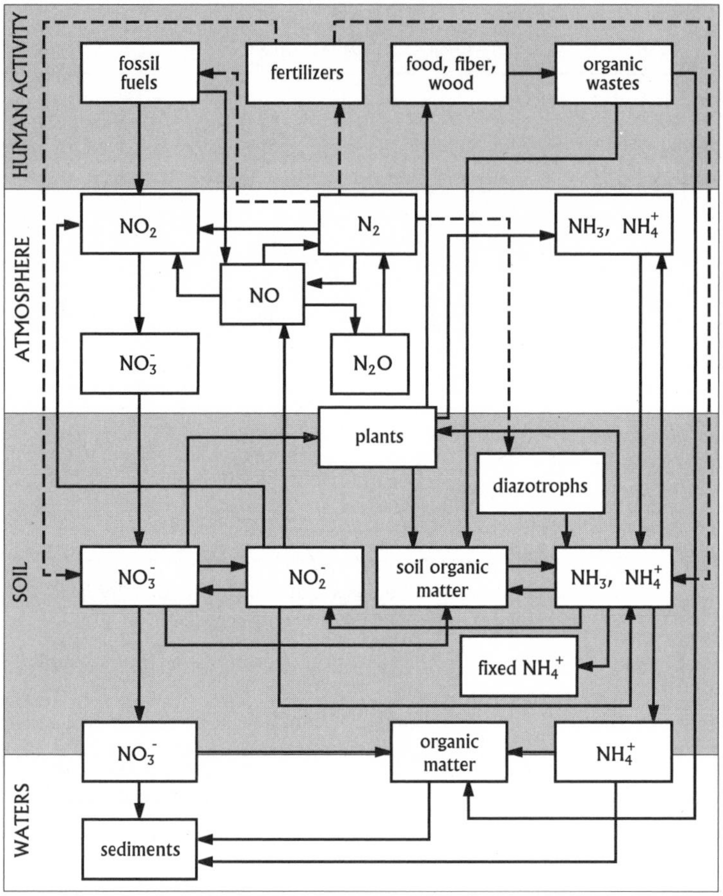

5.10 Principal reservoirs and flows of the biospheric nitrogen cycle (Smil 2000a).

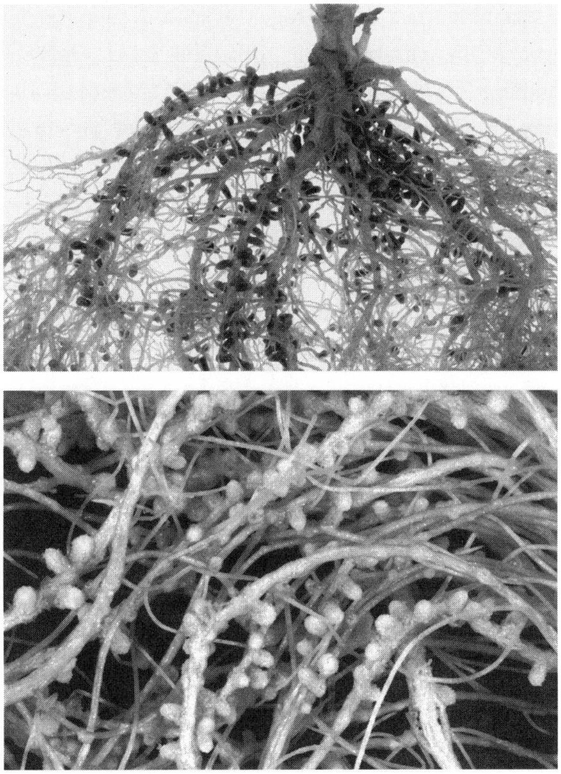

5.11 Roots of a pea (*Pisum sativum*) plant with numerous nodules containing *Rhizobium* bacteria. Images courtesy of Bert Luit, Department of Plant Science, University of Manitoba, Winnipeg.

that posses nitrogenase. This specialized enzyme that can cleave N_2, acquired during the almost anoxic conditions of the early Archean by copying a short-chained ferredoxin, was essential for Archean microbes, but biofixation is costly in energy terms, and no eukaryotes resort to it.

Diazotrophs (microbial fixers) can be free-living species, endophytes, or symbionts (Broughton and Puhler 1986; Smil 2000a). Free-living diazotrophs, including such common soil bacteria as *Azotobacter* and *Clostridium,* usually fix less than 0.5 kg N/ha per year, but cyanobacteria can add substantial amounts of nitrogen both to soils and surface waters. *Anabaena, Nostoc,* and *Calothrix* are by far the most important genera of cyanobacteria, enriching soils by as much as 20–30 kg N/ha per year. To protect the O_2-intolerant nitrogenase, they fix nitrogen in special heterocysts formed at regular intervals along their filaments. Common anaerobic fixers, such as *Oscillatoria* and *Plectonema,* do not need heterocysts. Colonial filamentous *Trichodesmium* was seen as the dominant oceanic diazotroph ever since its N-fixing ability was discovered in 1961, but new research indicates that unicellular cyanobacterial nanoplankton may be a significant source of reactive nitrogen (Zehr et al. 2001).

Endophytic diazotrophs (*Gluconacetobacter, Herbaspirillum*) live inside plant roots, stems, and leaves and can fix at least 50 kg N/ha annually; their presence explains high yields of unfertilized sugar cane and other tropical grasses (Boddey et al. 1995).

Some nitrogen-fixing organisms live in association with plants: *Frankia,* an actinomycete genus, forms nodules on roots of some 170 plants, including alder (*Alnus*) and *Myrica,* species common in bogs and on eroded slopes; *Anabaena* is associated with a water fern, *Azolla,* and *Beijerinckia* with sugar cane. But by far the most productive diazotrophs belong to just six genera of rhizobial bacteria: *Rhizobium, Bradyrhizobium, Sinorhizobium, Mesorhizobium, Azorhizobium,* and *Allorhizobium* (Spaink 2000). These chemoheterotrophs have the unique capacity to induce the formation of root nodules and are symbiotic solely with plants of Leguminoseae family, both large tropical trees and small temperate-climate annuals (fig. 5.11). These symbionts are concentrated in distinct root nodules, and they obtain carbohydrates in a rather energy-expensive exchange for their fixed nitrogen, which they supply in the form of ammonia or alanine: the typical cost is between 3 and 6 g C/g N fixed. Depending on the site and season, there may be anywhere between 10^0–10^6 rhizobia per gram of soil.

Not surprisingly, natural biofixation has strong positive relationships with evapotranspiration and NPP (for details on the latter, see chapter 7). Rhizobia can add annually more than 500 kg N/ha with such highly productive cover legumes as alfalfa, but more typical fixation rates in forests and grasslands are well below 100 kg N/ha. Because of large natural variability of biofixation rates observed in individual leguminous plants and in legume-rich ecosystems, it is difficult to offer a tightly constrained estimate of global biofixation. The most comprehensive recent evaluation of natural terrestrial biofixation, dominated by the contribution of tropical forests, ended up with a rather wide range of 100–290 Mt N, with 150–190 Mt N per year being the most likely value (Cleveland et al. 1999). This means that without any recycling, biofixation would exhaust the atmospheric reservoir of N_2 (3.9 Pt) in about twenty million years.

Plants can use the ammonia produced by diazotrophs as a source for nitrogen but prefer to assimilate the much more soluble nitrate. As previously explained (in chapter 3), the conversion of NH_3 to NO_3 is strictly microbial; nitrifying bacteria are common in both soils and waters, and the process is fairly rapid, particularly in warm and well-aerated soils. Assimilated nitrogen is embedded mostly in the amino acids that form plant proteins. Heterotrophs must ingest preformed amino acids in feed and food to synthesize their proteins. After plants and heterotrophs die, enzymatic decomposition (ammonification) moves nitrogen from dead biomass to NH_3, which is again oxidized by nitrifiers. Finally, denitrification returns the element from NO_3^- via NO_2^- to atmospheric N_2. This closing arm of the nitrogen cycle is also almost entirely microbial (abiotic denitrification, when soil organic matter reacts with nitrites, is of minor importance). But unlike nitrification, which depends on just a few bacterial genera, many common aerobic bacteria—including *Pseudomonas, Bacillus,* and *Alcaligenes,* whose enzymes use successively more reduced nitrogen compounds as electron acceptors—can perform denitrification.

Because not all of these bacteria have the requisite reductases for carrying the process all the way to N_2, denitrification also produces appreciable amounts of nitric oxide (NO) and N_2O. N_2O is unreactive in the troposphere, but in the stratosphere it reacts with O* and produces NO, whose release sets off the catalytic destruction cycle of O_3. This sequence, whose discovery by Crutzen (1970) was rewarded with the 1995 Nobel Prize in chemistry, is the most important natural cause of O_3 decomposition—and the reason for concerns about the long-term effects of increasing N_2O emissions from intensifying fertilization. In addition, N_2O is also a greenhouse gas considerably more potent than CO_2. Denitrification proceeds faster in soils with high levels of nitrate, organic matter, and moisture levels and low levels of O_2, and in fairly high temperatures, but its natural rates range over two orders of magnitude. Not surprisingly, global estimates of denitrification have more than a threefold range.

There are also many leaks, detours, and backtrackings along this main cyclical route from fixation to denitrification. Volatilization from soils, plants, and heterotrophic wastes adds NH_3 to the atmosphere, and the gas is rather rapidly redeposited in dry form or in precipitation. Both nitrification and denitrification release NO_x and N_2O. NO_x nitrogen is soon redeposited, mostly after oxidation to NO_3. In contrast, N_2O is basically inert in the troposphere, but as noted above, it is a powerful greenhouse gas. Highly soluble nitrates leak readily into ground and surface waters, and both organic and inorganic nitrogen in soils is moved to waters by soil erosion.

Compared to the huge amounts of atmospheric storage—besides the dominant N_2, there are also traces of NO, NO_2, N_2O, NO_3^-, and NH_3—other biospheric nitrogen reservoirs are small. Because most of the Earth's phytomass is composed of nitrogen-poor polymers (see chapter 6), the amount of the element stored in preagricultural vegetation added up to no more than 10 Gt. Soils contain about 100 Gt N, mostly in long- and short-lived humus. Average values for larger areas are not very meaningful, as nitrogen content varies by more than an order of magnitude among various soil types—ranging from less than 1,000 kg N/ha in poor glacial soils to well over 10,000 kg N/ha in excellent chernozems—and by a considerable margin even within a single field. The uncontaminated waters of the preindustrial era stored very little nitrogen: ammonia is not very soluble, and nitrate concentrations in preindustrial streams were very low (less than 0.1 mg NO_3-N/L), and even today clean rivers generally carry less than 1 mg N/L. Today's levels of aquatic nitrogen, as well as many other biospheric reservoirs and flows, have been much affected by the growing human interference in the element's cycle (chapter 9).

The Sulfur Cycle

Sulfur's critical role in life (as noted in chapter 2) is to keep proteins three-dimensional. Only two of the twenty amino acids providing building blocks for proteins (methionine and cysteine) have sulfur in their molecules, but when amino acids form polypeptide chains, disulfide bridges link them to maintain their complex folded structure, which is necessary for engaging proteins in biochemical reactions. Photosynthesis incorporates roughly one atom of sulfur for every 1,000 atoms of carbon. The initial uptake is in the form of soil sulfate, which is then reduced intracellularly to H_2S and incorporated mostly in cysteine and methionine. Nonprotein sulfur is in biotin (vitamin H), thiamine (vitamin B_1), glutathione, and coenzyme A. Phytomass contains between 0.08 and 0.5% S, which means that terrestrial ecosystems now assimilate annually about 150 Mt S, and standing phytomass contains less than 500 Mt S.

Degradation of sulfur-containing organic compounds by many common bacteria (*Pseudomonas, Escherichia*) and fungi (*Aspergillus, Streptomyces*) promptly returns most of the assimilated sulfur to soils. Organic sulfur, nearly equally split between amino acids and nonprotein compounds, accounts for more than 90% of total soil stores which, with carbon-to-sulfur ratios of 60 to 130, amount to about 15 Gt S; most of the rest is in highly soluble sulfates. The rapid sulfur cycle, consisting of assimilatory reduction of sulfate by plants, decomposition, mineralization, and plant uptake, has no major leak through which the element can be moved to a more persistent reservoir (fig. 5.12).

In contrast, dissimilatory bacterial sulfate reduction, which generates H_2S as well as more complex compounds including $(CH_3)_2S$ (methyl sulfide), COS (carbonyl sulfide) and CS_2 (carbon disulfide), has moved large masses of sulfur into long-term mineral stores. The process is carried on by thiopneuts, more than a dozen bacterial genera that reduce sulfate (or elemental sulfur) in a way analogical to denitrification. *Desulfutomaculum* and *Desulfovibrio* are particularly common in tidal and marine sediments, which combine an anoxic environment with abundant carbon substrate for metabolism. The reduction takes place in the topmost mud layers, with the required sulfate coming from the interstitial water. The portion of H_2S that does

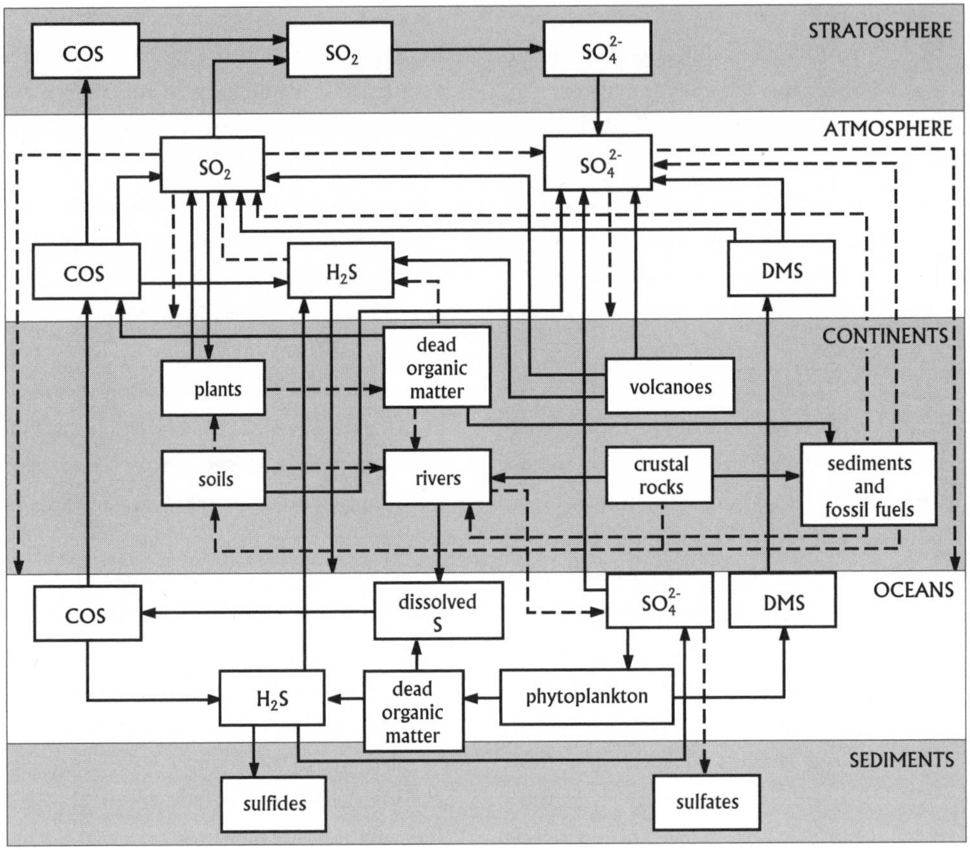

5.12 Principal reservoirs and flows of the biospheric sulfur cycle (Smil 2000a).

not escape into the water and the atmosphere diffuses within the mud and reduces the oxidized ferric ion (Fe^{3+}) to the ferrous ion (Fe^{2+}). Further reactions of Fe^{2+} with H_2S produce first FeS (frequently imparting black pigmentation to tidal muds) and then FeS_2 (pyrite).

Together with the formation of stromatolites (see chapter 2), this conversion is one of the two oldest biogenic processes of mineral formation, as the precursors of today's thiopneuts were among the earliest organisms. Ohmoto et al. (1993) described a pyrite deposit in South Africa's Barberton Greenstone Belt that was formed unmistakably by bacterial reduction of seawater sulfate about 3.4 Ga ago. Shen et al. (2001) found unequivocal evidence of microscopic sulfides in approximately 3.47-Ga-old barites from North Pole, Australia. And sulfate reduction remains at the core of long-term cycling of the element: best estimates of current marine FeS_2 formation indicate sequestration of at least 40 Mt S per year, but the actual flux may be much higher. The other process that has sequestered large amounts of oceanic sulfur is abiotic: episodic evaporation of seawater in shallow basins in arid climates precipitates carbonates first, and sulfates (mostly $CaSO_4{\cdot}2H_2O$) form only once the water volume is reduced by 80%.

Major natural sources of sulfur entering the biosphere include both nonvolatile and volatile compounds originating in sea spray, wind erosion, volcanoes, and a variety of biogenic emissions (Bates et al. 1992). Sea salt sulfates, lofted into the air primarily not by spectacularly breaking waves, but by incessant microbursts of countless bubbles at the ocean's surface, are by far the largest natural input of sulfur into the atmosphere. SO_4^{2-} is, after Cl^-, the second most abundant ion in seawater, but most of its huge mass (about 1.3 Pt) is beyond the reach of the near-surface turbulence. Global annual estimates of the annual sulfate flux have ranged from a mere 44 Mt to 315 Mt, with 140–180 Mt being the most likely input. In any case, only about a tenth of this mass settles on the continents; gravitation promptly returns the rest to the ocean.

Volcanic emissions are, on average, a relatively large source of sulfur compounds, mainly of SO_2, but their contributions are highly variable. For example, Mount St. Helens' spectacular plume and mud flows of May 18, 1980, contained no more than about 1,000 t S (Decker and Decker 1981). In contrast, the largest stratospheric cloud of SO_2 ever measured, emitted by Mount Pinatubo's eruption in June 1991, carried at least 17 Mt of the gas, or nearly 9 Mt S (Kress 1997). Reconstructions of total sulfur emissions based on sulfate deposits in Antarctic ice indicate that the 1815 Tambora eruption, the largest in modern history, spewed out as much as 50 Mt S.

Recent long-term annual global emission means, made up of quiescent degassing and explosive ejections originating mostly in the East Pacific arc between Kamchatka and Indonesia, add up to 14 ± 6 Mt S (Graf et al. 1997). Submarine volcanism, concentrated along the spreading zones of midocean ridges, is a source of identical magnitude, but we do not know what share of that flux enters the atmosphere. After reaching the stratosphere, sulfur-containing dust from volcanic emissions may contribute to global cooling not primarily, as originally thought, because it blocks the incoming radiation, but because atmospheric reactions produce sulfuric acid (H_2SO_4) aerosols, which are very efficient in back-scattering and absorbing incident radiation and have longer residence times than dust particles. Moreover, studies of polar ice cores show that these aerosols are dispersed globally, and hence it may be the sulfur emitted in an eruption rather than the total amount of dust that determines the degree of posteruption cooling. Estimates of sulfur in airborne dust, mainly desert gypsum ($CaSO_4{\cdot}2H_2O$) are also highly uncertain, ranging from 8 to 20 Mt S per year, with most of this burden promptly redeposited downwind. On the other hand, as already noted, major dust storms can carry considerable masses of particulate matter across continents or oceans, but some particulate matter crosses the oceans constantly.

Biogenic emissions of sulfur come from the decomposition of biomass by a variety of specialized bacteria that use the energy liberated by oxidation or reduction of elemental sulfur or sulfur compounds, including H_2S from the decomposition of cysteine and dimethyl sulfide (DMS), CH_2SH (methyl mercaptan), and $(C_3H_7)_2S$ (propyl sulfide) from the transformation of methionine. Terrestrial emissions come mostly from aquatic ecosystems, river and lake muds, and hot springs. Biogenic sulfur gases include H_2S (the dominant product from wetlands, lakes, and

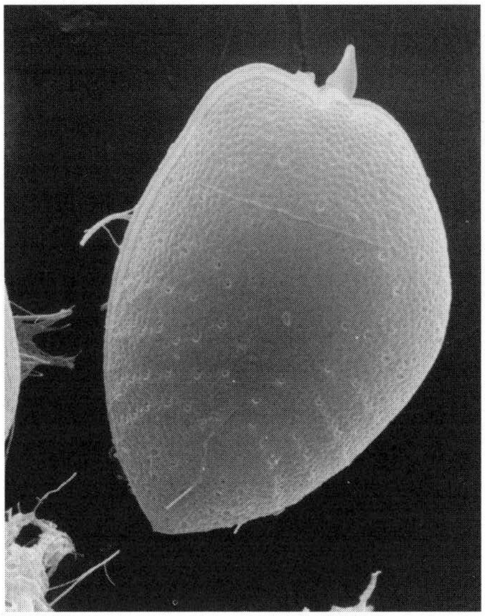

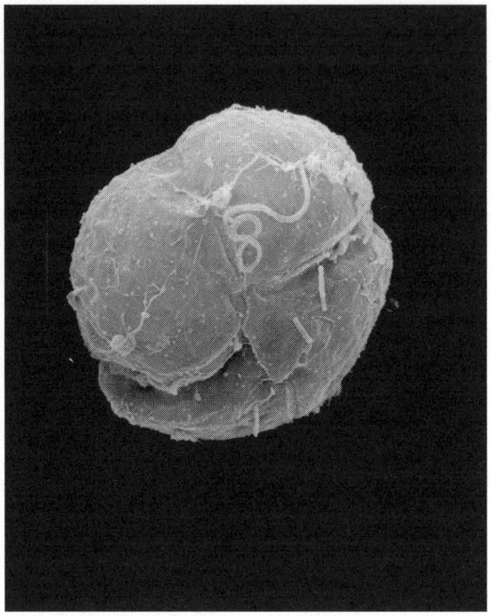

5.13 Two species of dinoflagellates, an armored cell of *Prorocentrum micans* and an unarmored *Gymnodinium mikimoti*. Both images courtesy of E.G. Vrieling and J. Zagers, Rijksuniversiteit Groningen.

anoxic soils), DMS (CH_3SCH_3), methyl mercaptan, and propyl sulfide. Terrestrial emissions peak during summer and are generally most intensive in the tropics; their total flux, about 14 Mt S per year, is small compared to the biogenic sulfur flux from the ocean.

H_2S was once assumed also to be the most important biogenic gas released from the ocean to the atmosphere, but we now know that DMS makes up at least four-fifths, if not nine-tenths, of that flux. The gas comes, via dimethylsulfonium propionate, from the decomposition of methionine in marine plants. Prymnesiophytes (brown algae) and coccolithophorids and dinoflagellates (fig. 5.13) are especially prolific producers. Global DMS flux is now estimated between 16–22 Mt S/year, with summer peaks produced in oceans at latitudes 50–65 °S and 50–75 °N (Bates et al. 1992). Products of its oxidation are either methanesulfonic acid (CH_3SO_3H, MSA) or SO_2 which is fairly rapidly oxidized to sulfate. As the MSA is not produced in any other way, its presence is an unambiguous indicator of biogenic sulfur emissions.

A possible feedback loop was postulated between DMS generation and received solar radiation (Lovelock et al. 1972; Charlson et al. 1987): more DMS would produce more condensation nuclei thus increasing cloud albedo,

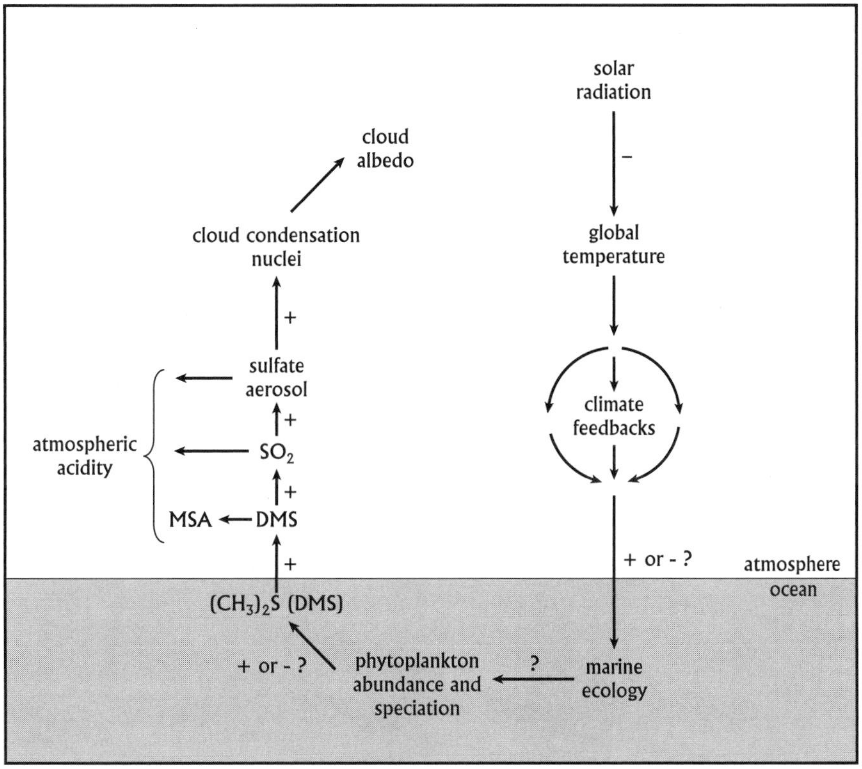

5.14 The mechanism proposed by Charlson et al. (1987) to explain the effect of DMS emissions on climate contains several uncertain links. Based on Watson and Liss (1998). *Note:* MSA = methanesulfonic acid

and the reduced radiation would lower planktonic photosynthesis, generating fewer condensation nuclei and letting in more insolation (fig. 5.14). This postulate of a homeostatic control of the Earth's climate by the biosphere was used for some time as an argument for the existence of Lovelock's planetary superorganism, a self-regulating Gaia (see chapter 9), but the magnitude, and indeed the very direction, of the feedback remains questionable (Watson and Liss 1998). In any case, the DMS hypothesis stimulated a great deal of research into interactions between ocean biota and the atmosphere.

In contrast to those produced by the carbon and nitrogen cycles, sulfur compounds have no long-lasting presence in the atmosphere. Primordial SO_2 and H_2S degassed into the Archean atmosphere were rapidly oxidized and deposited on continents and oceans. Sulfates became a permanent ingredient of ocean water, from which they were gradually incorporated into sediments or bacterially reduced to metal sulfides. Once the atmosphere became sufficiently O_2-rich, all reduced gaseous sulfur compounds were rapidly oxidized to sulfates. The residence time of SO_2 in humid air may be just a few minutes, and the global mean is only about a day; H_2S does not stay much longer, and marine DMS remains aloft up to three days, but commonly it takes less than 10 hours before it reacts with OH radical.

Consequently, the global burden of these gases, and of sulfates generated by their oxidation, staying in the lowermost troposphere usually no longer than three to four days, is low, only about 5–6 Mt S. This is five orders of magnitude less than the atmospheric CO_2 and nine orders of magnitude behind nitrogen's huge atmospheric stores. Carbonyl sulfide (COS), produced directly by sulfur-reducing bacteria, by photochemical reactions with dissolved organic matter in the ocean, by biomass burning, and by atmospheric oxidation of carbon disulfide (CS_2), has the highest background levels. At around 500 ppt, they are an order of magnitude above the levels of either H_2S or SO_2, but the atmospheric throughput of these two gases, as well as that of DMS, is much higher.

Short residence times limit long-distance atmospheric transport to a few hundred kilometers for SO_2 and H_2S, and between 1,000 and 2,000 km for sulfates: as a result, atmospheric sulfur does not have a true global cycle as does atmospheric carbon. Most of it (about 80%) is removed by wet deposition; the remainder is dry-deposited, and SO_2 is absorbed directly by soils and plants. Preindustrial rates of atmospheric deposition added to more than a few kilograms of sulfur per hectare of terrestrial ecosystems. Since 1850 human actions have also made a significant difference in the global sulfur cycle, with anthropogenic emissions coming largely from the combustion of fossil fuels and the smelting of color metals. I will examine these anthropogenic interferences in some detail in the book's last chapter.

Cycles of Mineral Elements

The grand and ponderous sedimentary-tectonic cycle begins with weathering of the Earth's surfaces, a mixture of physical and chemical processes that is also strongly influenced by vegetation. Physical changes in these surfaces are due largely to frost action (differential expansion of mineral grains by freezing and thawing of water in tiny rock crevices), pressure, thermal stresses, and crystal growth. Chemical reactions, involving dissolution, hydrolysis, carbonation, and oxidation and promoted by higher temperatures and water, break rocks down to soluble ions and insoluble residues. Silicate weathering releases dissolved silica (SiO_2), Ca^{2+}, Na^+, K^+, and Mg^+, dissolution of carbonates yields Ca^{2+} and bicarbonate (HCO_3^-) ions, and sulfide weathering produces H_2SO_4, which causes additional silicate weathering. Plants speed up weathering both by the mechanical action of penetrating roots and by introducing respired CO_2 into soils.

On the average, weathering strips less than 0.1 mm from the Earth's crust every year, with extremes ranging from just 1 μm in lowlands to nearly 10 mm in parts of the Himalayas (Selby 1985). Weathered material is removed by water, ice, and wind erosion. Rivers carry the bulk of the mass, in ionic solutions and as suspended matter, and their transport capacities are governed primarily by water flow, gradient, and sediment load itself. Most of the eroded material is temporarily deposited in river valleys and lowlands, and the sediment reaching the ocean, estimated at 9–10 Gt per year during the preagricultural era, is perhaps only a tenth of all annually eroded material. Human activities have markedly increased the rate of soil erosion, above all through deforestation and careless crop cultivation, but in many regions the construction of dams during the twentieth century actually reduced the transport of sediments to the ocean (Hay 1998).

Holeman's (1968) subsequently much cited estimate put sediment transport in the world's rivers at 18.3 Gt per year, and the latest detailed account by Milliman and Syvitski (1992) is little changed at 20 Gt per year. Glacial transport is difficult to estimate, but it may be as low as 1–2 Gt per year, and highly variable wind erosion carries away most likely between 0.5 and 1.0 Gt per year. The total (natural and anthropogenic) annual rate of denudation may be thus on the order of 25 Gt, implying the removal of about 0.17 kg of weathered material per square meter of the Earth's surface. If this eroded material had a composition corresponding to the average makeup of the crust, it would contain about 7 Gt Si, 1.25 Gt Fe, 900 Mt Ca, and 20 Mt P.

An annual sedimentation rate of 25 Gt would denude the continents to the sea level in less than 15 Ma, but tectonic forces keep rejuvenating the Earth's surfaces. Marine sediments can reenter the biosphere in a number of ways. They can reemerge, either still unconsolidated or after they have undergone dehydration, lithification, and metamorphosis, after a simple coastline regression or because of a tectonic uplift during a new mountain-building period. Or they can be carried by a subducting oceanic plate deeper into the mantle, where they are reconstituted into new igneous rocks creating a young spreading ocean ridge or vigorous magmatic hot spots. The transit between a subducting trench and a spreading ridge can take just 10^7 years, and a typical supercontinental cycle takes on the order of 10^8 years (for more on the process, see chapter 4).

Among the scores of mineral elements whose cycling we observe on the civilizational time scale merely as an oceanward flux of eroded sediments, four stand out: phosphorus, because of its fundamental role in metabolism and because of its frequently limiting role in plant growth; calcium and silicon, because of their relatively large uptakes by biota; and iron, because of its controlling effect on marine productivity.

Phosphorus

Phosphorus is rare in the biosphere: in mass terms it does not rank among the first ten elements either on land or in water. Although it is, besides nitrogen and potassium, one of the three macronutrients needed by all plants for vigorous growth, it is entirely absent in cellulose, lignin, and proteins. Phosphorus averages a mere 0.025% in the above-ground forest phytomass, and the biosphere's entire above-ground phytomass stores no more than 500 Mt P. But neither proteins nor carbohydrate polymers can be made without phosphorus. Phosphodiester bonds link mononucleotide units forming long chains of DNA and RNA, synthesis of all complex molecules of life is powered by energy released by the phosphate bond reversibly moving between ADP and ATP, and no life would be possible without at least one atom of phosphorus per molecule of adenosine (Deevey 1970).

The element is relatively abundant in vertebrate bodies, because bones and teeth are composite materials comprised mostly of the phosphorus-rich ceramic constituent: hydroxyapatite, $Ca_{10}(PO_4)_6(OH)_2$, containing 18.5% P and making up almost 60% of bone and 70% of teeth (Marieb 1998). The global anthropomass contains approximately 2.5 Mt P, the reservoir less than half as massive as that of the anthropomass nitrogen (Smil 1999b). Unlike other micronutrients (Ca, Fe, I, Mg, Zn), P is almost never in short dietary supply, with dairy foods, meat, and cereals being its main sources. Crustal apatites—$Ca_{10}(PO_4)_6X_2$, with X being F in fluorapatite, OH in hydroxyapatite, or Cl in chlorapatite—store most of the element. Soluble phosphates are released by the weathering of these apatites: with an average lithosperic content of 0.1% P and a mean global denudation rate of around 750 kg/ha (Froehlich et al. 1982), about 10 Mt P are released annually. Soils store about 40 Gt P, with no more than 15% of this total bound in organic matter.

Unfortunately, these phosphates are usually rapidly immobilized by reactions with aluminum and calcium into insoluble forms (Khasawneh et al. 1980). As a result, only a tiny fraction of phosphorus present in soils is available to plants as a dissolved oxy-anion (PO_3^{-4}), and the element is commonly the growth-limiting nutrient in terrestrial ecosystems in general and in tropical soils in particular. This forces plants to absorb the element from very dilute solutions and concentrate it up to 1,000-fold. The average land phytomass carbon-to-phosphorus mass ratio of around 700 to 1 implies annual assimilation of close to 100 Mt P. The marine phytomass stores only some 75 Mt P, but because of its rapid turnover, it absorbs about 1 Gt P annually from surface water, a flux an order of magnitude higher than in terrestrial photosynthesis.

Rapid recycling of the element released by the decomposition of biomass is essential in both terrestrial and aquatic ecosystems, with turnover times of just 10^{-2} to 10^{0} years in either case. As with carbon, nitrogen, and sulfur, biota are essential in this process. Decomposition of dead biomass, solubilization of otherwise unavailable soil phosphates by several species of bacteria, and enhancement of the release of phosphorus from soil apatites by oxalic acid produced by mycorrhizal fungi are especially critical during later stages of soil development, when primary minerals have weathered away (Smil 2000c). This cycling must be highly efficient. As there is neither any biotic mobilization of the element (akin to nitrogen fixation) nor any substantial input from atmospheric deposition, the nutrient inevitably lost from the rapid soil-plant cycling can be naturally replaced only by slow weathering of phosphorus-bearing rocks (fig. 5.15). Shortages of phosphorus in terrestrial ecosystems are thus common.

The nutrient's scarcity is usually even greater in aquatic ecosystems. Only in shallow waters can phosphates circulate easily between sediments and aquatic biota; in deep oceans phosphorus is relatively abundant only in regions of vigorous upwelling, and dissolved phosphorus is often nearly undetectable in surface waters of the open ocean (Jahnke 1992). Scarcity of the nutrient is a key factor limiting photosynthesis in many freshwater bodies, and external phosphorus inputs control longer-term primary production in the global ocean (Tyrrell 1999). But as with all mineral cycles, the long-term phosphorus cycle is not dominated by biota; moreover, the element's transfers within the biosphere are greatly limited, because it does not form any long-lived gaseous compounds. As a result, its atmospheric reservoir is negligible, its flows have no airborne link from ocean to land, and particulate phosphorus that sinks into marine sediments—the rate of phosphorus burial in ocean sediments may have recently added over 30 Mt P/year (Smil 2000c)—becomes available to terrestrial biota only after the tectonic cycle introduces new crystalline rock into the biosphere (Guidry et al. 2000). As with the three doubly mobile cycles, there has been considerable human interference in biospheric phosphorus flows (see chapter 9).

Calcium

Calcium has both structural and dynamic roles in biota. As noted in the previous section, hydroxyapatite is the dominant constituent of bones and teeth, and myriads of aquatic invertebrates, protists, and autotrophs build their support structures from bioprecipitated $CaCO_3$, the compound also used by reptiles and birds to build

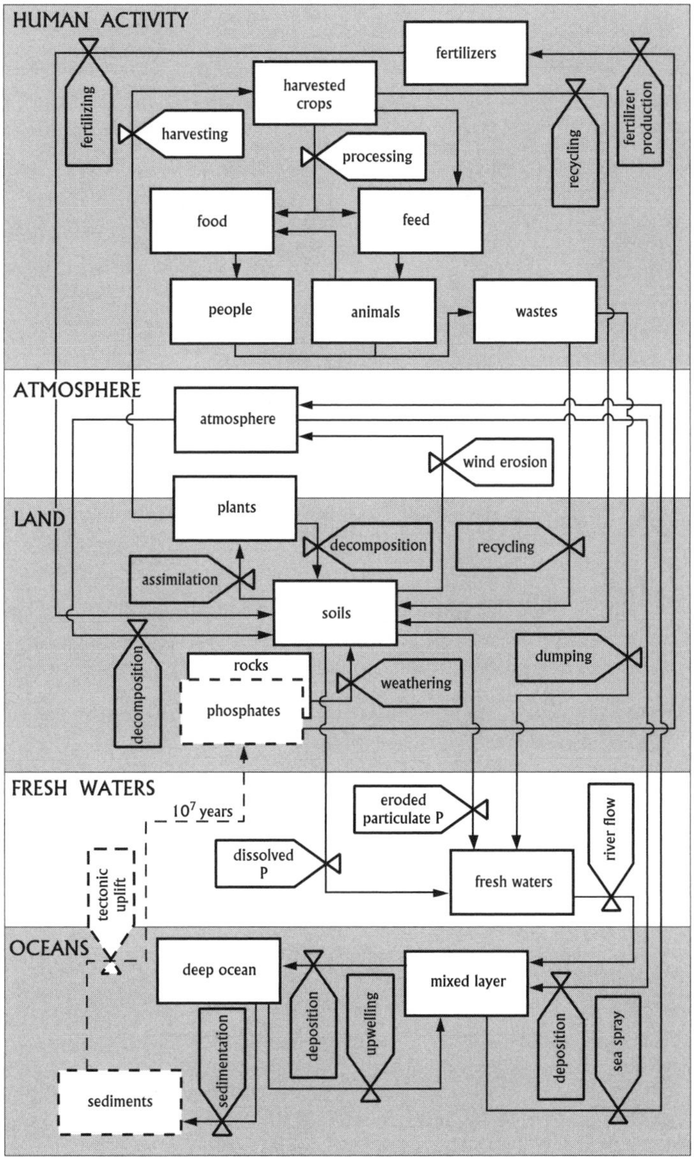

5.15 Principal reservoirs and flows of the biospheric phosphorus cycle (Smil 2000c).

their eggs. Calcium's most important functional involvements are in excitation-contraction coupling in skeletal and heart muscles, in neurotransmission at synapses, in intracellular signaling, and in mitotic cell division. In plants the element is in salts of pectic acid, which make up most of the central lamella that binds adjacent cells. Its adequate supply boosts the availability of phosphorus and micronutrients and helps increase the rate of biofixation; most of the element taken up by perennials is recycled in leaf fall.

Long-term cycling of calcium is to a high degree coincidental with the slow carbon cycle, and it also entails the hydrothermal exchange in the oceanic crust. Although there are substantial amounts of calcium locked in apatites (see the preceding section) and in gypsum ($CaSO_4$, produced by evaporation in shallow waters in arid climates), the element's largest sedimentary repository is $CaCO_3$ forming limestones, followed by another carbonate, dolomite ($CaMg(CO_3)_2$), which makes up the bulk of dolostones. Garrels and Mackenzie (1971) estimated the total mass of limestones at 350 Pt, with about a quarter of this huge mass deposited during the Precambrian era. The ready solubility of carbonates in slightly acidic precipitation makes Ca^{2+} one of the most abundant cations in river water: about two-thirds of it originates in carbonate weathering, roughly a fifth comes from silicates, and the rest from gypsum.

Deposition of carbonates can take place only in continental shelves and in the shallower parts of the deep ocean containing warmer water. In the colder waters of the deeper layers—below the carbonate compensation depth (CCD), which is about 5000 m at the equator, close to 3000 m in high latitudes—dissolution is greater than precipitation, and no carbonates can accumulate at the ocean's bottom. Most inorganically precipitated carbonates have thus been laid down in shallow waters, as must have been the case with biomineralized calcium, whose producers either are photosynthesizers or feed on these autotrophs. In the early lifeless ocean or in waters containing just prokaryotic organisms, formation of carbonates could proceed only after the two constituent ions reached critical concentrations in the seawater.

Only the emergence of marine biomineralizers using dissolved $CaCO_3$ to make calcite or aragonite shells (differing only in crystal structure) greatly accelerated the rate of calcium sedimentation in the ocean. Reef-building corals are certainly the most spectacular communal biomineralizers, but coccolithophorids, golden motile algae which surround themselves with intricate disclike calcitic microstructures, and foraminiferal tests (pore-studded microshells of protists) are the main contributors to the rain of dead $CaCO_3$ shells on the sea floor (fig. 5.16). Mollusks, often intricately shaped and patterned, are only minor contributors. Where the detritus rain is heavy, the remains of organisms can be buried quickly to form calcium deposits below the CCD. The eventual return of calcium to subaerial weathering through tectonic uplift is abundantly demonstrated by the extensive (and often very thick) layers of limestone and dolomite encountered in all of the world's major mountain ranges.

Silicon

Silicon is best known today as a semiconductor, but to produce it in the purest elemental crystalline form required for microchips takes a great deal of ingenuity and energy. In contrast, in its various oxidized forms—mostly as silica (SiO_2, pure sand) and various silicate minerals—the element is, with 27% of the total, the second most abundant presence (after oxygen) in the Earth's crust. As it is the second element in the carbon group, with four valences forming a variety of polymers—silicates, whose

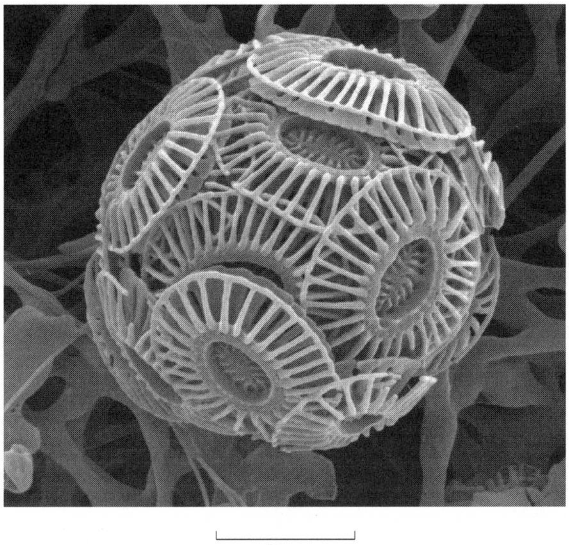

5.16 The coccolithophore *Emiliania huxleyi,* a unicellular planktonic protist containing chloroplasts and covered by coccoliths formed inside the cell and extruded to form a protective armor. Cretaceous and Tertiary chalks are composed almost exclusively of the remnants of these tiny (diameter less than 0.01 mm) organisms. This and other excellent images of coccolithopores are available at <http://www.soc.soton.ac.uk/SOES/STAFF/tt/eh/pics/>

common property is Si atoms at the center of tetrahedrons with O at the corners—it has been suggested that silicon might be the base of noncarbon life, possibly extant deep in the crust (Gold 1999). Although plant physiologists do not class the element as an essential micronutrient, it is absorbed from soil, as H_4SiO_4 (silicic acid), in quantities surpassing several fold those of other minerals, and in some instances even rivaling the uptakes of macronutrients. For example, rice (*Oryza*) takes up about as much silicon as nitrogen in order to stiffen its stems and leaves, which contain as much as 5% silicon (Ponnamperuma 1984). The aggregate annual uptake of the element by the terrestrial phytomass is nearly 500 Mt Si.

Silicon is carried to the ocean mostly in suspended sediments, with dissolved silicic acids amounting to only about 5% of the net flux transported by rivers (fig. 5.17). Silicon is dissolved in the ocean largely as undissociated monomeric silicic acid, with concentrations being very low in surface waters of central ocean gyres and high in the Antarctic ocean as well as in bottom layers almost everywhere. Silicon is assimilated vigorously by a variety of marine organisms, including diatoms, silicoflagellates, and radiolarians, which take up silicic acid from seawater to build their intricate opal (hydrated, amorphous biogenic silica, $SiO_2{\cdot}0.4H_2O$) structures (fig. 5.18). Tréguer et al. (1995) estimated that these organisms' annual silicon uptake amounts to about 7 Gt Si; this silicon dissolves after these organisms die, and about half of it is promptly recycled within the mixed layer. The remainder settles into the deeper ocean, but all but a small share of it is returned to the mixed layer by upwelling, with net coastal and abyssal sediments of biogenic silicon accounting for less than 3% of the annual uptake by biota.

After burial, these deposits are rapidly recrystallized, primarily as chert. Schieber et al. (2000) suggest that quartz silt in mudrocks comes from the siliceous skeletons of marine plankton rather than, as previously believed, from eroded continental sediments.

Iron

Iron's biospheric importance is due to its common presence in the metalloproteins that activate or catalyze just about every essential life process (Butler 1998). Iron is the essential metal in chlorophyll, and it is present both in the cofactor and phosphorus-cluster of nitrogenase. The

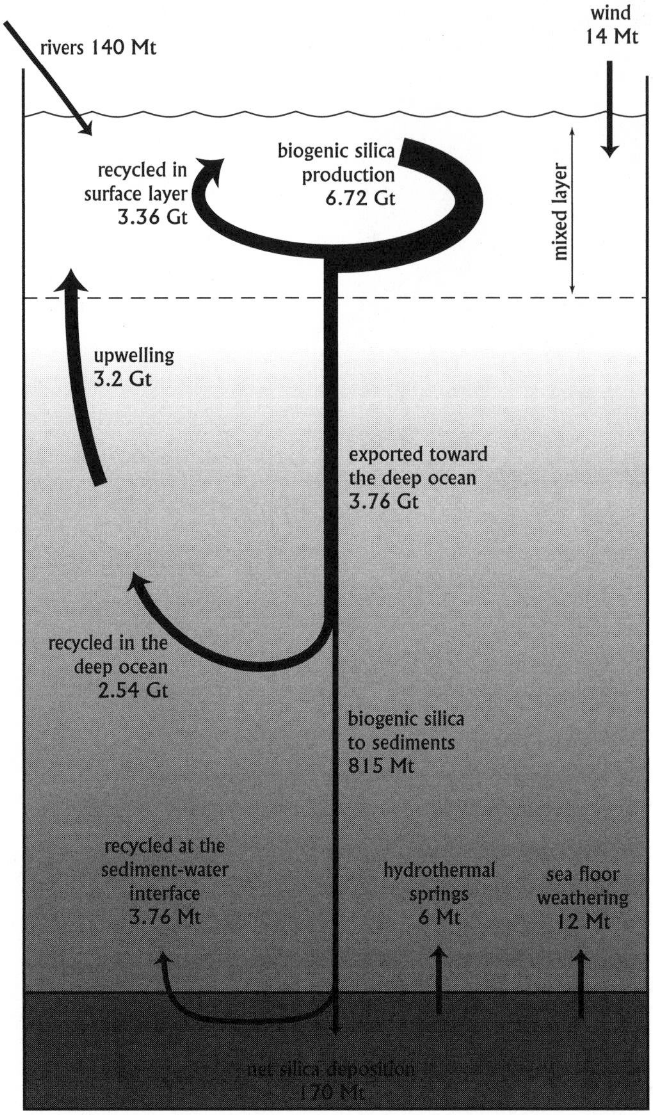

5.17 Silicon cycle in the ocean. Based on Tréguer et al. 1995.

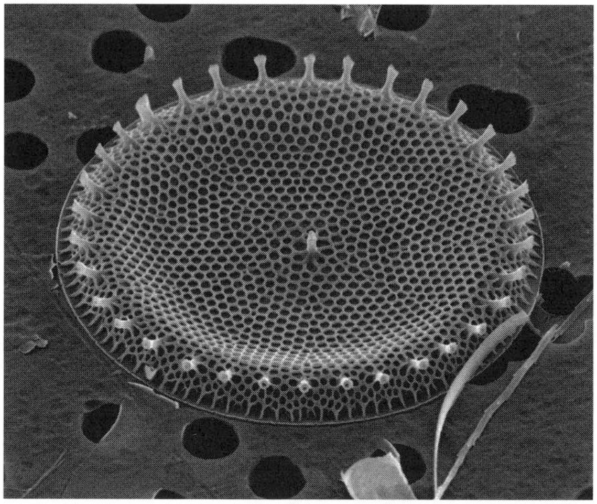

5.18 *Thalassiosira aestivalis* is an excellent example of a centric diatom with a near-perfect radial symmetry. Electron microscope image of its SiO_2-impregnated test courtesy of James M. Ehrman, Digital Microscopy Facility, Mount Allison University, Sackville, New Brunswick.

element's importance is magnified because of its insolubility under the neutral pH that prevails in living organisms. Iron's special place in oceanic photosynthesis was pointed out for the first time by Martin and Fitzwater (1988), who demonstrated that, in spite of a relative abundance of macronutrients, phytoplankton growth in the subarctic Pacific is not more vigorous largely because of the scarcity of iron. The element's poor supply also limits phytoplankton productivity in the equatorial Pacific and in the Southern Ocean. Most of the element assimilated by phytoplankton comes from aeolian deposition of poorly soluble dust iron (Fung et al. 2000).

Consequently, it has been suggested that iron enrichment could explain why atmospheric CO_2 levels during the last glacial maximum were 40% lower than today. According to this explanation, additional iron transported to the ocean in dusts blown by stronger winds across drier lands stimulated ocean photosynthesis and hence removed more carbon from the atmosphere in sedimented organic matter. Transfers of iron in surface ocean waters have been assumed to move in tandem with nitrogen flows, and grazing of heterotorphic bacteria and phytoplankton is assumed to balance iron uptake rates (Tortell et al. 1996). In any case, phytoplankton in shallow coastal waters depends primarily on iron in continental-shelf sediments (Johnson et al. 1999), and in deeper waters it is entirely dependent on aeolian deposition.

Iron's controlling role in phytoplankton production was experimentally confirmed during the IronEx II test in 1995, when nearly half a ton of iron was added to 72 km^2 of the ocean west of the Galapagos Islands and its effects were tracked for 18 days (Coale et al. 1996). In a matter of days, the specific growth rate of phytoplankton doubled and its abundance increased more than twenty times, with larger sizes (mostly diatoms) dominant, changes clearly supporting the conclusion about iron's growth-limiting role. In addition, atmospheric levels of DMS increased 3.5-fold during the experiment. Similar results were obtained during a mesoscale iron fertilization experiment in the polar Southern Ocean (61° S), when phytoplankton bloom (mostly due to diatoms) stimulated by iron caused a large drawdown of CO_2 and macronutrients and elevated DMS levels after 13 days (Boyd et al. 2000).

As the IronEx experiment was conducted in waters that are representative of roughly one-fifth of the ocean's surface, where nitrate levels are high but chlorophyll levels, and hence photosynthetic rates, are low, the experiment also led to some unjustified expectations that iron fertilization of seas could significantly counteract a future rise in CO_2 levels. That is not the case, however: iron may boost photosynthesis in surface waters, but this may not necessarily intensify the removal of carbon into the abyss. Moreover, the unknown effects of continuous massive additions of iron to surface seas (they might stimulate toxic algal blooms rather than a desired detrital rain to the abyss) would dictate a very cautionary approach in any case. And as Tréguer and Pondaven (2000) suggest, it might be silica rather than iron that is the ultimate controller of CO_2 sequestered by ocean phytoplankton. These uncertainties illustrate perfectly the interdependence of various biospheric processes and hence the necessity of avoiding simplified single-variable explanations and of striving for more complex understanding, untidy and full of spatial and temporal exceptions and variations as it may be.

6

The Biosphere's Extent

The Moveable Boundaries

This pressure is apparent in the ubiquity of life. *There are no regions which have always been devoid of life. . . . Life has always tended to become master of apparently lifeless regions, adapting itself to ambient conditions.*
Vladimir Ivanovich Vernadsky, *Biosfera*

. . . we now gazed at the most wondrous phenomenon which the secret seas have hitherto revealed to mankind. A vast pulpy mass, furlongs in length and breadth. . . . So rarely it is beheld, that though one and all of them declare it to be the largest animated thing in the ocean, yet very few of them have any but the most vague ideas concerning its true nature and form; notwithstanding, they believe it to furnish to the sperm whale his only food.
Herman Melville, *Moby Dick*

Three distinct realms of the Earth's environment are available for invasion by living organisms: the gaseous shroud, the atmosphere; the hydrosphere, deepest in the Pacific; and the planet's solid crust. The extent of these three spheres is well defined by obvious physical discontinuities. The presence of the gases that form the atmosphere can be detected for thousands of kilometers above sea level, but all but a tiny fraction of the atmospheric mass is within the first 30 km above the ground: the rest resembles more the vacuum of interplanetary space that the relatively dense mixture of dinitrogen, oxygen, water vapor, and trace gases that form the troposphere. The troposphere encompasses the turbulent part of the atmosphere, and its seasonally variable thickness ranges from about 15 km above the tropics to about 10 km above the polar regions. The stratosphere extends to about 50 km above the ground and contains the ozone layer (the highest ozone concentrations are at an altitude of about 30 km) that screens nearly all of the incoming UV radiation.

These two layers contain about 99% of the total mass of the atmosphere; the overlying mesosphere and thermosphere are marked by sharp temperature reversals, but they have an extremely low density of atmospheric gases (fig.

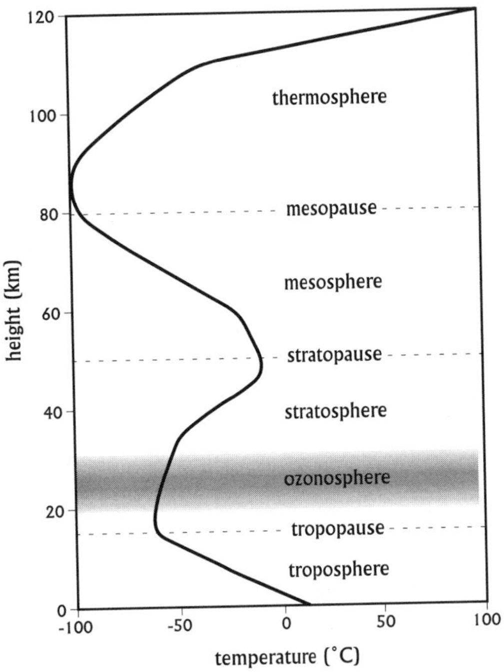

6.1 Atmospheric stratification, with a typical temperature profile.

6.1). The average depth of the world's oceans is about 3.7 km, and the deepest trenches, the sites of the Pacific plate subducting underneath the Asian continental plate, plunge to almost 11 km below the sea surface. The Earth's crust, the uppermost layer of the lithosphere, has a thickness ranging from about 5 km below the oceans to as much as 80 km below the continents.

To what extent have these three distinct realms of the Earth's physical environment become permeated by life? Answering this question may seem to be just an obligatory chore of mapping the biosphere's boundaries, a simple descriptive matter of outlining how high life has penetrated into the atmosphere and how deep it has advanced into the hydrosphere and into the solid surfaces of the lithosphere. In reality, tracing the extent of the biosphere is an intriguing task of dealing with elusive markers and amazing life forms, with variable edges and moveable limits, and unequivocal answers cannot always be given even to very narrowly framed questions.

Should viruses count when delimiting the extent of the biosphere? And if the marking begins only with prokaryotes, should it consider only metabolizing archaeons and bacteria, or should it take into account their cryptobiotic forms, which enable these microorganisms to spend extended periods of time, even many millions of years (see chapter 3), in a deathlike state? And if we are to look only at microbes that are actually metabolizing, do we draw the extreme lines wherever we might encounter the first solitary cells, or are we to insist on a reasonably contiguous occupation of an extreme niche? And if the latter is the case, at what cellular density should that contiguity begin?

Moreover, until recently we have delimited the biosphere with so little imagination, and hence in such a restrictive manner, that we should be very cautious about claiming that we now have the right answers. Just two generations ago nobody imagined that surprisingly complex autotrophic ecosystems energized by oxidation of sulfides could thrive, without sunlight and without chlorophyll, at the very bottom of deep oceans. Twenty years ago it was generally believed that microorganisms, unless carried deeper in percolating water, could prosper only in the few uppermost meters of the ground: since then we have found them in the few uppermost kilometers of the Earth's crust, correcting a long-standing belief that was mistaken, in terms of distance, by three orders of magnitude. And even today when asked how deep the soil is, nearly everybody will still say half a meter or so, a total that according to a revised definition may be least an order of magnitude too small.

A necessary reiteration before I start looking at these extreme niches of life. As stressed in chapter 2, we know that outer space is full of the essential building blocks of life: complex organic molecules appear to be abundant in dark parts of interstellar clouds (Bernstein et al. 1999), and they have been repeatedly found in meteorites. But as also previously noted, we still lack any convincing evidence for panspermia, of the simplest forms of life transmitted to the Earth from elsewhere in the universe, and hence I will not assume that life extends beyond the stray organisms found in the upper atmosphere.

Life Aloft

For life that originated on the Earth, the altitude limit in the atmosphere seems to be the easiest border to describe, but the task is actually complicated, and the successive limits are the functions of particular definitions. Passive drifters will be able to travel far: the smallest windborne organisms can rise far beyond the troposphere and stay aloft indefinitely at no energy cost to themselves. But microorganisms in the stratosphere are just defenseless hostages of extreme conditions because of the high levels of UV radiation there and temperatures close to –60 °C.

At those altitudes bioaerosols—microorganisms that can exist aloft on their own or can be part of larger (although in absolute terms still very tiny) solid or liquid agglomerations—are, at best, cryptobiotic survivors. And whereas myriads of viruses, bacteria, fungi, spores, and pollens that travel to such heights will eventually return to multiply in the gentler conditions that prevail far below the stratosphere's forbidding environment, myriads will not survive the experience. The only exceptions are some strains of radiation-resistant bacteria: spores of *Clostridium perfringens* and *Bacillus subtilis* (fig. 2.18) are resistant not just to far-UV but also to X-rays, and the family Deinobacteriaceae contains species that are extremely resistant to far-UV radiation as well as to even deadlier gamma rays (M. D. Smith et al. 1992).

In contrast to cryptobiotic prokaryotes, migrating birds never metabolize more vigorously than during their high-altitude flights, which are the most energy-intensive activities per unit of time any animal can choose, and their trajectories are flight lines admirably charted with the aid of sunsets, stars, and the Earth's geomagnetic field aimed at taking them most efficiently from their overwintering areas to their summer breeding grounds (Terres 1991; Berthold and Terrill 1991). But birds are also only temporary visitors to the higher layers of the troposphere: they do not feed there, and they spend typically no more than a few weeks there every year. There are exceptions: the Arctic tern (*Sterna paradisea*) has to spend about eight months aloft in order to make the round trip of about 40,000 km between its Arctic breeding grounds and circum-Antarctic seas (Baker 1981).

Only in its lowermost few hundred meters is the troposphere full of macroscopic life, thanks to insects and birds, many of which spend a large part of their active life aloft. The European swift (*Apus apus*) flies easily 900 km a day during its breeding season, performing just about any function, including some sleeping, on the wing (Weitnauer 1956). And foraging bees, zipping along at 25 km/h, may fly as far as 6 km from their hives, an equivalent of 400,000 body lengths (15 mm), a one-way trip comparable to humans walking daily some 600 km, or from Boston to Washington (Seeley 1995).

Bioaerosols

Anything with a diameter smaller than 1 μm can become almost indefinitely airborne, and that category includes an enormous number of bioaerosols (Lighthart and Mohr

1994; Cox and Wathes 1995). A 1 µm particle, the size of small, but far from the smallest, bacteria, will have a settling velocity of just 0.2 cm a minute, or about 12 cm/hour; such a bacterium shed from a migrating bird's wing at 3 km above sea level would need, in a perfectly still air, nearly 3 years to hit the ground; in reality, it will be almost constantly buffeted by air turbulence, either drifting randomly without any great change of altitude, lofted much higher, or carried earthward. Still, that bacterium's still-air descent is a rapid one compared to a virus with a diameter of 0.01 µm (or 10 nm), which has no discernible terminal settling velocity as it falls by mere 10^{-9} cm/s. The actual atmospheric residence times of small aerosols are much shorter, as they are removed by collisions with other particulates and by incorporation into water droplets (fig. 6.2).

In contrast, a bioaerosol with 100-µm (or 0.1 mm) diameter—either a very large grain of pollen or a mist particle containing a microcosm of bacteria and fungi—descends at 25 cm/s, and in still air, it would make it from a very tall treetop or from a low-lying cloud to the ground in just about five minutes. Bioaerosols encountered in the atmosphere are thus both solid and liquid, their sizes ranging from the smallest single viruses (0.003 µm or 3 nm) through fungal spores (mostly smaller than 4 µm) and bacteria (most of the airborne bacteria have aerodynamic particle size below 10 µm) to large droplets, plant debris, or soil particles (all larger than 1 mm or 1000 µm) with embedded or attached microbes or pollen grains. They enter the atmosphere not only as wind-driven particles—dust storms, tornadoes, and hurricanes are naturally their best mobilizers—but also with falling plant litter, ocean spray, and bursting and splashing rain bubbles, and they are also dispersed from animal and human bodies by activities ranging from moving, rubbing, and shedding skin to sneezing and voiding wastes.

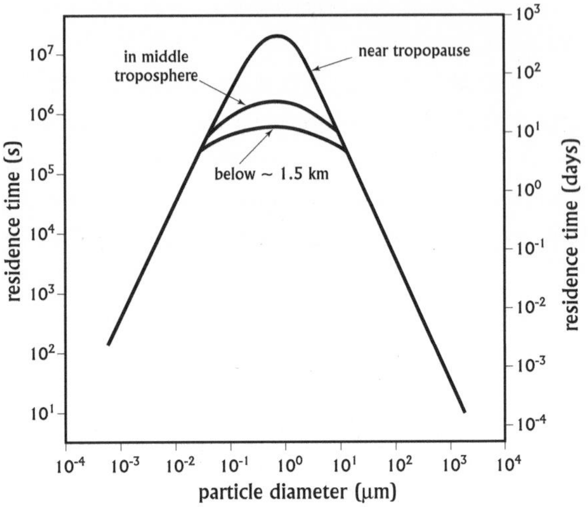

6.2 Average residence times of aerosols in the troposphere range from minutes to more than a year. Based on Hobbs (1999).

Although their densities change with seasons, airborne viruses and pollens generally abound; bacteria cultured from the atmosphere range from pleomorphic rods to clustered cocci; and commonly encountered airborne fungi include species of the two most important antibiotic-producing genera, *Penicillium* and *Streptomyces,* as well as *Saccharomyces,* the bread- and wine-making yeast (fig. 3.10). An infected wound exposed to the air would not be healed by airborne fungi, but a baker who is after a truly natural *pain de levain* (traditional leavened bread) does not need to start a *chef,* the seed the sourdough is built upon, with a pinch of commercial yeast. All she has to do is just to mix up some flour and water, and airborne *Saccharomyces* will alight on the mixture and it will start fermenting it in a matter of minutes or hours (Leader and Blahnik 1993). But airborne microbes also pose health risks as potentially hazardous bacteria and fungi are carried between continents in dust plumes. The best example of this phenomenon is the seasonal (usually February to April) transport of the Saharan dust to the western Atlantic region (Griffin et al. 2001). Microbes hidden in the tiny cracks of dust particles are shielded against UV radiation as they are carried at altitudes of 2–4 km, and the eventual inhalation and deposition of them causes increased respiratory problems (including asthma attacks) in humans and diseases in domestic animals and the Caribbean corals (a fungus, *Aspergillus sydowii,* being the cause).

The aerial motions and survival of bioaerosols are complex functions of their size, shape, and density, as well as of air viscosity, composition, pressure, and temperature, and the intensity of UV radiation. Throughout the first few hundred meters of the atmosphere, bioaerosols' densities generally decrease with altitude, except when thermal inversions trap warmer air above a colder layer near the ground. Such inverted stratifications aside, bioaerosols are rather uniformly mixed through the first few kilometers of the troposphere. They are much rarer in the stratosphere: only massive volcanic eruptions that entrain myriads of airborne microorganisms in enormous ash and tephra clouds can inject large numbers of microorganisms as high as 25 or 30 km above sea level. And although the highest recorded occurrence of bacteria has been between 55 and 77 km above the surface, microorganisms are exceedingly rare at altitudes above 50 km.

Of course, bioaerosols, living or dead, are eventually removed from the atmosphere. Swirlings of atmospheric turbulence fling them against water or solid surfaces to which they may stick, or, more likely, they will coagulate or conglomerate to form larger particles that are heavy enough to settle gradually or are scavenged by droplets of rain or fog or by snow and deposited on land or water.

Birds

Descending into more hospitable levels we encounter the uppermost layers of the atmosphere periodically frequented by migrating flocks or by a few soaring birds. There have been many bird sightings, and bird-aircraft encounters, at very high altitudes. The highest ever encounter happened on November 23, 1973, when a commercial jet collided with a vulture — Rüppel's griffon (*Gyps rueppelli*), a bird weighing up to 8 kg, with a wingspan of as much as 2.4 m — at an altitude of 11,100 m above the Ivory Coast (Laybourne 1974).

Another often-cited and visually irresistibly appealing image is that of a high-altitude migration: George Lowe told Lawrence Swan how he watched from the slopes of Mount Everest a flock of bar-headed geese (*Eulabeia indica*) flying in an echelon directly over the mountain's summit, a feat implying level flight at about 9 km (Swan 1961). The fact that the birds would not skirt the peak and fly nearly 3 km lower in a layer that contains about 40% more oxygen (the pass on the mountains western flank, Lho La, is only 6,026 m above sea level) attests to their admirable adaptation for high-altitude flight and to their preference for a straight and speedy migration route.

Geese share with all other avian species a respiratory system unique among vertebrates (Sturkie 1986; Norberg 1990; fig. 6.3). Also, like many other long-distance migrants, they can store relatively high deposits of fat that enable them to energize a nonstop flight from Indian lowlands to the lakes of Tibet. In addition, these geese, and other bird species, have been able to minimize the effects of the high-altitude oxygen shortage to an amazingly high degree by evolving a remarkable two- or three-stage cascade of hemoglobin with graded oxygen affinities that provides a sufficient supply of the gas over a wide range of altitudes (Berthold and Terrill 1991). Consequently, birds

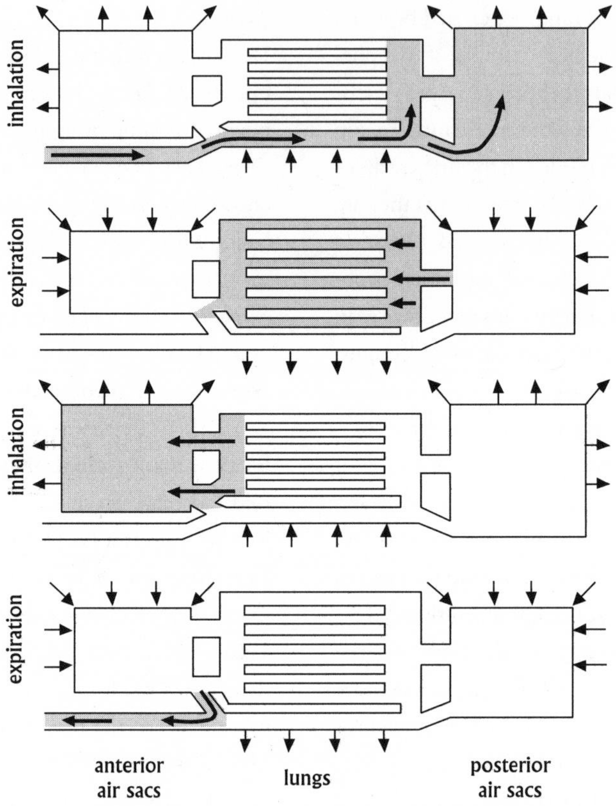

6.3 One-way air flow through avian lungs supports superior performance in oxygen-poor air at high altitudes (Smil 1999d).

may actually prefer to fly in higher altitudes whose thin air requires higher speeds: they will accomplish their migration in a shorter time.

Naturally, most birds stay most of the time far below the altitude of even low Himalayan passes. Most migratory birds accomplish their journeys at altitudes of just 100–1,500 m. Radar scans of high-flying flocks on transcontinental or transoceanic journeys put their altitudes no higher than 6–7 km, with flight profiles often changing along the journey. For example, shorebirds (sandpipers, plovers) and small songbirds (warblers) of Eastern North America make the first leg of their journey to Venezuela, between Massachusetts and Bermuda, at around 2 km above sea level, but by the time they arrive over Antigua, they are flying at 3–6 km, many even a bit higher (Williams and Williams 1978). Only then do they begin their gradual descent toward Tobago and the South American coast.

The highest nesting elevation is probably that of Alpine choughs (*Pyrrhocorax graculus*) living near the snow line in the Himalayas (between 4 and 5 km) and seen in aerial acrobatics at about 6.5 km (Gilliard 1958). The zone of everyday patrolling—for birds of prey scouting small rodents or for carrion eaters spotting a newly abandoned large kill—lies typically below 500 m. But both raptors and vultures, as well as many other species, can soar much higher: the ill-fated 1921 British expedition to Mount Everest saw a bearded vulture (*Gypaetus barbatus*) circling at 7,500 m. And slope-soaring auks or gannets can keep rising to high altitudes for hours even in gale-force winds (Pennycuick 1972). Storks use the lines of thermals even during their migration from Europe to Africa by alternating soaring and gliding. Insectivorous birds feeding on airborne species flit mostly below 200 m, and fructivorous birds have no good reason to fly much higher than the tops of trees that yield their juicy meals.

And very large birds have a difficult time rising even that high: they cannot generate enough power for sustained flying, because the power needed for flight rises faster (with the exponent 1.0) than the power that can be delivered by pectoral muscles (with the exponent 0.72). As a result there are few frequent fliers heavier than 10 kg, and the heaviest Kori bustards (*Ardeotis kori*) in East Africa, whose mass is between 13 and 19 kg, are nearly as earthbound as the mammals around them: they get airborne only rarely, executing just extended hops low above the ground.

Insects

Most of the nearly one million identified insect species spend their lives within the few lowermost meters of the troposphere or up to a few tens of meters above the ground in tree canopies. But the class Insecta includes some fliers whose feats are comparable, even in absolute terms, to some great bird migrations. Locusts, above all the migratory locust (*Locusta migratoria*) and desert locust (*Schistocerca gregaria*), are perhaps the most impressive, and most unwelcome, long-distance insect fliers (Uvarov 1977; Krall et al. 1997). Their swarms, containing billions of individuals, may cover hundreds of square kilometers and move with a ground speed of up to 30–40 km/h. Their swarms may be shallow and stratiform, moving no higher than 20 m above the ground, as well as cumuliform, extending to more than 1000 m upward, and they can reach much higher altitudes when confronted with physical barriers or when caught up in intense convection currents. *Schistocerca* swarms were recorded more than 3,000 m aloft when crossing the snowy ranges of the High Atlas in Morocco, and they were found embedded in the ice of the Mount Kenya glacier at an altitude of 5,000 m above the sea (Uvarov 1977).

Small insects cannot fly against even moderate winds, and their presence at higher altitudes is almost always a matter of passive entrainment in convective air currents. That is clearly the case with aphids whose bodies get smashed against the leading wing edges of gliders: they rise to high altitudes by being entrained in powerful summer thermals that are used by the glider pilots to ascend into the midtroposphere; once the effect wears off, the aphids may drop, en masse and manna-like, far from their places of entrainment (Pennycuick 1972).

Active long-distance insect flying goes on at much lower altitudes. What is certainly the best-known North American migration of an insect, the flight of Monarch butterflies (*Danaus plexippus*) from the Great Lakes region to small patches of coniferous forests in Mexico's Michoacán, takes place at speeds of about 20 km/h and usually at altitudes of no more than 20 m, although individual butterflies have been occasionally observed at around 100 m above the ground (Scott 1986). Large subtropical and tropical nocturnal moths can climb rapidly to between 400 and 1,500 meters and then gradually descend after several hours of migration (Baker 1981). And numerous nocturnal radar observations have shown swarms of unidentified migrating insects flying between 100 and 500 m above the ground.

Life in the Ocean

Oceans remove what is commonly the greatest limitation for the life of plants on the land—the scarcity of water—but they add two immense challenges to life based on photosynthesis: even the cleanest water does not allow sunlight to penetrate beyond a hundred or so meters, whereas particulate matter containing nutrients indispensable for photosynthesis (whether nitrogen and phosphorus, the two key macronutrients, or iron, an essential micronutrient) will sink rapidly beyond that some limit. Not surprisingly, the surfaces of open oceans are among the least productive environments on the Earth.

Of course, zoomass abundance is not limited by the reach of sunlight: countless zooplankters, as well as invertebrates and the fishes and mammals that feed on them, profit from the constant, and in lower depths fairly dense, marine snow of decaying biomass reaching the midocean layers and the floor of the shallower parts of the sea. But conditions for life on the deep-sea floor are extremely taxing (Tyler 1995; Sokolova 1997). Temperatures at no more than 4 °C (fig. 5.3), not even the faintest detectable trace of light below 1,000 m, only small amounts of nutrients reaching the bottom, and very high pressures (there is an increase of 10^5 Pa for every 10 m of descent, so the pressure in Pacific trenches is more than 100 MPa, or a thousand times more than at the ocean's surface) make life unbearable for but a few briefly visiting large vertebrates.

Still, even the deepest waters—abyssal (below 3,000 m) and hadal (below 6,000 m; yes, the adjective is from the mythical Greek Hades) zones—are not lifeless. Although their biomass densities drop sharply with the increasing depth, there are no less 700 species of Metazoa, mostly invertebrates, living at depths exceeding 6,000 m (Vinogradova 1997), and, as in every other extreme niche, there are numerous prokaryotes and viruses. The existence of bacteria in sediments from one of the world's deepest ocean trenches, at 10,400 m below the sea level near the Philippines, was first proved by ZoBell (1952). Recently a research program led by Koki Horikoshi of the Japan Marine Science and Technology Center collected thousands of microbial strains from benthic organisms and deep-sea mud, including the first scoop from the Earth's deepest sea floor. On March 2, 1996, their unmanned submersible *Kaiko* scooped out mud at a depth of 10,897 m (the deepest spot is now given as 10,924 m) in the Challenger Deep in the Mariana Trench (Horikoshi et al. 1997).

Distribution of marine viruses follows the same general pattern as the abundance of bacteria, decreasing with the depth from about 10^{10}/L in nutrient-rich coastal surface waters (five to twenty-five times the bacterial abundance) to an order of magnitude less in the open ocean and another one or two orders of magnitude decline with the descent to abyssal seas (Fuhrman 1999). Deep-dwelling animals were first brought from depths up to 5,500 m by the *Challenger* expedition between 1872 and 1876, proving that the abyssal seas are not devoid of macroscopic life. Only between 1950 and 1952 did the Danish *Galathea* catch in its bottom dredges samples from more than 10 km below the surface (Menzies et al. 1973). Since that time the existence of live animals at these depths has been documented with direct observations from the bathysphere *Trieste,* which descended to 10,916 m in 1960, and with photographs and benthic traps (Yayanos 1998). And the late 1970s brought an unexpected discovery of unique autotrophic bacteria that support complex ecosystems in the vicinity of hydrothermal vents on the ocean floor.

Shallow Waters

Obviously, no photosynthesizing organism can live deeper than the reach of solar radiation. The depth of the euphotic zone, the uppermost layer of water that receives sufficient radiation to support photosynthesis, in a particular area is determined by the degree of turbidity. Euphotic depth is measured in mere millimeters where the Huang He discharges its slurry (a more correct term than water, as the mixture often carries an equal amount of silt) to the North China Sea (Smil 1984). In contrast, enough sunlight to enable photosynthesis to take place may be available in such clear waters as the Caribbean Sea offshore the Cayman Islands at a depth of 100 m, and the typical thickness of the euphotic zone in clean seas is between 50 and 60 m. This means that less than 2% of the ocean's water volume can support photosynthesis.

The tiny sizes of phytoplankters are an effective adaptation to the dual challenge of staying close to the Sun and dealing with the oceanic nutrient scarcity: these organisms have high surface-to-volume ratios that maximize the absorption of nutrients present at very low concentrations and minimize the rate of sinking through the euphotic layer. Even so, phytoplankton produce only small amounts of phytomass in the nutrient-starved open ocean, and euphotic zones burst with rapidly dividing cells only near the coasts, where the continental runoff brings the nutrients released by the natural denudation of land, farming, cities, and industries—or off the western coasts of continents, where the upwelling of cool, nutrient-laden water enriches the surface layers with nutrients released by the decay of sinking biomass.

Although some new research suggests that symbiotic algae may actually be repressing, rather than enhancing, the calcification rates of reef-building corals (Marshall 1996), there is no doubt that the tiny polyps living in symbiosis with large numbers of yellow-brown unicellular algae are incomparably more successful builders of reefs than are their algaeless counterparts (Gattuso et al. 1999). Obviously, photosynthesizing symbiotic algae cannot grow beyond the euphotic zone, and coral colonies in deeper waters that are shaded by the growth in shallower layers maximize their light-gathering capacity by growing more branches or horizontal plate-like structures.

Zooplankton embraces an enormous variety of herbivorous, carnivorous and detritivorous species ranging from very common and very tiny flagellates (so-called nanoplankton) and dinoflagellates (fig. 5.13) that produce huge amounts of seasonal blooms to less common but spectacularly shaped foraminifers (Laybourn-Parry 1992). Naturally, the depths of zooplankton existence overlap with the euphotic zone, but they go also well below it: for example, in the Kuroshio stream, a considerable amount of zooplankton biomass can be found during the day at depths between 300 and 600 m (Hirota 1995), because many zooplankters spend days in darker and cooler waters below the euphotic zone, where their metabolic rates and their exposure to predators are lower. They migrate toward the surface during the night, much like, in John Isaac's (1969) apt comparison, "timid rabbits emerging from the thicket to graze the nighttime fields" (p. 155). The vertical distribution of *Calanus euxinus,* a common herbivorous planktonic copepod in the central part of the

Black Sea, is an excellent illustration of this daily migration (Vinogradov 1997; fig. 6.4).

Deep Ocean

The abundance of zoomass declines with the descent from the dimly lit mesopelagic zone (1,200–2,000 m) to the perpetual midnight of abyssal seas. The first layer is frequented by large squids and octopuses; the second is home to bizarre swallowers and gulpers, fishes with oversize and disjoinable mouths, a most useful adaptation for not letting pass any opportunity for seizing a meal, albeit one bigger than the predator itself, in those largely empty waters, where one can go for very long periods without any food. The other shared attribute among many bathypelagic fishes is their use of bioluminescence, of steadily glowing or suddenly flashing lights (sometime produced by symbiotic luminescent bacteria) to lure or scare, confuse or identify (Robinson 1995). Perhaps the most intriguing use of this device is by anglerfishes (Lophiiformes), which deploy miniature lighted lanterns just above their mouths to swallow any unsuspecting prey that comes close enough to investigate the source of light.

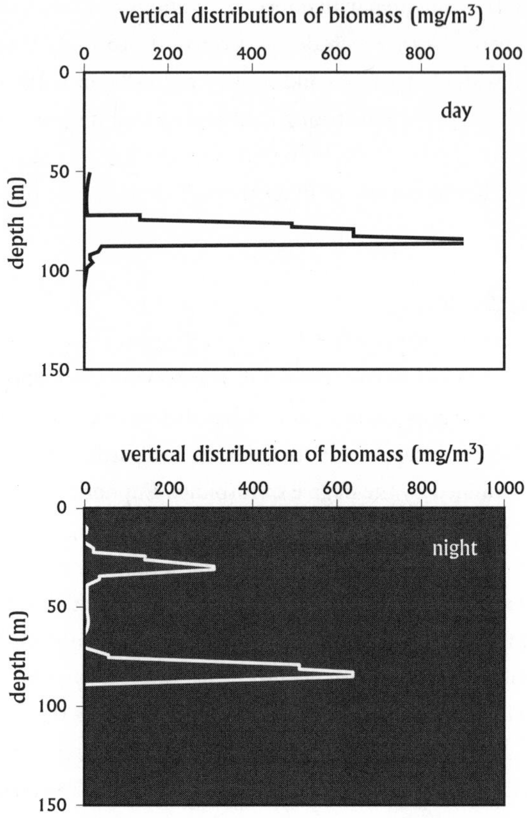

6.4 Nocturnal vertical migration of a planktonic copepod (*Calanus euxinus*) in the Black Sea. Based on Vinogradov (1997).

The typical density of life at the bottoms of deep oceans is exceedingly low. Although more than half of the ocean's zoomass is bottom fauna (benthos), more than 80% of this total is found in shallow (less than 200 m deep) coastal waters, and less than 1% is at the bottom of abyssal seas, below 3,000 m, which account for three-quarters of the world ocean's floor (Menzies et al. 1973). Omnipresent but invisible bacteria aside, the most common thinly distributed organisms at this depth are brittle stars, glassy sponges, crinoids (sea lilies), brachiopods, holothurians, bivalves, and polychaetes that can be seen as solitary species or in tiny clumps separated by huge expanses of floor that is devoid of macrobiota but that is in many places overlaid by rich deposits of manganese nodules (they also contain iron and nickel).

But there are three notable exceptions that induce much denser and sometime stunningly diverse assemblages of life on the ocean's floor. The first one is brought from above as an astonishing variety if organisms are attracted to large chunks of dead zoomass, particularly to whale carcasses that settle on the ocean's bottom. The other two are brought from underneath the ocean's floor as natural crude oil and gas seeps provide hydrocarbon substrate for bacterial growth that attracts other heterotrophs, and as hydrothermal vents spew hydrogen sulfide, a substrate for chemolitotrophic bacteria whose metabolism sustains diverse mini-ecosystems.

The skeleton of a single whale can support a greater variety of organisms than are crowded around a rich hydrothermal vent (Smith et al. 1998). Anaerobic bacteria are the main decomposers of oil-rich whale bones, and their releases of H_2S support chemosynthetic bacteria that oxidize the gas and provide food for a variety of worms, mollusks, and crustaceans. When hauled up from their resting place, five vertebral bones from a whale off southern California with a surface area of less than one square meter harbored more than 5,000 animals from nearly 180 species, 50% more than the species count from the most fertile known hydrothermal sites and a community five times as diverse as the typical assemblage based on a hydrocarbon seep.

Hydrocarbon seeps in shallower seas have been known for decades to be bases for a denser benthic life, but the existence of thriving deep-ocean communities that exist totally without sunlight and are associated with often violent geotectonic events, extremely high temperatures, and copious releases of a toxic gas must surely rank among the most startling discoveries of twentieth-century biology (Tunnicliffe 1992). The first discoveries of warm (7–17 °C, compared to 2 °C for the surrounding water), gently flowing thermal springs on the ocean floor and the peculiar animals in their surroundings were made after an extensive search in February and March of 1977 during dives with the deep submersible *Alvin* to a depth of about 2,400 m in the Pacific near the Galápagos Islands (Corliss et al. 1979). Soon afterward came the discoveries of more spectacular hydrothermal discharges with maximum exit temperatures of 270°–380° C.

The genesis of hydrothermal flows is simple: old deep-ocean water percolates through fissures below the sea floor, is superheated by the underlying magmas to as much as 400 °C, and leaches various minerals from the surrounding basaltic rocks. H_2S in the hot water comes from the leaching of crustal basalts as well as from the conversion of seawater sulfates in the anoxic subsurface environment to sulfides. As the hot water spews, at rates of 1–2 m/s, from the ocean floor vents it cools rapidly, and the dissolved polymetallic sulfides precipitate instantly. A portion of these sulfides, together with the precipitated $CaSO_4$, piles up around the vent and builds bizarre, tall (as much as 45 m), heavy but brittle chimneys. As the sulfide precipitates are commonly dark-colored, these vents became known as black smokers; white smokers are characteristic of vents with lower exit temperatures.

Both gentle seafloor springs and gushing vents have been found in every tectonically active region studied around the world; they appear to be most common along the ocean ridges, the spreading zones where the new ocean floor is being created by upwelling of basaltic magmas, but they have also been found in areas where subducting plates create excessive heating that results in hydrothermal releases. The H_2S flowing from the vents is very poisonous to all higher organisms, but chemotrophic bacteria, the only aerobic autotrophs that can fix carbon in the permanently dark environment, oxidize it to sulfate in a reaction that releases energy for their growth and forms the basis of fairly complex trophic pyramids (Jannasch and Mottl 1985).

Various thiobacilli, including *Thiotrix* and *Thiomicrospira,* and *Beggiatoa,* the first chemotroph discovered by Sergei Winogradsky (see chapter 3), are the dominant H_2S-oxidizing bacteria. Other prokaryotes found near the vents include aerobic methanotrophs that oxidize CH_4, as well as hyperthermophilic (preferring temperatures above 80 °C) anaerobic methanogens. Among the latter species is *Methanococcus jannaschii,* the first archaeon to have its genome completely sequenced (Bult et al. 1996); it was discovered at a depth of 2,600 m at the base of a white smoker chimney on the East Pacific Rise in 1983, and its optimal growth temperature is 85 °C.

About 450 different invertebrate species have been found in these unique, and by all indications very ancient, circumvental ecosystems, and all but a few are endemic to hydrothermal habitats (Tunnicliffe et al. 1998; Van Dover 2000). The ecosystem's most prominent inhabitants are giant tube worms, *Riftia pachyptila,* a previously unknown genus of the class Vestimentifera. These worms dwell in slender white tubes up to 3 m long; their red-coated bodies, often 1 m long, have no mouths or digestive organs, just a body cavity whose tissues contain dense concentrations of symbiotic sulfide-oxidizing bacteria. Sulfide is carried to them on special binding sites on the worm's hemoglobin molecule, and blood vessels within the body cavity absorb the products of bacterial metabolism (Jannasch 1989).

Commonly encountered circumvental organisms include another species of vestimentiferan tube worms (*Tevnia jerichonana*), white clams (*Calyptogena magnifica*), mussels (*Bathymodiolus thermophilus*), small crabs (*Bythograea thermydon*), small fishes, and polychaete worms. The total biomass of these vent communities is often 500–1,000 times denser than the mean prevailing at the deep ocean floor. An obvious puzzle is how these highly specialized and closely related circumvental organisms disperse over great distances that separate many active hydrothermal vents.

Mammals in the Ocean

Mammals in deeper waters face the dual challenge of oxygen supply and rising pressure, but many of them have evolved adaptations that allow them to make relatively prolonged stay in pelagic seas. Humans, of course, are not in that category: we are poorly adapted for anything but very brief underwater stays (Ashcroft 2000). In 1913 a Greek sponge diver, Georghios Haggi Statti, used a 7-kg rock do go down to 75 m to retrieve the anchor from a sunken Italian ship; his feat was replicated and his record was surpassed in 1998 by Umberto Pelizzari, who reached a depth of 100 m in 2 minutes and 43 seconds (Pelizzari 1998). In 1991 Pelizzari went 16 m deeper while holding on to a weighted, rope-guided sled.

Pelizzari also holds the world record, 71 m, for constant-weight dives; these are achieved without any weight, but divers are equipped with fins nearly a meter long to aid their ascent (Beavan 1997). These are extremely hazardous feats, however, accomplished by exceptionally endowed and highly risk-tolerant individuals. Cuban diver Alejandro Ravelo set the world record of holding his breath for 6 minutes and 41 seconds when lying motionless at the bottom of a hotel pool; most of the readers of these lines will have a great difficulty holding their breath for much more than a minute while sitting in their armchair. Sponge and pearl divers—Japan's female *amas* are perhaps the most famous practitioners of this ancient art—go usually no deeper than 30 m, and professional divers who go unaided about 20 m below the surface and can stay under water for about 3 minutes belong to an expert group.

The world's largest mammals, blue whales (*Balaenoptera musculus*), have no reason to go very deep: their usual swimming depths of less than 100 m are determined by the occurrence of krill (*Euphausia superba*), their dominant

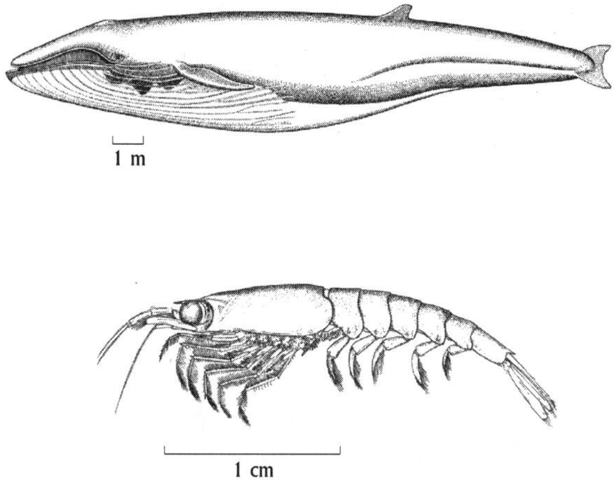

6.5 The world's second largest animal, the fin whale (*Balaenoptera physalus*), can grow to a length of 20–26 m and weigh in excess of 50 t; its main food, Antarctic krill (*Euphausia superba*), is 2–6 cm long and weighs less than 1 g.

feed, which they filter through their baleen bones, comb-like plates hanging from their upper jaw instead of teeth (fig. 6.5). "Krill" (the name means "small fish" in Norwegian) is a misnomer: the animals are actually small, yellow- or orange-hued crustaceans resembling shrimp and feeding on diatoms (Nicol and de la Mare 1993). Their swarms in Antarctic waters can contain up to a billion individuals, and their annual zoomass production totals perhaps as much as 1.3 billion t (Nicol and de la Mare 1993). Echosounders show by far the greatest densities of euphausiids in the top 50 m, and so the whales often feed right on the surface.

Another baleen whale, the right whale (*Eubalaena*), also skim-feeds with its mouth open, and Melville's (1851) memorable description of this spectacle is worth quoting at length:

On the second day, numbers of Right Whales were seen, who . . . with open jaws sluggishly swam through the brit, which, adhering to the fringing fibres of that wondrous Venetian blind in their mouths, was in that manner separated from the water that escaped at the lip. As morning mowers, who side by side slowly and seethingly advance their scythes through the long wet grass of marshy meads; even so these monsters swam, making a strange, grassy, cutting sound; and leaving behind them endless swaths of blue upon the yellow sea.

But baleen whales also feed at the bottom, usually in depths less than 60 m (Würsig 1988). Dolphins and porpoises go deeper, to 80–300 m, in dives that last typically only 2–4 minutes. Using free-ranging animals fitted with time-depth recorders monitoring the depth and duration of their dives, researchers found that many seals go much deeper (fig. 6.6). To feed on Antarctic cod, the Weddell seal (*Leptonychotes weddelli*) dives to more than 500 m and stays submerged for more than 70 minutes (Zapol 1987). These are not, of course, usual feeding depths, as record dives go normally two or three times deeper than the average dives of each species.

The record diver among the seals is the northern elephant seal (*Mirounga anguistirostris*), whose females were repeatedly measured descending to about 1,000 m and whose maximum diving depths are around 1,500 m, the distance that the animals cover in about 20 minutes in either direction without any signs of exhaustion. High oxygen storage in blood (70% as opposed to about 50% in humans), collapsible lungs, and a large spleen that injects red blood cells into the circulation after the animal submerges make it possible for seals to hold their breath longer, and hence go deeper, than any mammal except sperm whales (Kooyman and Ponganis 1997).

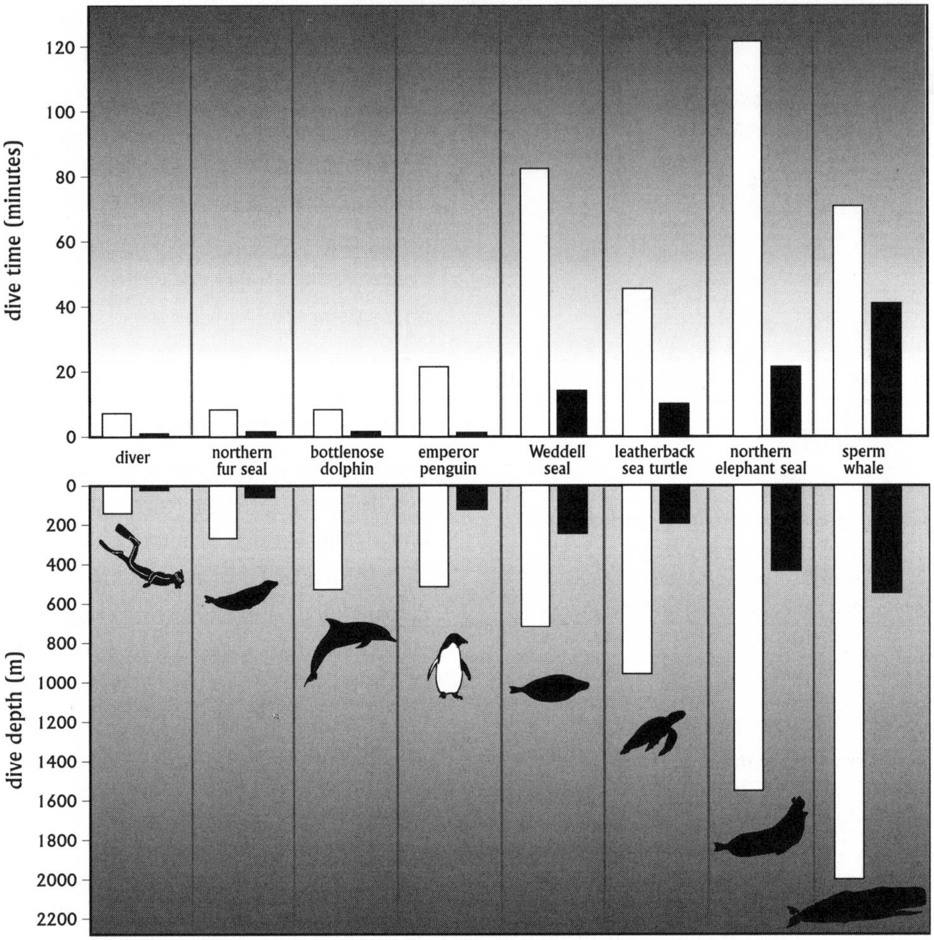

6.6 Average (black) and maximum (white) diving times and depths for various animals. Based on Kooyman and Ponganis (1997).

Sperm whales (*Physeter macrocephalus*), the world's largest toothed animals, feed almost entirely on a mixture of various squid species, and hence they must dive to the depths preferred by those pelagic creatures and be able to stay there for long periods of time. Examinations of whale stomachs indicate that they get about three-fourths of their food by simply swimming through luminous shoals of small, neutrally buoyant cephalopods (Clarke et al. 1993). The rest of the whale's diet comes from chasing faster-swimming, larger species, including the elusive *Architeuthis,* the giant squid so vividly described by Melville (Roper and Boss 1982; Melville 1851). Dives to 500 m or more are common, sonar observations of sperm whales have recorded maximum dives to more than 2,000 m, and maximum dives to over 3,000 m are likely.

Subsurface Life

Groundwater has always carried microorganisms far below the surface, and deep wells and boreholes yield a great variety of dark-adapted subterranean microbes (Gilbert et al. 1994). Where the bedrock is cracked by tectonic stresses or earthquakes, the fissures allow for an even speedier and deeper percolation of microorganism-laden water into deep underground aquifers. And where the precipitation reacts with soluble limestone, water can carve intricate and bizarre karst formations: many are tens or hundreds meters deep, and the deepest ones are just over 1,500 m below the surface. Deep sinkholes, convoluted tubes, and sometimes enormous underground caverns with mazelike passages and pools of water shelter troglobites—peculiar-looking animals including various invertebrates (flatworms, snails, arachnids, crustaceans, insects) as well as fishes and salamanders—that dispense with body pigments and with sight (Holsinger 1988).

And Sarbu et al. (1996) found the first limestone cave ecosystem that is not dependent on the inputs of new organic material of photosynthetic origin from the surface. The diverse fauna (forty-eight cave-adapted terrestrial and aquatic invertebrate species, including thirty-three endemic organisms) of the Movile Cave near the Black Sea in Romania is based, like that of the vent communities in the deep sea, on chemosynthesis in bacterial mats that use H_2S as their energy source. Leaving these obvious waterborne penetrations aside, until a generation ago the depth limit of life on land seemed to correspond to that apparently impervious barrier, the presence of solid bedrock. Of course, weathering would open cracks in even the hardest igneous minerals, and many of these would be quickly filled by opportunistic roots, tufts of grass, mosses, lichens, or bacterial mats, but such penetrations are generally shallow, and hence biologists considered all but a thin sliver of the solid lithosphere as devoid of life. Agriculturists also used to subscribe to the idea of shallowly embedded life: they saw soils ending at a plow depth, or a few spade depths, below the surface. Both of these perceptions were wrong.

The concept of soil as a component of ecosystems has been including progressively deeper layers ever since the beginnings of modern soil science during the late nineteenth century (Simonson 1968). Depths of many meters, or many tens of meters, are now seen as functionally justifiable. And this deepening of the biosphere goes far beyond the redefined soils. As remarkable as the deep-ocean black smokers are, I think it is the profundity of the biosphere on land, or below the sea floor, that is the most fascinating case of the receding boundaries of life. The adjective "receding" must be applied not just to the progressively greater depth at which we find peculiar assemblages of bacteria, but also to the lineage of these remarkable species: they may be direct descendants of very first organisms that inhabited the Archean Earth.

Soils

Soils are complex assemblages of living and abiotic components formed through prolonged interactions of organisms and minerals that are critically influenced by climate and geomorphic processes. The nascent soil science of the nineteenth century paid attention only to the two thin topmost layers of these formations that are now known as O and A horizons. They refer, respectively, to organic matter in various stages of decomposition on the soil surface and to the uppermost soil layer, usually darkened with decomposed organic matter. Their combined depth is most commonly no more than 10–30 cm, and they are densely permeated with life. Photosynthesizing bacteria, cyanobacteria, and algae are obviously limited to the uppermost layer of soil, where there is enough direct or transmitted solar radiation for photosynthesis to take place. Bacteria are invariably the most abundant organisms in topsoils, with typical counts of 10^9–10^{10}/g. Fungi come next (10^4–10^6/g), but their much larger size and their often extensive networks of filamentous mycelia mean that they dominate microbial biomass in some ecosystems, particularly in forests.

Unicellular protozoa (amoebas, ciliates) may be as abundant as fungi, and they are prayed upon by soil microarthropods, above all by mites (*Acari*) and springtails (*Collembola*). Common eelworms (*Nematoda*), which often cause great damage to crops, are barely visible, and most of the macroscopic soil fauna consists of millipedes and centipedes (*Myriopoda*), earthworms, and, particularly in many forest and grassland soils, ants and termites. But as we go deeper, soil macroorganisms get noticeably rarer. Earthworms, who can every year drag more than 10 t/ha of surface litter into the soil (Brown 1995), do most of this vertical redistribution of organic matter within the A horizon. In sandy soils ants can excavate vertical corridors more than 3 m deep, but most ant species usually do not penetrate that deep; ant hills of *Formica polyctena,* the European wood ant, extend only about 50 cm below the ground (Hölldobler and Wilson 1990). And excavations of termite colonies show that, on the average, only 7% of these insects are found at depths of 1.75–2 m (Brian 1978).

As the concept of soil evolved, the B horizon, underlying the O and A, was recognized as an essential part of soils. This layer has much less organic matter but a much higher clay content, as well as more aluminum and iron, than the layers above it. Inclusion of the B horizon increased the depth of most soils to at least 50 cm, but commonly to as much as 2.5 m. Until recently the C horizon, the soil's parent material below the B layer, has been generally seen as a part of the underlying bedrock rather than as belonging to the soil as such. And yet this layer is not only subject to the same chemical influences that create what might perhaps best be termed the "farmer's soil" of the overlaying horizons, it is also clearly affected by life.

Studies of deep soil profiles have revealed a remarkable uniformity of the total bacterial mass with increasing depth: as expected, the count is highest in surficial horizons, but it decreases by less than an order of magnitude at a depth of 8 m (Richter and Markewitz 1995). The presence of respiring biomass elevates the CO_2 level much above the average measured in the ambient air, up to nearly 3% in the C horizon compared to 0.036% in the atmosphere. Obviously, higher biogenic CO_2 levels contribute to faster weathering, and that process is also affected by biogenic acidification, which solubilizes and releases rockbound elements. Living organisms thus strongly influence, directly and indirectly, the entire C horizon; its addition extends the thickness of many U.S. soils to 50 m, and maxima elsewhere may be up to 100 m.

Hot Subterranean Biosphere

When Edson Bastin, a University of Chicago geologist, suggested in the 1920s that the sulfate-reducing bacteria found in groundwater samples extracted from crude oil deposits hundreds of meters below the surface are descendants of organisms buried when the sediment was initially formed more than 300 million years ago, his claims were dismissed, and contamination of his samples by surface bacteria was offered as the logical explanation. And how would bacteria survive deep underground, where temperatures are in excess of 50 °C and where there are no organic substrates to feed on? The answer we now know is obvious: by being both thermophilic and autotrophic. The first indubitable proof of subsurface bacteria living in deep formations came only during the 1980s as a result of a research program established by the U.S. Department of Energy in 1985. The first three boreholes were completed to the depth of 289 m at the Savannah River nuclear facility in South Carolina (Fliermans and Balkwill 1989).

The ubiquity of subsurface microorganisms has been subsequently demonstrated by sampling in several countries and in different substrates (Pedersen 1993; Frederickson and Onstott 1996). Special drilling and sampling methods are obviously needed for obtaining such samples to avoid contamination by surface microorganisms (Griffin et al. 1997). Tracers added to the drilling liquid will clearly indicate whether the liquid has penetrated the core samples that are to be tested for the presence of bacteria. Some studies have merely confirmed the presence of unspecified subsurface bacteria; others have gone on to culture and to identify dominant species. *Thermoanaerobacter* and *Archaeoglobus,* both sulfate reducers, have been found at 1,670 m below the surface in a continental crude oil reservoir in the East Paris Basin (L'Haridon et al. 1995).

Sulfate-reducing bacteria have also been detected in marine sediments more than 500 m below the Pacific sea floor (Parkes et al. 1994), and bacteria related to *Thermoanaerobacter, Thermoanaerobium,* and *Clostriduim hydrosulfuricum* were obtained from water samples collected at 3,900–4,000 m in Gravberg 1 borehole, drilled to a depth of 6,779 m in the complete absence of any sediments in the granitic rock of the interior of the Siljan Ring, an ancient meteorite impact site in central Sweden (Szewzyk et al. 1994). Many subsurface bacterial communities include such common surface- and soil-dwelling species as *Pseudomonas* and *Arthrobacter.*

Predictably, bacterial densities decline with increasing depth, the viability of cells found in subsurface environments has extreme range (0–100%), and the metabolic activity in deep-rock communities, or in consolidated sediments, is orders of magnitude slower than in other environments penetrated by bacteria (Kieft and Phelps 1997). The heating of organic matter during marine burial may be the best explanation of its availability to subsurface bacteria: experiments have demonstrated that heating sediments to 10–60 °C notably increased acetate generation from organic matter (Wellsbury et al. 1997). There can be no doubt not only that deep-living bacteria are common in porous sedimentary formations that may contain relatively oxygen-rich zones or comparatively large amounts of organic matter, but that they can also be found in the pores between interlocking mineral grains of igneous rocks. Indeed, they appear to be everywhere there is fluid and temperatures no higher than those tolerated by microorganisms found in deep-sea vents, that is, about 110 °C.

This means that there should be bacterial communities as deep as 7 km below the seafloor (where temperature rises by about 15 °C/km) and up to almost 5 km below the continental surface (where temperature goes up by as much as 25 °C/km). Because of its large crack porosity, the top few kilometers of the ocean's basaltic ocean crust should be permeated with microorganisms. This conclusion has been confirmed by findings of widespread alteration of oceanic volcanic glass in ocean crustal rocks by microbial activity: glass brought from boreholes in the ocean crust shows an accelerated rate of weathering, including deep pits and etched feathery patters suggestive of microbial action (Fisk et al. 1998). High underground pressures have no effect on bacteria lodged in tiny rock or clay interstices.

There is no shortage of fundamental questions concerning these deep finds. Are the microorganisms in oil fields, as Bastin proposed, direct descendants of bacteria that were trapped during the formation of sediments tens or hundreds of millions of years ago? If they are of a much more recent origin, their downward penetration could have been accomplished at, geologically speaking, very rapid rates: bacteria migrating vertically 0.1–1 m a year would reach a 1-km depth in a mere 1,000–10,000 years. To what extent is subsurface life dependent on the photosynthesis? Bacteria feeding on 100-million-year-old organic matter in buried sediments are clearly dependent on ancient photosynthates, but not the recently discovered microbial ecosystems that may be entirely hydrogen-based (Stevens and McKinley 1995).

Other Extreme Environments

Hydrothermal vents at the bottom of the ocean or tiny interstices in igneous rocks several kilometers below the ground are just two among many kinds of extreme environments on the Earth that are colonized by extremophiles, archaea and bacteria preferring niches that at first sight seem quite unsuited for life (Madigan and Marrs 1997; Horikoshi and Grant 1998). There is no shortage of archaean or bacterial thermal extremists, both on land and in the ocean, on the either end of the temperature spectrum. Thermophilic species have received much more attention than cold-loving, psychrophilic microbes, largely because of the often spectacular nature of their habitats, which include hot springs, geysers, solfataras, and volcanic muds, but the often unnoticed presence of psychrophiles is also widespread.

The short-term heat tolerances of arthropods and vertebrates are remarkable. Certainly no other demonstration of the human and canine ability in this respect has been more convincing than the classical 1775 experiment of Charles Blagden, at that time the Secretary of the Royal Society, and his friends. They entered a room heated to 126 °C and stayed for 45 minutes: a steak they took with them was cooked, but the men, and a dog kept in a basket to protect its feet from getting burned, did fine (Blagden 1775). Heat dissipation through evaporative cooling—humans are particularly good at producing copious volumes of sweat—is the key to this admirable thermoregulation, but vertebrates cannot tolerate body temperatures less than half as high for extended periods of time.

Terrestrial and marine environments also contain a variety of niches that are chemically extreme: very salty, highly acidic or alkaline, burdened with very high concentrations of heavy metals or radioactivity. Specialized prokaryotes are associated with all of these extreme niches, and in some cases there is no need to go to geothermal sites, soda lakes, or other unique locations remote from densely inhabited areas to see the work of these extremophilic species. Although most travelers are not aware of the cause, spectacular color displays created by a halophilic (salt-loving) archaean are known to anybody who has looked out of the airplane window just before landing at the San Francisco International Airport or out of a car window when crossing the Dumbarton bridge across the San Francisco Bay. Salt ponds in the southern end of the Bay contain large quantities of *Halobacterium,* whose red pigment, bacteriorhodopsin (similar to the light-detecting rhodopsin in vertebrate retina), colors the plots in beautiful shades of purple.

Temperature Extremes

Cryptobiotic forms can survive the greatest temperature extremes, and this ability is not limited just to prokaryotes. Phylum Tardigrada contains small (50 μm to 1.2 mm) invertebrates related to arthropods as well as to nematodes; these tiny *Echiniscus* water bears, common in thin water films and on mosses, lichens, and algae and also in sand, form cryptobiotic, barrel-shaped tuns able to survive exposure to as much as 151 °C and to as little as –270 °C, very close to absolute zero (Greven 1980; fig. 6.7). High temperature survival extremes for some metabolizing prokaryotes are, remarkably, not far below such cryptobiotic records.

Before the 1960s no bacterium was considered more heat tolerant than *Bacillus stearothermophilus,* which is able to grow between 37 and 65 °C; during the 1960s discoveries of hyperthermophilic strains of *Bacillus* and *Sulfolobus* raised the extreme to 85 °C (Cowan 1992). Then during the 1980s, marine hydrothermal thermophiles raised the optimal growth extremes to 95–105 °C and the tolerated temperature extreme to 113 °C. This record belongs to *Pyrolobus fumarii,* an archaeon that grows in the walls of deep-sea vent chimneys and stops growing at temperatures below 90 °C (Blöchl et al. 1997; Stetter 1996, 1998). *Methanopyrus* is another hyperthermophilic archaeon found in deep-sea chimneys.

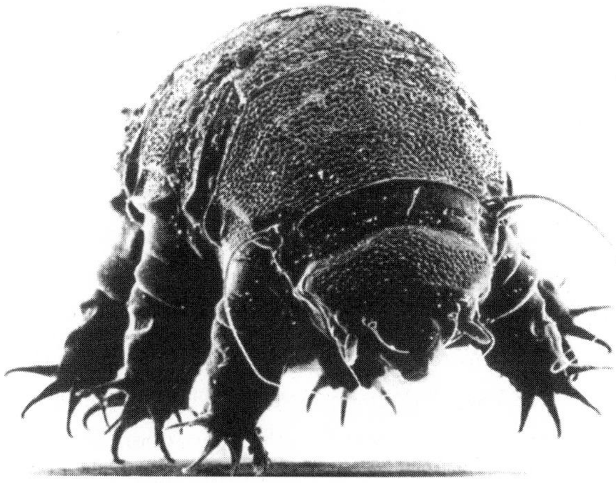

6.7 Tardigrades (water bears or moss piglets) form a separate phylum of microscopic, arthropod-like invertebrates and are best known for their extraordinary survival capabilities. The species shown here, *Echiniscus arctomys,* grows usually to less than 0.5 mm.

Terrestrial surface or near-surface temperatures above 70 °C are limited to geothermal habitats, to thermal springs, geysers, hot pools, and fumaroles; in contrast, habitats with temperatures repeatedly rising to 40–70 °C are quite extensive, including most desert, subtropical, and tropical soils and phytomass decomposing in thicker layers, as well as such anthropogenic environments as waters receiving discharges of industrial heat, coal tips, and ore tailings. The extreme buffering compounds—H_2SO_4 and sodium carbonate (Na_2CO_3) or bicarbonate ($NaHCO_3$), make many hot habitats either very acidic, with pH 1.0–2.5, or highly alkaline, with pH above 8.0 (Cowan 1992).

Two hyperthermophiles were among the first organisms with completely sequenced genomes: *Aquifex aeolicus,* a marine bacterium with growth temperature maxima near

95 °C (Deckert et al. 1998), and *Thermotoga maritima,* belonging to one of the deepest and most slowly evolving lineages of bacteria and isolated from geothermal marine sediment at Italy's Vulcano, where it grows at about 80 °C (Nelson et al. 1999). Some hyperthermophillic enzymes degrading sugars and proteins can be effective up to 140–150 °C, and experiments have proven that autotrophic synthesis of all twenty protein-forming amino acids is energetically favored in 100 °C submarine hydrothermal solutions compared to syntheses in 18 °C warm surface seawater. We still do not understand why most of the hyperthermophilic enzymes are able to have optimal catalytic activity above 100 °C, because they contain the same twenty amino acids as enzymes of other organisms and there are no gross structural differences between them and the compounds found in mesophilic prokaryotes (Zierenberg et al. 2000).

Eukaryotes are much less heat tolerant. When body temperatures approach 50 °C mammalian enzymes pass their optima, lipids change, and cell membranes become increasingly permeable (Spotila and Gates 1975). Even the most thermotolerant insects are not active above 60 °C: the Sahara desert ants of genus *Cataglyphis* that forage during the midday heat have critical thermal maxima at 53–55 °C (Gehring and Wehner 1995; fig. 6.8). Some species of both ectothermic (cold-blooded) and endothermic (warm-blooded) vertebrates can be active in deserts whose midday surface temperature is in excess of 70 °C, and mammals in hot environments have evolved a variety of remarkable thermoregulative adaptations (Schmidt-Nielsen 1964). But neither desert lizards and snakes nor foxes and rodents can be active in such temperatures throughout the day: resting in shade, hiding in burrows, and venturing out mostly during the night are common adaptations.

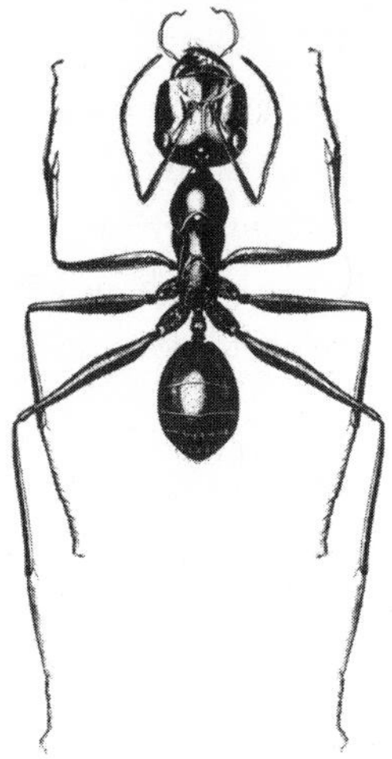

6.8 The *Cataglyphis* ant can forage on Saharan sands when the air temperature is in excess of 50 °C. Based on a drawing reproduced in Hölldobler and Wilson (1990).

As far as we know, only one marine eukaryotic organism comes close to the extreme heat tolerances of microbes. The Pompeii worm (*Alvinella pompejana*) is a 6-cm-long, shaggy-looking creature that colonizes the sides of active deep-sea hydrothermal vents (Desbruyères and Laubier 1980; fig. 6.9). Recent measurements on the East Pacific Rise showed that the temperatures within the worm tubes averaged 69 °C, and frequent spikes exceeded 81 °C (Cary et al. 1998). With temperatures at the tube opening averaging 22 °C, there is an astonishing gradient of up to 60 °C as the worm transfers heat from the hot to the cold end of its small body.

Inhabited cold environments are much more extensive on the Earth than are extremely hot niches. The vast Arctic and Antarctic cold deserts come first to mind, and, as already noted, some birds and many mammals can remain

6.9 The polychaete worm *Alvinella pompejana* lives on the sides of black smokers and tolerates extraordinary temperature gradients in its small (up to 9-cm-long, up to 2-cm-diameter) body. Color image, courtesy of the University of Delaware Graduate College of Marine Studies, available at <http://www.udel.edu/PR/NewsReleases/Worms/worm.jpg>

active in temperatures below –40 or even –50 °C. But in terms of the total volume of cold environment, the ocean is far ahead: except for the topmost 1,200 m, all of it remains constantly below 5 °C; deeper waters are just 2–3 °C, and the underlying sediments are permanently at about 3 °C. In addition, all bacteria living in these environments must tolerate high pressure, and some of them actually prefer it: these piezophilic species cannot grow in low-pressure environments (Yayanos 1998). Most of the continental surfaces experience seasonal or diurnal drops of temperature to close to, or below, 0 °C. And modern civilizations have created a multitude of cold niches with their ubiquitous use of refrigeration. Microorganisms that prefer temperatures lower than 15 °C are considered psychrophilic, but many prefer even lower temperatures, close to 0 °C (Russell and Hamamoto 1998).

Bacteria responsible for food spoilage in refrigerators—*Pseudomonas,* forming a slime layer on meat; *Lactobacillus,* attacking both meat and dairy products; and *Flavobacterium*—are the most commonly encountered genera thriving in chilly temperatures (Russell 1992). Because the lowest temperatures compatible with active life must still allow for the existence of liquid water, metabolizing microbes will be found in supercooled cloud droplets—Psenner and Sattler (1998) reported their occurrence at altitudes above 3,000 m in the Alps at around –5 °C—or in brine solutions, which remain liquid at well below the freezing point. Consequently, bacteria, algae, diatoms, crustaceans, and fishes can be found in ice-laden Antarctic waters whose temperature is as low as –2 °C. Notothenioids, Antarctic bony fishes resembling perches, survive in these waters by synthesizing at least eight different antifreeze compounds (glycopeptids) whose adsorption to minute ice crystals inhibits growth of such crystals in body tissues (Eastman and DeVries 1986; fig. 6.10).

Other marine organisms protecting their cells by secreting ice nucleators into the extracellular fluids are mollusks, both bivalves and gastropods, living in intertidal zones that are temporarily exposed to low temperatures (Loomis 1995). *Littorina littorea,* a snail from the northeastern United States, may survive with more than 70% of its total water content frozen when it is exposed to temperatures as low as –30 °C; barnacles (*Balanus balanoides*) survive even with 80% ice in their extracellular tissues. Many nonmarine organisms, including some turtles, amphibians, and caterpillars, also use antifreeze proteins, together with

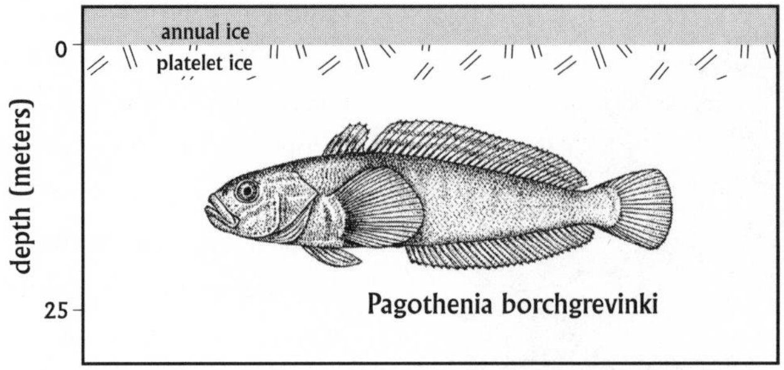

6.10 An Antarctic fish (*Pagothenia borchgrevinki*) adapted to platelet ice habitat. Based on Eastman and De Vries (1986).

ice-nucleating proteins initiating the formation of tiny extracellular ice crystals, to freeze solid while preserving their vital intracellular structures (Storey and Storey 1988). And recent discoveries describe ice-binding proteins not only in an insect (the spruce budworm) but also in a grass (Graether et al. 2000; Sidebottom et al. 2000).

So far the most extreme known psychrophiles are microorganisms living, literally, within ice. Slush layers on Alpine lakes (mixtures of ice and snow crystals with lake water, meltwater, and rainwater) contain rich communities of autotrophic organisms whose productivity has been found to be up to an order of magnitude higher at 0 °C under the snow cover than at 3–4 °C in the pelagic water of lakes (Psenner and Sattler 1998; fig. 6.11). An even more extreme psychroecosystem has been discovered in the perennial Antarctic lake ice in the McMurdo Dry Valleys (Priscu et al. 1998). The ice there is 3–6 m thick, and it would seem impossible that it could harbor any active life; however, miniature pockets of liquid water develop because of solar heating of the sediment layer containing sand and microbes blown from the surrounding cold deserts and embedded in the ice (fig. 6.12).

The maximum concentration of these pockets is at a depth of 2 m, where the sinking and ice growth are balanced, and they contain numerous species of bacteria and cyanobacteria (198 total in Lake Bonney). Most cyanobacteria found in these pockets are filamentous species of *Phormidium,* but there are also nitrogen-fixing *Nostoc* and *Chamaesiphon.* These miniature ecosystems are active for only about five months during the Austral summer, when eventually up to 40% of the total ice cover can be liquid water, but its temperature still never exceeds 0 °C. The most incredible Antarctic find, however, is that of microbes in subglacial ice above Lake Vostok that lies more than 3.5 km below the continent's surface (Priscu et al. 1999). Organisms found in the ice at a depth of 3,590 m were closely related to extant species of Proteobacteria and the Actinomycetes, and it is plausible that the lake itself, once explored, may support microbial populations in spite of more than a million years of isolation from the atmosphere.

In contrast to their relatively limited tolerance of high temperatures, the portable microenvironment of endotherms makes them unrivaled masters of the cold. As long

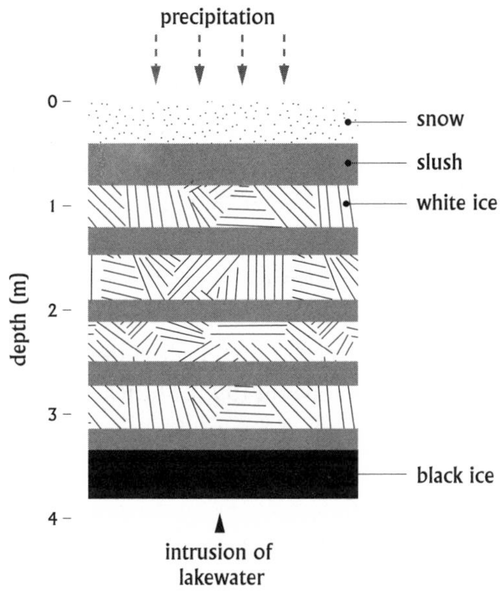

6.11 Cross-section through ice and slush layers of Alpine lakes, a habitat harboring numerous autotrophic microorganisms. Based on Psenner and Sattler (1998).

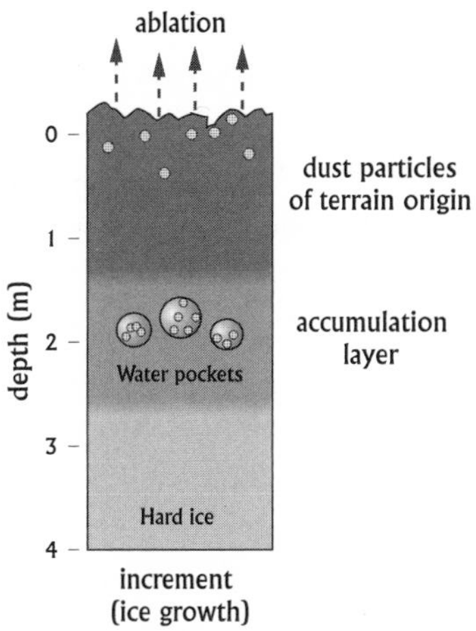

6.12 Location of miniature water pockets harboring microbes within Antarctic ice. Based on Priscu et al. (1998).

as food supplies are adequate, mammals (maintaining core temperatures between 36 and 40 °C) and birds (thermoregulating at between 38 and 42 °C) can survive in environments that may remain for most of the year well below the freezing point and that can plunge repeatedly to below –40 °C in absolute terms, and to chill factor equivalents of more than –60 °C, conditions that prokaryotic organisms could survive only by becoming cryptobiotic. Musk oxen (*Ovibos*) and polar bears (*Thalarctos maritimus*) prove that both herbivores and carnivores can thrive in extremely cold environments, but their thermoregulatory achievement pales in comparison with that of sparrows that live through cold Canadian winters by maintaining, during the coldest days, a temperature gradient of about 80 °C across the less than 5 cm between the core of their tiny bodies, warmed to just above 40 °C, and the outside air.

Chemical Extremes

Acidophilic life is almost entirely restricted to microorganisms, including a relatively large number of eukaryotic species: scores of strains of yeast and more than 300 strains of filamentous fungi, algae, and protozoa tolerate environments with acidity as high as pH 3 (Norris and Johnson 1998). But bacterial acidophiles have evolved not merely to tolerate highly acid environments or to be able to grow and reproduce in them: some of them are actively producing dilute sulfuric acid, an ability that has been put to a widespread commercial use.

Because both the key biochemical processes as well as essential cellular structures tolerate only slight departures from the neutral pH, acidophilic bacteria must prevent their internal, cytoplasmic pH from falling much below

6.0. The lowermost limit of external tolerance belongs to *Picrophilus oshimae,* whose growth optimum is at pH 0.7 (more than million times more acid than the neutral environment!) and which can metabolize at a pH approaching 0. The enormous pH gradient that is needed to maintain a relatively neutral cellular pH in such external conditions is astounding! The acid in question is virtually always H_2SO_4, as only SO_4^{2-} or its close, but much rarer, analog SeO_4^{2-} can sustain metabolism at extreme acidities (Ingledew 1990; Norris and Ingledew 1992). At the other end the delimitation of acidophily is arbitrary, with a pH no higher than 4.0 being a reasonable choice.

There is no shortage of environments containing pyrites or reduced sulfur: besides regions naturally endowed with metallic sulfides, they include geothermal areas and waterlogged acid soils. Thiobacilli are the largest group of acidophiles, and *Thiobacillus ferrooxidans* is certainly the most studied species of the genus; it is able to oxidize ferrous iron (Fe^{2+}) in pyritic (FeS_2) ores to ferric iron (Fe^{3+}), which subsequently oxidizes the reduced sulfur. The process yields a solution of iron in H_2SO_4 that can be used to dissolve other heavy metals (Cu, Zn, Pb, Sb, Ni, Ga).

Extraction of copper from low-grade ores is by far the most common commercial use of this acidophily. Enormous dumps of crushed poor-grade ore are flooded or sprinkled with acidified water and naturally invaded by *Thiobacillus ferrooxidans* and *thiooxidans;* after the dissolved metal is extracted, the percolate may be recycled. Currently about 25% of the world's copper output, worth more than U.S.$ 1 billion, is produced this way (Bosecker 1997). The very same process is quite unwelcome when thiobacilli work in an uncontrolled manner on pyritic materials and cause acid runoff from operating or abandoned coal or ore mines. Needless to say, *Thiobacillus* must be able to tolerate high concentrations of cations in the leached heavy metal. The bacterium can also fix enough nitrogen for its generally slow growth. Species of *Sulfolobus* are perhaps the most common acidophilic archaea, and like other Crenarchaeota, they are also hyperthermophilic: *Sulfolobus acidocaldarius* grows in waters at 75 °C and pH 2.

The opposite of acidophiles are alkaliphiles: they cannot grow even in neutral environments, the optimal pH for their metabolism is between 9 and 10, and some bacteria can tolerate levels a bit over pH 11 (Kroll 1990; Grant and Horikoshi 1992; Horikoshi 1998). Much like acidophiles they have the challenge of maintaining near-neutral pH in their cytoplasm despite extremely alkaline external conditions. Alkaline environments are much more extensive on the Earth than highly acid niches. Calcium hydroxide–dominated groundwaters are found in some rock formations, but soda lakes and deserts are much more common. California's Mono and Owens Lakes, Turkey's Lake Van, Kenya's Lake Nakuru and Tanzania's Lake Natron are among the world's best-known soda lakes. Soda deserts cover large areas in Central Asia, North and Northeast China, and North Africa. Some uncontrolled releases from industrial operations (cement production, electroplating, Kraft pulp process) have also created highly alkaline environments.

Bacillus (species *pasteurii, firmus,* and *alcalophilus*) is the common alkaliphilic genus; *Halobacterium* is an extreme haloalkaliphile, as are *Natronobacterium* and *Natronococcus,* the two archaeons that dominate in such highly saline alkaline environments as Lake Owens or Egypt's famous Wadi Natrun and can withstand pH levels between 8.5 and 11. And whereas there are very few eukaryotes that tolerate high acidities, there are many alkalitolerant fungi as well as some green algae (including the very common

Chlorella), protozoa, and crustaceans that can survive in pH up to 10.

Halophiles, organisms that require sodium chloride (NaCl) for their growth, thrive in hypersaline environments, that is, those with more than 3.4% salt, the normal concentration in seawater. Although there are a few halophilic eukaryotes—flagellates of genus *Dunaliella,* the brine shrimp (*Artemia salina*), and the brine fly—life in the saltiest environments is solely the domain of halophilic bacteria and archaea (Grant et al. 1998). They can be found in such diverse habitats as the Great Salt Lake sediments, the waters of the Dead Sea, the soils of Death Valley, and Antarctic saline lakes. They are also common in soils salinized by excessive irrigation, in some hot springs, and in salted fish. The already noted archaean *Halobacterium* belongs to the order of the most halophilic organisms; its red pigmentation promotes rapid precipitation of sea salt and has been known for millennia from saturated salterns.

Before closing this section a few more details on radiation-resistant Deinobacteriaceae. *Deinococcus radiodurans,* red with carotenoid pigment and smelling of rotten cabbage, is found in soils, animal feces, and sewage and can tolerate extreme doses of radiation as well as exposure to solar-flux UV radiation, desiccation, and freeze-drying. The sequencing of its genome (White et al. 1999) showed that its resistance to radiation is due to an extraordinary capacity to repair any radiation-damaged DNA. The polyextremophilic nature of *Deinococcus radiodurans* has led some researchers to think of it as a perfect candidate for the initial terraforming of Mars, gradually altering the planet's environment to make it eventually habitable for humans (Richmond et al. 1999). The most obvious problem with such a suggestion is that *Deinococcus* is not heat tolerant.

The Biosphere's Extent

New understanding amassed during the second half of the twentieth century makes it much easier to outline the extremes of the biosphere. Although their densities normally decline rapidly with altitude and although most bioaerosols high above the ground are cryptobiotic, microorganisms can be found not only in the entire troposphere but also up to, and even above, the highest layers of the stratosphere, that is, throughout a vertical distance of some 50 km or more. Extreme bird flights go up to about 9 km above sea level, and long-distance migrants, including small songbirds, often fly higher than 5 km. The troposphere below 3 km teems with microorganisms, and below 500 m with birds and insects.

The entire water column, which is in the deepest oceans just short of 11 km, is inhabited by eukaryotic organisms. There is the expected decline of total zooplankton, invertebrate, and vertebrate biomass with increasing depth, but large carcasses of whales and hydrothermal vents give rise to surprisingly rich micro-ecosystems on deep-ocean floor. Viruses and prokaryotes can be found in enormous counts throughout the water column. On land, the traditional, highly restrictive concept of soil should be replaced by a new one that recognizes that the activity of soil-forming organisms extends commonly to tens of meters, rather than tens of centimeters, below the surface. We have yet to glimpse the limit of the subterranean biosphere: most likely it is somewhere between 5 and 7 km inside the Earth's crust.

A few specialized prokaryotes can metabolize in waters whose temperature is in excess of 110 °C, and many actually prefer temperatures above 80 °C; in contrast, eukaryotes are generally much less heat tolerant, and vertebrates try to avoid any long exposures to heat above 50 °C. But eukaryotes can handle cold much better than microorganisms. Even some insects can be briefly active in temperatures well below 0 °C by raising their body temperatures through vigorous shivering, and well-adapted birds and mammals tolerate long periods of temperatures below –30 °C, and some even below –50 °C. The extreme range of ambient temperatures compatible with active life is thus in excess of 150 °C. Extreme tolerance of acidity goes to below pH 1 and that of alkalinity to pH 11; both records belong to prokaryotes: eukaryotes prefer overwhelmingly to stay within a much narrower range between pH 4 and 8.

And in closing, a couple of paragraphs on *Homo sapiens.* Unable to survive for more than a few minutes underwater, incapable of becoming airborne, and uninclined to live underground, the only two extremes mastered by our species unaided by modern technical support and huge energy subsidies were those of temperature and altitude. Excellent human thermoregulation, thanks largely to an unmatched capacity for sweating, has made it possible to inhabit all but the most extreme desert environments (Hanna and Brown 1983). Where adequate surface or underground water supplies are present, humans can lively indefinitely in areas where temperatures in shade commonly surpass 40 °C. And Arctic populations have been able to survive in environments where the recurrent temperature plunges to below –40 °C are made even less tolerable by months of polar darkness.

The highest traditional permanent human settlements are in the Himalayas: Kargyak village in the Indian Ladakh at 4,060 m above sea level, and Dingboche, the highest Buddhist monastery, in the Mount Everest region of the Nepali Himalayas at 4,358 m above sea level. Wenchuan, a road post established by the Chinese army during the construction of the Qinghai-Tibet highway north of the Tanggula range, is at 5,100 m. Himalayan passes repeatedly crossed by pastoralists, pilgrims, and refugees lie above 5,000 and even 6,000 m. The major high-altitude cities are Lhasa and La Paz, both above 3,600 m. Gradual adaptations allow most people to cope with these altitudes, but research has shown that the natives have clear genetic advantages (West and Sukhamay 1984).

7

The Biosphere's Mass and Productivity

Quantifying Life's Presence and Performance

The amount of living matter in the biosphere . . . does not seem excessively large, when its power of multiplication and geochemical energy are considered.
Vladimir Ivanovich, Vernadsky *Biosfera*

As soon as the sun begins to diffuse its warmth over the surface of the earth in the spring . . . the trees display in a few days the most wonderful scene that can be imagined . . . they all at once increase, perhaps more than a thousand times, their surface by displaying those kind of numberless fans which we call leaves.
Jan Ingenhousz, *Experiments upon Vegetables*

What is the total mass of living matter on the Earth? How much new biomass is produced in a year? Attempts to answer these fascinating questions long predate the recent interest in studies of the global environment and of its change. The first estimate of worldwide primary production is about 140 years old (Liebig 1862), and one of the worst errors in Vernadsky's seminal work on the biosphere was his gross overestimate of the total mass of the Earth's living matter (Vernadskii 1926). The range of extreme estimates of primary production began to narrow during the second half of the twentieth century, when the unprecedented availability of remotely sensed information and a growing interdisciplinary interest in the global environment considerably improved our understanding of many natural processes on the planetary scale.

Although the estimates of global phytomass and of its annual productivity have profited from the advances offered by remotely sensed data, there has been no need for drastic revisions of values that were calculated during the 1970s and the early 1980s without the benefit of these advances. A clearer consensus regarding annual plant and phytoplankton photosynthesis has emerged from the latest set of primary production models, but considerable uncertainties remain as far as the standing biomasses of prokaryotes and eukaryotic heterotrophs and their productivities are concerned. This uncertainty is greatest in estimating the contributions of extremophilic archaea and bacteria, whose ubiquity and abundance have been properly recognized only since the late 1970s (Herbert and Sharp 1992; Horikoshi and Grant 1998).

Narrowing these uncertainties is not merely a matter of accounting curiosity: a better understanding of the biosphere's stores of living matter and of its annual fluxes is important for assessing the degree of human impact on the global environment in general and on biodiversity in particular. And it is also essential for more realistic predictions of the most likely changes in the biosphere's key biogeochemical cycles, especially for the three doubly mobile elements, carbon, nitrogen and sulfur (Smil 2000a). Such fundamental unresolved questions of global environmental change as the fate of missing carbon, nitrogen eutrophication of the biosphere, and effects of biogenic sulfur require the best possible accounts of global biomass and of its annual production.

Basic Concepts

Biomass, as usually defined, encompasses all living organisms on the Earth, including their nonmetabolizing tissues. Substantial differences in the moisture content of fresh biomass (from less than 5% in tooth enamel to more than 90% in the youngest plant shoots or in a newborn's brain) require the use of a common denominator. Totals are usually expressed either as dry mass or as carbon, the main constituent of living matter. The carbon content of biomass is much less variable than is its water share—extremes are around 20% C for some species of marine phytoplankton and 49% for many fungi—and the average share of 45% C works well both for plants and animals (Bowen 1966). Most of the Earth's above-ground biomass is stored in phytomass, whose abundance and partitioning in trees and grasses of major ecosystems became better understood only as a result of the International Biological Programme (Breymeyer and Van Dyne 1980; Reichle 1981).

Destructive sampling is the most reliable—but obviously also a time-consuming, labor-intensive, and spatially limited—way of quantifying the standing terrestrial phytomass. Consequently, some storage estimates still refer only to the above-ground biomass. Better allometric estimates of both above- and below-ground phytomass and advances in our understanding of fine-root biomass (Waisel et al. 1996; Jackson et al. 1997) have helped, but reliable information on above-ground phytomass is still patchy even in affluent countries and very poor elsewhere. Below-ground phytomass is well known from only a few one-time measurements at a small number of sites around the world. Short life spans and large periodic fluctuations, sometimes in massive blooms, are the main challenges in quantifying aquatic phytomass. Estimates of microbial and fungal biomass on ecosystemic scales are still quite rare, as are large-scale quantifications of invertebrate and vertebrate zoomass.

Gross primary production (GPP) is the amount of carbon fixed in a year by photosynthesizing bacteria and plants as well as chemotrophic prokaryotes. Subtraction of autotrophic respiration (R_a; carbon returned to the atmosphere by primary producers) from this total yields net primary production (NPP), the preferred measure of photosynthetic (or chemotrophic) performance. Biomass created by NPP is available for heterotrophic consumption, which ranges from rapid microbial decomposition to seasonal grazing by large ungulates. Subtraction of heterotrophic respiration (R_h; carbon released by consumers) from the NPP of a particular ecosystem shows the net ecosystem production (NEP; the annual rate of change of carbon storage), whose ultimate aggregate is the net primary biospheric production.

The production and respiration rates of organisms are, much like their biomass, usually expressed either in terms of dry organic matter per unit area, or, more frequently, as the amount of assimilated or respired carbon. As with the standing phytomass, the above-ground component of the NPP of plants is much easier to estimate than the below-ground share, and biomass production by extremophilic prokaryotes is particularly uncertain. And although heterotrophic productivities have been studied in considerable local detail for many microbial, invertebrate, and vertebrate species, it is not easy to extrapolate these results to regional, continental, and global scales. These uncertainties make little overall difference as far as inherently limited vertebrate productivities are concerned, but they can lead to disparities of several orders of magnitude when estimating activities of subterranean and subsea microbes.

The Earth's Biomass

In 1926 Vladimir Vernadsky concluded in the first edition of his pioneering *Biosfera* that the "estimates of the weight of green matter, alone, give values of 10^{20} to 10^{21} grams, which are the same in order of magnitude as estimates for living matter *in toto.*" He believed that this aggregate "does not seem excessively large, when its power of multiplication and geochemical energy are considered." In fact, this range, corresponding to 10^{13}–10^{14} t of organic carbon, proved to be an enormous exaggeration. During the 1930s Vernadsky kept revising the total downward, and his last published estimate was 10^{13} t of organic matter (Vernadskii 1940). We now know that this total is still more than an order of magnitude above the actual mass of even the most generous estimate of all living matter. In contrast, Noddack's (1937) estimate of just 270 Gt of carbon was undoubtedly far too low (Appendix E).

Phytomass

The first systematic attempt to calculate the total mass of plants was made during the late 1960s. N. I. Bazilevich, L. Y. Rodin, and N. N. Rozov (1971) classified the world's plant formations into 106 types and came up with a continental phytomass total of 2.4 Tt, or almost 1.1 Tt C. As this account considered the world's potential vegetation (the reconstructed plant cover of the preagricultural era before the conversion of forests and grasslands), it was clear that it greatly overestimated the recent total.

Robert Whittaker and Gene Likens (1975) applied the average biomasses of 14 ecosystem types to vegetation areas that existed around the year 1950 to calculate the continental phytomass at 1.837 Tt, or 827 Gt C. Olson et al. (1983) tried to improve the accuracy of the global total by subdividing continents into $0.5° \times 0.5°$ cells and collecting the best available data on climatic factors, vegetated areas, and phytomass ranges on that scale. Their range came to 460–660 (mean of 560) Gt C. They also argued that unless the storage in tropical forests is much more massive than is generally known, any estimates in excess of 800 Gt C would be unrealistically high, but they acknowledged that totals well below 560 Gt C would be less surprising.

Models of the global carbon cycle published during the 1990s offered total continental phytomass values as low as 486 Gt C (Amthor et al. 1998) and as high as 780 Gt C (Post et al. 1997; Appendix E). Two principal reasons for this large divergence are substantial disparities in the categorization of major land covers and differences in assumed typical phytomass densities. Land cover assessments published since 1980 have used values as low as about 25 and as high as 75 million km^2 for the total area of the Earth's forests (Emanuel et al. 1985; Solomon et al. 1993; Cramer et al. 1999). Oceanic phytomass is negligible in comparison with huge terrestrial stores, and values between 1 and 3 Gt C have been used in recent global carbon cycle models (appendix E).

And although cultivated land was expanded by a third during the twentieth century—to roughly 1.5 billion hectares, accounting for about 12% of the ice-free land—and average crop productivity rose more than fourfold, agricultural phytomass remains a small share of all plant matter. The recent global harvest has been about 1.25 Gt C in crops and 1.7 Gt C in their residues (Smil 1999c). To derive the maximum standing phytomass, we must enlarge that total by preharvest field losses (at least 20%) and add roots and unharvested above-ground tissues (another 20%). But because the global multicropping ratio is approaching 1.5, we must reduce the total by at least a third (an adjustment neglected in most published global crop biomass estimates). Excluding orchard and plantation trees and shrubs, this process yields about 3 Gt C, or less than 1% of all terrestrial phytomass.

Prokaryotic and Heterotrophic Biomass

An unknown number of species, their highly variable densities, and their occupation of many extreme niches are the main factors precluding any reliable assessments of prokaryotic biomass. Enormous uncertainty regarding the overall diversity of arthropods (Rosenzweig 1995), and the variety, mobility, and variability of other consumer species pose similar difficulties for quantification of other heterotrophic biomass. Extrapolation of this information and its aggregation into global estimates may carry large errors. Fundamental energetic limits make it impossible for endotherms to dominate the planetary consumer biomass: prokaryotes and invertebrates must form its bulk.

Perhaps the best-studied segment of the prokaryotic presence is soil microorganisms. As noted in chapter 7, proper accounting must recognize that soils are deeper than has been traditionally assumed, as several meters, rather than just 0.5–2 m, are filled with considerable numbers of bacteria, as well as with fungi (Richter and Markewitz 1995). Given the wide range of bacterial presence in soils—10^1–10^3 g/m^2 (Paul and Clark 1989; Coleman and Crossley 1996)—global estimates can differ substantially. Whitman et al. (1998), assuming an average of about 450 g/m^2, estimated about 26 Gt C in soil bacteria. A more conservative assumption of 250 g/m^2 would yield about 15 Gt C (Appendix F). But these are minor uncertainties compared to those we face in estimating the mass of subterranean prokaryotes.

Gold's (1992) assumptions (3% porosity in the first 5 km, and 1% of that space filled by bacteria) would result in 200 Tt of subterranean prokaryotes, a total roughly 200 times larger than all of the above-ground phytomass. But Gold correctly acknowledged that we do not know how to make a realistic estimate of the subterranean biomass, and the same caveat must apply to its more recent estimates by Whitman et al. (1998). They also used 3% porosity for the uppermost 4 km, but they assumed that the total pore space taken up by prokaryotes is just 0.016%. Their estimate of 2.2×10^{30} subsurface cells would be equivalent to some 200 Gt C, but they offer a range of 22–215 Gt C instead. Pedersen's (1993) review makes it clear that even this span may be much too narrow: total bacterial counts at depths between 10 and 2000 m at 12 drill sites on three continents ranged between 10^1–10^8 per gram of groundwater, and we have no representative information on actual counts in deeper layers.

Large uncertainties also surround the estimates of numbers of prokaryotes in subsea sediments. The best evidence comes from just five Pacific Ocean sites from depths up to 518 m below the sea floor (Parkes et al. 1994). A depth integration of numbers encountered (rapidly declining with depth) results in a total mass of about 300 Gt C in the top 500 meters (average depth of oceanic sediments). The actual total may be a fraction of that mass, but even one-third of it would still surpass the biomass in all boreal forests! In mass comparison, prokaryotes in waters, and even more so in rumens, colons, and hindguts of vertebrates and termites, are insignificant (Whitman et al. 1998). The fungal biomass in soils also ranges widely, from 10^0–10^2 g/m^2 (Bowen 1966; Brown 1995; Reagan and Waide 1996). A fairly conservative global aggregate would be 3 Gt C, but a mass twice as large, implying about 100 g/m^2 of ice-free surface, may not be excessive (appendix F).

Earthworms are the most conspicuous soil invertebrates. Darwin (1881) thought that "it may be doubted whether there are many other animals which have played so important a part in the history of the world, as have these lowly organized creatures" (p. 316), but their mass is usually just around 5 g/m^2 (fig. 7.1). Only in cultivated soils may there be well over 10 g/m^2 (Edwards and Lofty 1972; Hartenstein 1986). Inconspicuous or microscopic nematodes average about a tenth of annelid biomass, and ants and termites each add typically no more than 0.1 g/m^2 (Brian 1978). The most common range for invertebrates may be 7–10 g/m^2: their global biomass, dominated by annelids, nematodes, and microarthropods, would then be between 400 and 550 Mt C (appendix F).

Reptiles and amphibians often dominate vertebrate zoomass in tropical forests, where their biomass density may rival that of invertebrates (Reagan and Waide 1996). Only in the richest savannas supporting large numbers of ungulates will the average density of mammals surpass 2 g/m^2; in contrast, the biomass of animals in equatorial and montane rainforests and in temperate woodlands is below 1 g/m^2 (Prins and Reitsma 1989; Plumptree and Harris 1995). Small mammals, mostly rodents, add generally less than 0.2 g/m^2, and the zoomasses of insectivorous mammals are mostly below 0.05 g/m^2 (Golley et al. 1975). The total for all carnivores in the Ngorongoro Crater, one of the world's best habitats for large predators to hunt ungulates, is less than 0.03 g/m^2 (Schaller 1972). Similarly, total avifaunas usually do not surpass 0.05 g/m^2 (Edmonds 1974; Reagan and Waide 1996).

In aggregate, vertebrate zoomass rarely sums to more than 1 g/m^2 over large areas, and uncertainties in its quantification thus hardly matter in the overall biospheric

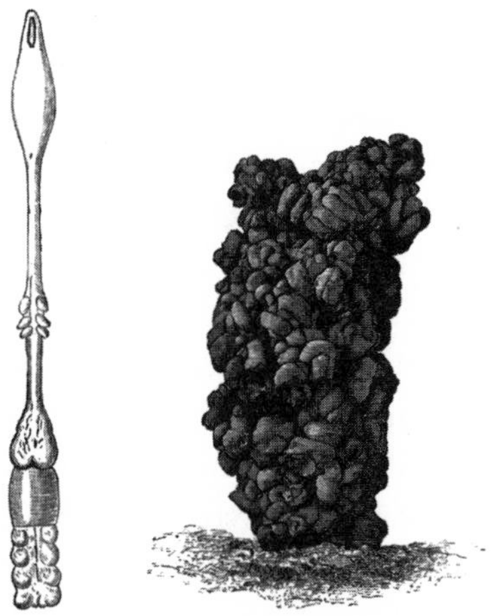

7.1 Images from Darwin's last book (1881): a diagram of the alimentary canal of an earthworm (*Lumbricus*) and a tower-like earth casting voided by a species of *Perichaeta*.

count: errors inherent in estimating prokaryotic biomass are easily an order of magnitude larger. Published estimates of terrestrial invertebrate and vertebrate biomass range between 500 and 1,000 Mt C, with wild mammals contributing less than 5 Mt C (Bowen 1966; Whittaker and Likens 1973; Smil 1991). Vertebrate zoomass also includes domestic animals, whose biomass is dominated by bovines (about 1.5 billion heads of cattle and water buffaloes), and calculations based on Food and Agriculture Organization (FAO 2000) animal counts and on conservative averages of their live weights result in 100–120 Mt C, or at least twenty times the wild mammalian total (appendix F; fig. 7.2).

The great depth of the inhabited medium and the extraordinary patchiness and mobility of many oceanic heterotrophs make the quantification of marine zoomass exceedingly difficult. Published estimates range from 300 to 500 Mt C, with invertebrates dominant and mammals contributing no more than 5% of the total (appendix F). Mann (1984) estimated total fish biomass at 300 Mt of fresh weight, or less than 40 Mt C. I have used conservative population estimates for all major whale species (IWC 2001) to put their global live mass at no more than 80 Mt, or about 15 Mt C. Even the highest combined estimate of terrestrial and oceanic values means that the global zoomass adds up to less than 0.3% of standing phytomass.

The lower average body mass and a higher share of children in populations of low-income countries mean that that the weighted global mean of human body mass is only about 45 kg per capita. Consequently, at the beginning of 2000, the global anthropomass of just over six billion people amounted to about 270 Mt of live weight containing approximately 40 Mt C (appendix F). My best estimate is that at the beginning of the twentieth century, the zoomass of wild mammals was at least as large as the anthropomass of 1.6 billion humans (Smil 1991), but by 2000, human biomass was an order of magnitude larger!

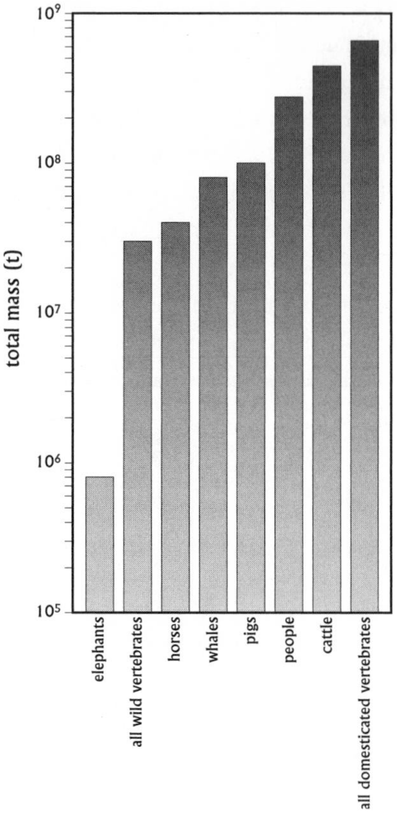

7.2 The zoomass of domesticated land animals, dominated by cattle, is now at least twenty times larger than the zoomass of all wild vertebrates. Based on data in the text.

The global average of human density for ice-free land is not a very meaningful measure. Cultivated area is the proper denominator: it now supplies about 85% of all food, and the global anthropomass (live weight) now amounts to almost 200 kg/ha of arable land and permanent plantations; China's mean is almost 500 kg/ha, and

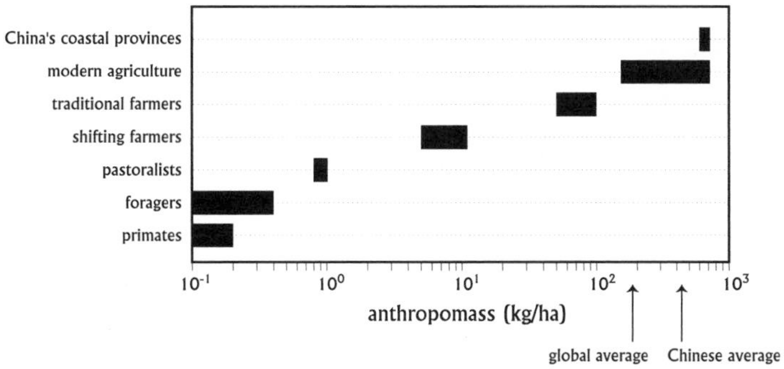

7.3 Common population densities among humans supported by foraging differed little from those achieved by other primates. Traditional agricultures raised average population densities by two orders of magnitude, and modern farming, heavily subsidized by energy and material inputs, can feed ten times as many people per hectare of arable land.

the country's most intensively cultivated provinces support 600–700 kg of humanity per hectare of arable land. This means that in densely populated regions, human biomass is now more abundant than that of all soil invertebrates. In contrast, the average densities of the two large African primates, chimpanzees and gorillas, are mostly less than 1 kg/ha of their now so limited (and disappearing) habitats (Bernstein and Smith 1979; Prins and Reitsma 1989; Harcourt 1996). These comparisons demonstrate impressively the relentless ascent of the most adaptable as well as the most destructive of all heterotrophic species (fig. 7.3).

Primary and Secondary Production

Primary production is, of course, the most important natural flux of the biospheric carbon cycle. Consequently, it has recently received a great deal of research attention as a part of increasingly detailed modeling of past and future perturbations of the carbon cycle due to the anthropogenic releases of CO_2 caused by conversions of natural ecosystems and by fossil fuel combustion. However, estimates of the Earth's primary production have a surprisingly long and interesting history. Inescapable losses during the energy transfer to the primary consumer level (see chapter 8 for details) mean that even the biomass of obligatory herbivores are only a small fraction of the phytomass they feed on, and the aggregate biomasses diminish rapidly in higher trophic levels. Even so, as we have seen in chapter 6, the bulk of the biosphere's diversity can be found among primary and higher-level consumers—not among plants!

History of Productivity Estimates

In 1862 Justus von Liebig, who pioneered in so many aspects of the new sciences of plant physiology and agronomy, published the first estimate of global primary

7.4 Justus von Liebig (1803–1873) was one of the most influential scientists of the nineteenth century. His rough estimate of the global phytomass was surprisingly close to the best current totals. Photo from author's collection.

productivity (fig. 7.4). He based his estimate on the seemingly unrealistic assumption of all land being a green meadow and yielding annually 5 t/ha (Liebig 1862). The result, about 63 Gt C, falls within the range of the best calculations published since the 1970s, which implies that the Earth's mean primary productivity resembles that of a rich grassland. Most of the values offered during the intervening generations were either serious under- or overestimates of the most likely total. Two decades after Liebig's pioneering attempt, Ebermayer (1882) used average productivities for woodland, grassland, cropland, and barren areas, based on yields of Bavarian forests and crops, to estimate the annual flux of carbon from the atmosphere to plants at just 24.5 Gt C.

Svanté Arrhenius offered a very similar estimate in 1908. Even lower values (16 and 15 Gt C, respectively) were published by Schroeder (1919) and Noddack (1937). In contrast, Edward Deevey's (1960) estimate of 82 Gt C of the terrestrial NPP erred on the high side. A clearer consensus began emerging only during the 1970s, when a number of American and European appraisals, many of them based on regression models using empirically derived relationships between climatic variables and NPP, ended up with totals between 45 and 60 Gt C; calculations resulting from better NPP models of the 1980s and 1990s have remained mostly within the same range (appendix G; fig. 7.5).

The first estimate of global phytoplankton production—25.4 Gt C per year for the open ocean and 3.2 Gt C per year for coastal seas, published by Noddack (1937)—was not so unrealistic, especially given the fact that it was based on just a few values for relatively unproductive areas. Gordon Riley of the Woods Hole Oceanographic Institution tried to derive a more realistic total by using productivity values for seven Western Atlantic sites, but we now know that only the lowest extreme of his range of 44–208 (mean 126) Gt C was on the target (Riley 1944). The first well-supported estimates of the productivity of open oceans were prepared thanks to Russian researchers led by O. J. Koblents-Mishke (Koblents-Mishke et al. 1968) and based on data from 7,000 sites around the world. Their

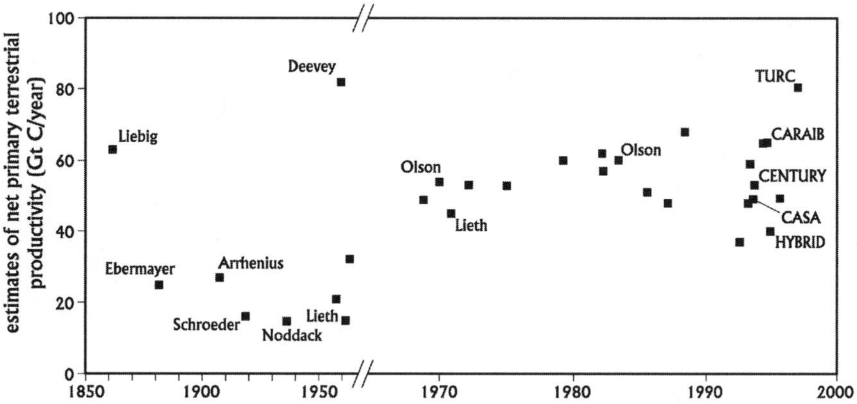

7.5 Nearly 150 years of estimates of global annual NPP show most of the totals within the range of 50–70 Gt C per year.

total of 23 Gt C a year, however, excluded all benthic production. Whittaker and Likens (1975) used 55 Gt C, and De Vooys (1979) estimated the total marine NPP at 46 Gt C (appendix G).

The latest estimates of global productivity are derived from satellite and surface data in combination with models of basic ecological processes. They have been made possible by remotely sensed observations on a planetary scale and by a better understanding of fundamental relationships determining GPP, NPP, and R_a. Naturally, models of ecosystem productivity are also useful for exploring the effects of rising atmospheric CO_2 concentrations as well as changes in other climatic variables.

Remote Sensing of Ecosystems

The first LANDSAT, launched in July 1972, produced land use data with a resolution of 80 m, good enough for a large-scale mapping of ecosystems and of their changes (fig. 1.10). Chlorophyll reflects less than 20% of the longest wavelengths of visible light but about 60% of near infrared radiation. These differences in reflectance can be used to distinguish between vegetated and barren areas, to do relatively detailed ecosystemic mapping, and, when backed up by good ground observations, to estimate total phytomass. The French SPOT offered even higher resolution (10–20 m), but high cost of both SPOT and LANDSAT images (particularly after LANDSAT's privatization in 1986) led to the use of the advanced very-high-resolution radiometer (AVHRR).

The AVHRR, installed on NOAA's polar-orbiting satellites, has a maximum resolution of only 1–4 km, but this is sufficient to appraise large-scale patterns of vegetation coverage and to detect seasonal and statistically significant year-to-year variabilities of global plant conditions (Gutman and Ignatov 1995). The normalized difference vegetation index (NDVI) compensates for changing

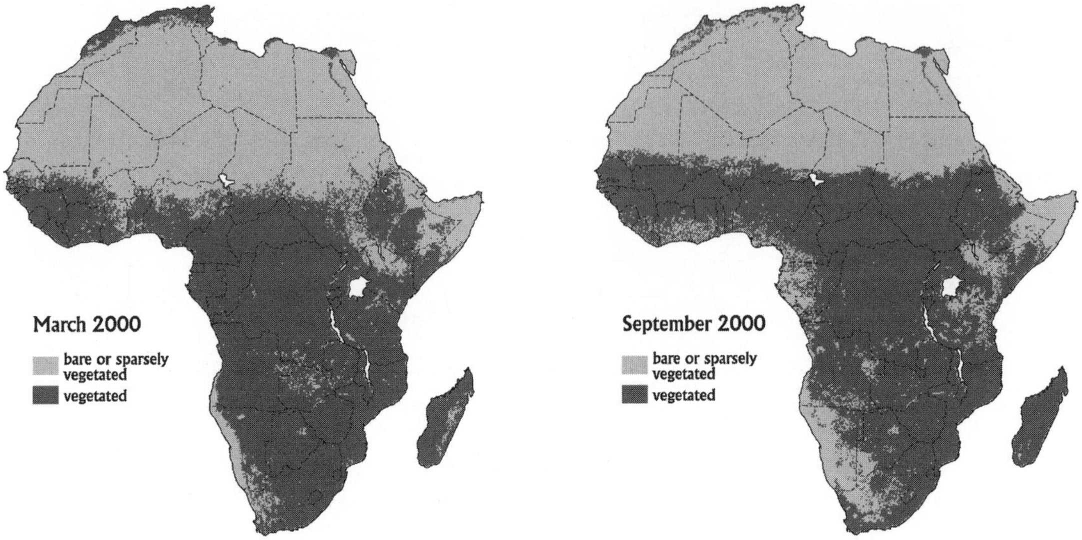

7.6 Sequential NDVI illustrates particularly well the shifting boundaries between barren and vegetated areas and the intensity of forest growth. A sharp border between Sahelian vegetation and the Sahara reaches its northern limit in September, when Central Africa's seasonally dry forests also grow more vigorously. Based on images available at <http://metart.fao.org/~~/gbr/E-VVGTGL.htm>

illumination conditions, surface slope, and viewing aspect. Monthly averaging and spatial bilinear interpolation eliminate cloudiness present in daily observations. The first NDVI calculations used reflectances in the visible band (0.58–0.68 μm) and in the near IR (0.73–1.1 μm); since April 1985 they have also included brightness temperatures in the thermal IR (11 and 12 μm) and the associated observation illumination geometry (Gutman et al. 1995).

Sequential monthly averages of NDVI show dramatically the seasonal ebb and flow of the Earth's photosynthesis, and the images provide an excellent record for monitoring long-term changes in principal ecosystem patterns as well as a highly reliable input to global models of photosynthetic productivity (fig. 7.6). Similarly, data from the Coastal Zone Color Scanner (CZCS) have revealed global patterns of phytoplankton growth, including major seasonal changes in relatively nutrient-rich coastal and upwelling regions (Gordon and Morel 1983; Antoine et al. 1996; Falkowski et al. 1998; fig. 7.7).

Recent Models of Global Primary Production

Annual and seasonal NPPs for the land biosphere produced by seventeen global models of terrestrial biogeochemistry were recently compared using standardized input variables

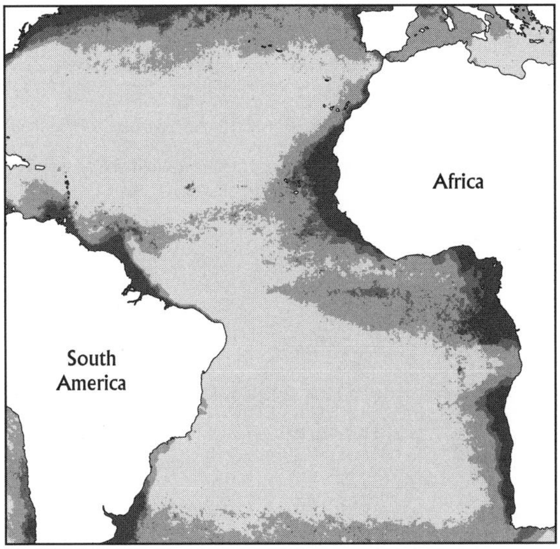

7.7 Monitoring by CZCS shows high phytoplankton productivity in relatively nutrient-rich regions—in shallow coastal seas, in regions of vigorous upwelling (shown here offshore from the western coasts of Africa), in waters receiving a high inflow of nutrients from major rivers (shown here from the Amazon)—and very low photosynthetic rates in central ocean gyres. This image is derived from CZCS data averaged for the period between November 1978 and June 1986. Color original is available at <http://seawifs.gsfc.nasa.gov/SEAWIFS/CZCS_DATA/global_full.html>

(Cramer et al. 1999). These models belong to three major categories. Nine models that simulate carbon fluxes using a prescribed vegetation structure, including carbon assimilation in the biosphere (CARAIB) and terrestrial ecosystem model (TEM) form the largest group; five satellite-based models, including the Carnegie, Ames, Stanford approach (CASA) and terrestrial uptake and release of carbon (TURC), used data from the NOAA/AVHRR sensor as their major input; and three models simulated both vegetation structure and carbon fluxes. After removing two outliers (which produced extreme results as artifacts due to the comparison) the simulations produced a range of 44.4–66.3 Gt C/year (Appendix G; fig. 7.5).

CARAIB, a mechanistic model, combines three submodels calculating leaf assimilation, radiative transfer through canopies, and wood respiration (Warnant et al. 1994). Its most likely result was 65 Gt C/year. The process-based TEM, built at the Marine Biological Laboratory in Woods Hole, calculates the global NPP in monthly steps as the difference between GPP and R_a, including both maintenance and construction respiration (Melillo et al. 1993). Its annual global NPP total with 355 ppm CO_2 was 53.2 Gt C. The CASA model calculates NPP as a product of absorbed photosynthetically active radiation (PAR) and the efficiency of its use (Potter et al. 1993; Field et al. 1995). Its latest version (Potter 1999) put the global terrestrial NPP at 56.4 Gt C/year. The TURC model combines a remotely sensed vegetation index to estimate the fraction of solar radiation absorbed by canopies with parametrization of the relationship between absorbed solar radiation and GPP based on measurements of CO_2 fluxes above plant canopies (Ruimy et al. 1996). Because of the model's high R_a-to-GPP ratios for the tropical rain forests (70%) and for temperate forests (50%), its highest predicted NPPs are in tropical savannas of Africa and South America and in temperate, intensively cultivated zones of North America rather than, as is in other cases, in forested equatorial regions. The model's best global annual NPP rate was 62.3 Gt C.

The Global Primary Production Data Initiative was launched in 1994 as a part of the International Geosphere-Biosphere Programme to make NPP measurements readily available in a standardized format. The world's most comprehensive NPP database is managed by the Oak Ridge National Laboratory and complements detailed field measurements of NPP for about fifty intensive terrestrial study sites (mainly grasslands and tropical and boreal forests) with geo-referenced climate and site-characteristics data (ORNL 2000). Three recent satellite-based models of marine NPP came up with very similar results: 37–46 Gt C per year by Antoine et al. (1996), 49 Gt C per year by Field et al. (1998), and 50.8 Gt C per year by Yoder et al. (1998). With an oceanic phytomass of 1–2 Gt C, this implies an average turnover rate of one to two weeks; in contrast, a continental NPP of 50–60 Gt C translates to mean turnover rates of one decade for terrestrial phytomass. Although the Southern Hemisphere's ocean area is about a third larger than that of the Northern Hemisphere, the marine NPP of the two hemispheres is about equal (41 vs. 38%, with the remainder produced in equatorial waters) because of the much higher (about 60%) output per unit area in northern oceans.

Secondary Producers

Bioenergetic imperatives constrain the amount of biomass that can be produced annually by heterotrophs. The simplest estimate can be limited to the overall production by herbivores, because the inevitable decline in energy transfers means that values for subsequent trophic levels will all be smaller than the estimation error in the first instance. Shares of NPP consumed by land herbivores range from just 1–3% in deciduous temperate forests to 5–10% in most forest ecosystems; values in excess of 25% in temperate meadows and wetlands and 30% in rich tropical grasslands are exceptions (Crawley 1983; Valiela 1984). Rates between 5 and 10% (average close to 8%) are most common, implying an annual consumption of about 4 (3–5) Gt C.

Endothermy exacts its high metabolic cost, as birds and mammals convert only 1–3% of ingested phytomass to new zoomass; in contrast, many ectotherms convert an order of magnitude more (10–25%) of their energy intake (Golley 1968; Humphreys 1979). With respective means at 2% and 20%, an even split of the consumed phytomass would result in net herbivore production of some 450 Mt C per year. Whittaker and Likens (1973) thought that 370 Mt C per year was a high estimate. But ratios of net secondary production to zoomass show a more complex picture. They are around or below 0.1 for the largest mammals, but they increase to 2–3 for larger rodents and to as much as 5–6 for mice and shrews, rising predictably with falling body mass. In contrast, there is no orderly production-to-biomass pattern for ectotherms: for example, many arthropods are prolific producers, with production-to-biomass ratios above five, but some ant species have production rates lower than that of hares (fig. 7.8). With average ratios of 0.5 for endotherms and at least 1 for ectotherms, the total zoomass production would be on the order of 400 Mt C a year, confirming the total derived in the previous paragraph.

Domestic animals do not increase this total as much as they do the zoomass aggregate. Annual meat production now amounts to about 200 Mt, and reverse application of typical live weight–to–dressed carcass factors (Smil 2000b) yields about 450 Mt of slaughtered live weight, or at least 60 Mt C. With perhaps a fifth to a quarter of all terrestrial zoomass, domestic animals thus account for less than 15% of net secondary production. This is hardly surprising given that nearly a third of all meat comes from bovines, whose production-to-zoomass ratio is inherently very low, just around 0.1.

With terrestrial and oceanic NPP almost equal, secondary marine productivity must be higher, as aquatic herbivores consume a larger share of available phytomass—commonly 30–50% (Valiela 1984)—and, being

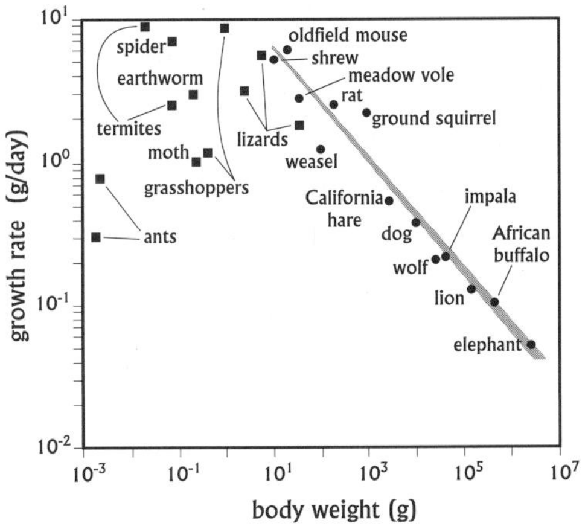

7.8 An excellent double-log fit of production-to-biomass ratios for mammals contrasts with the lack of a similar relationship among invertebrates (Smil 1999d).

overwhelmingly ectothermic, convert it to new biomass with relatively high efficiencies. Whittaker and Likens (1973) estimated total marine production at nearly 1.4 Gt C per year, three times their total of oceanic zoomass. Assuming, conservatively, a 20% consumption and 15% conversion rate would yield annually between 1.1 and 1.5 Gt C. Fish appear to account for only about a tenth of this net biomass production. Two estimates of fish production, published a generation apart, are fairly close: Schaefer's (1965) total was 1.08 Gt, whereas Houde and Rutherford (1993) estimated a live weight of 1.359 Gt. These values reduce to about 130–160 Mt C, roughly twice as much as a recently published estimate of annual production of mesopelagic cephalopods (Nesis 1997).

No aggregate totals of heterotrophic microbial production—continental, marine, or extremophile—will be offered here: they would be just guesses rather than uncertain, but reasonably bounded and hence acceptable, approximations. Microbial productivities are notoriously hard to measure in situ, the majority of removed organisms will not be culturable (although many nonculturable cells may be viable), and rates of CO_2 production based on cultures range widely (by at least three orders of magnitude for lab-incubated subsurface samples) and may grossly misrepresent the natural processes. Available estimates of large-scale bacterial production vary so widely that, for example, its total in various marine ecosystems has been put equal to anywhere between less than 1 and as much as 50% of phytoplanktonic NPP (Valiela 1984).

The best estimates of average turnover rates for soil prokaryotes are between 180 and 900 days, and applying these rates to the store of 26 Gt C estimated by Whitman et al. (1998) would result in annual production of 10–52 Gt C, an equivalent of between just 20% and 100% of terrestrial phytomass NPP. Ducklow and Carlson (1992) give a similar difference (26–70 Gt C per year) for all marine bacterial production. And we have virtually no reliable information regarding long-term productivities of most extremophiles: are their average turnover times just considerably slower than those of other nonextreme prokaryotes, or are most of their cells nonviable? Even if we had what might be considered a fairly representative sample of subterranean prokaryotic productivities, we could use it to estimate the right order of magnitude of the global production of these organisms only after narrowing the huge uncertainty regarding their actual biomass.

The Biosphere's Ethereal Presence

By necessity, any attempt to quantify global biomass and its annual production must be flawed and incomplete. A great deal has to be assumed, some important components must be left out, and a number of the order-of-magnitude uncertainties cannot be narrowed with confidence. The greatest weaknesses of our understanding concern belowground phytomass and its productivity, biomasses and productivities of microorganisms, particularly those of subterranean and subsurface extremophiles, and zoomasses and production of invertebrates. And there are also omitted data whose inclusion may make little difference to the overall stores or rates, but whose contributions are critical in their respective niches. Biomass and NPP estimates do not include either the chemolithotrophic prokaryotes, which are found in environments ranging from anoxic ocean sediments to hot sulfurous springs, or the endophytic fungi, ubiquitous mutualists found in virtually all plant species (Kuenen and Bos 1989; Saikkonen et al. 1998).

We are becoming masters of the smallest scales, assembling complete genomes, following the progress of chemical reactions by a femtosecond, and constructing nanomachines, but our fundamental understanding on the global scale remains poor. Nevertheless, some interesting conclusions are possible, as long as they are stated circumspectly. The Earth's phytomass stores are most likely between 500 and 600 Gt C, but the difference of some 300 Gt C between the recently published extreme estimates is still unacceptably large. The most likely phytomass stores are now perhaps as little as one-half, and almost certainly no more than two-thirds, of their preagricultural total. A large part of this massive loss has taken place only during the past 150 years, first because of the expansion of temperate cropping, and more recently because of tropical deforestation (Houghton 1995, 1999). This rapid transformation inevitably induces some sobering thoughts about the prospects for the Earth's plant cover.

Whatever the actual continental phytomass, standing phytoplankton is less than 1% of that total, but its turnover is 160–200 times faster than the average for land plant tissues. Consequently, the NPPs of these two very disparate phytomasses may be approximately equal, or the terrestrial plants may produce 10–25% more. The global averages of these rates prorate to annual productivities of 400–500 g C/m^2 of ice-free land and 125–150 g C/m^2 of ice-free ocean. The restrictions placed on marine photosynthesis by availability of macro- and micronutrients are well known, but of no less importance for this difference between terrestrial and marine NPP is a large disparity in absorbed photosynthetically active radiation (PAR). Because of competition from water and dissolved organics, phytoplankton absorbs a mere 7% of incident PAR, compared to just over 30% for land plants (Field et al. 1998).

With 500–800 Gt C, the biosphere's phytomass binds an equivalent of 65–102% of the element's current atmospheric content (786 Gt C, corresponding to 369 ppm CO_2). In terms of penetration of the Earth's physical spheres, eukaryotic biomass constitutes a mere 7×10^{-9} of the ocean's volume and less than 0.01% of the all carbon in the ocean. Subterranean and subsea prokaryotes, even when credited with very high cell totals, amount to just 6×10^{-7} of the top 4 km of the crust, the generally accepted depth limit for the survival of extremophiles. A uniform distribution of dry terrestrial phytomass over ice-free land would produce a layer about 1 cm thick; the same process in the ocean would add a mere 0.03 mm of phytoplankton (in both cases I am assuming the average biomass density equal to 1 g/cm^3). I know of no better examples to illustrate the evanescent quality of life.

This is perhaps the best place to correct another of Vernadsky's exaggerations, his claim that "generally radiation cannot reach any locality of the Earth's surface without passing through a layer of living matter that has multiplied by *over one hundred times* the surface area that would otherwise be present if life were absent from the site" (Vernadsky 1998, p. 79). Vernadsky's conclusion was based on unrealistic assumptions about the area of leaves compared to the ground area they cover: he believed, for example, that for temperate meadows this area is 22–38 times larger. This may be true for a small patch but not on an ecosystem scale. Modern studies of the leaf area index (LAI), the upper area of foliage per unit area of ground, show that large-scale means for grasslands and cereal crops are no higher than 2–3 during the months of peak leaf growth; average LAI is 3–4 for needle forests, and the highest values, for heavily canopied broadleaf ecosystems, range between 4 and 6 (Myneni et al. 1997; Buermann et al. 2000). Moreover, Vernadsky assumed that places "poor in life" constitute only 5–6% of the Earth's surface, whereas satellite observations show half of the continental area with an average LAI below 1 (fig. 7.9). Consequently, even when assuming, very generously, the annual global mean LAI at 1.5 and then doubling it for the lower side of leaves, green surfaces would be only about three times as large as the entire ice-free area of the Earth's continents.

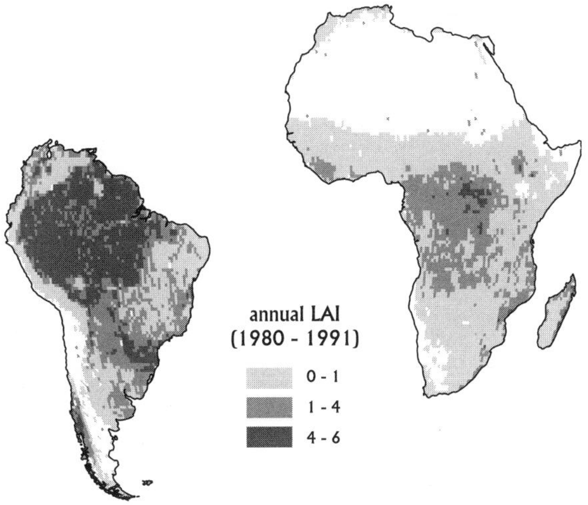

7.9 Comparison of average indices of annual leaf area in Latin America and Africa contrasts the dense Amazonian vegetation and less dense African tropical forests. Derived from a color image available at <http://earthobservatory.nasa.gov:81/Newsroom/Stories/Images/>

Accounting for the surface area of marine phytoplankton is much more uncertain. An acceptable order-of-magnitude estimate, assuming 3 Gt C of standing oceanic phytomass and, to err on the low side and hence to make the overall photosynthetic surface larger, an average cell diameter of just 10 μm, would give a total area of roughly 2×10^9 km^2, four times the Earth's surface. For comparison, Volk's (1998) estimate is 6 (range 3–8), and even that estimate is far smaller than Vernadsky's calculation of between 5.1×10^{10} and 2.55×10^{11} km^2. And so rather than being approximately equal, as Vernadsky estimated, to the surface of Jupiter or to as much as 4% of the Sun's surface, the aggregate area of the biosphere's marine phytoplankton equals perhaps only about 3% of Jupiter's and 0.03% of the Sun's surface. Although it is very difficult to estimate their totals, it will be the areas of fine roots and fungal hyphae and, naturally, the surfaces of prokaryotes that are, respectively, 10^1 and 10^2 times larger than the Earth's surface.

Redefining Biomass

Cogent arguments can be made for both drastically reducing and greatly expanding the definition of the Earth's biomass Restricting biomass to living protoplasm makes the biosphere even less substantial, because most of the terrestrial phytomass is locked in structural polymers and in cell walls, tissues that play essential supportive, protective, and conductive roles but are not alive. With forests containing roughly 90% of all phytomass, most of the necessary adjustment could be made by estimating the share of living cells in trees, but no simple generalizations are possible although it is safe to conclude that most of the mass in any mature living tree is not alive, it is difficult to suggest reliable large-scale corrective multipliers.

The radial extent of the cambial zone, the generator of tree growth, is difficult to define because of the gradual transition to differentiating xylem and phloem; most of the conducting tissue in trees is sapwood, with typically only 5–8% living cells in conifers and 10–20% living cells in hardwoods. Conversion of sapwood into nonconducting heartwood always involves death of the cytoplasm of all living cells in softwoods, but in hardwoods some cells in axial and radial parenchyma may remain alive not just for many months, but for years and decades. And there are substantial specific differences in the shares of total phytomass made up of fresh leaves, buds, young branches, and rapidly growing fine roots (Philipson et al. 1971; Reichle 1981; Shigo 1986; Gartner 1995; Waring and Running 1998).

In addition, trees and shrubs have dead branches and roots. Assuming that no more than 15% of all forest phytomass is alive is a generous allowance that would reduce the amount of terrestrial phytomass to less than 80 Gt C, and the real total may be less than 50 Gt C. In contrast, prokaryotic biomass needs much less adjusting for nonliving structures: cell walls take up no more than about 20% of the organism's volume in most bacteria (Cummins 1989). The global biomass estimate needs an unknown degree of reduction however, because of large numbers of nonviable cells, particularly for the abundant but often moribund subsurface prokaryotes. Available estimates of these organisms' viability show that even in shallow aquifers the percentage ranges from less than 1% to over 90% (Kieft and Phelps 1997). There is little to guide us in choosing the representative fraction needed to calculate the share of active microbial life. And even if we knew it, we ought to make some qualitative distinctions regarding the tempo of life.

Subterranean prokaryotes, be they in deep vadose zones, in rocks, or in consolidated sediments, may metabolize ten to fourteen orders of magnitude (!) more slowly than their counterparts in soils or shallow lake sediments (Kieft and Phelps 1997). Such organisms are masters of starvation survival, but it is hardly appropriate to combine their collectively large biomass with that of, in aggregate, much smaller but incomparably more active cells in root tips, leaves, or vascular cambium. But even a conservative estimate of all prokaryotic biomass (just a third of Whitman et al.'s [1998] more than 500 Gt C) and its subsequent halving would still mean that the protoplasm of prokaryotes is as large as that of plants. Kluyver's nearly fifty-year-old estimate of almost one-half of living protoplasm's being microbial is thus easily defensible, and it may actually err on the low side (Kluyver 1956).

On the other hand, the standard inclusion of all nonliving structural cells in total phytomass can be logically extended first to the addition of all dead biomass in standing trees, shrubs, and grasses, as well as in their roots, then to all accumulated surface litter and to huge stores of long-lived soil organic carbon. If structural polymers count because they provide indispensable services, then litter and soil organic matter should count no less as irreplaceable sources of recyclable nutrients. Undecomposed surface litter and standing dead plants may add close to 200 Gt C to the element's biospheric stores (Potter et al. 1993); soils contain at least 1,400 Gt of organic carbon (Schlesinger 1977; Zinke et al. 1986), and they may have in excess of 2,000 Gt of it (Amthor et al. 1998).

I will end my extension at this point, but a further progression along this enormous secular continuum of dead and transformed biomass would eventually end with all carbon in reduced compounds present in the crust: this extreme inclusion would push the total of organic carbon as high as 1.56×10^{16} t (Des Marais et al. 1992). Those who would wish to go yet another step further and start adding animal-built structures should consult the latest book on the extended organism (Turner 2000).

Another, much smaller extension of actual biomass can be constructed by adding viruses. They have, of course, no intrinsic metabolism, but they exert enormous influence on the survival and productivity of cellular organisms, especially in the ocean, where recent research have showed them to be the most abundant entities containing nucleic acids. The typical viral abundance of 10^{10} per liter in surface waters is five to twenty-five times the usual bacterial counts (Fuhrman 1999; Wilhelm and Suttle 1999). The size disparity (an average of 0.2 fg C per virus and 20 fg C per bacterial cell) between the two types of organisms means that the total mass of oceanic viruses is most likely less than 300 Mt C.

Broadly defined, biomass—an aggregate of living and dead cells, standing and circulating organic carbon, fresh and aged tissues, and entities ranging from viruses to metazoa—may thus amount to at least 2,200 and to as much as 4,000 Gt C. Of course, even this most liberal definition does not make biomass really abundant: when compared to the planet's other spheres, life on the Earth is, by any definition, ethereal—and yet so adaptable and persistent. The fact that we may be doing relatively little damage to the half (or more) of protoplasm of superabundant and admirably resilient prokaryotes is of little solace when we contemplate the consequences of tropical deforestation on the abundance and diversity of the eukaryotic life during the twenty-first century.

8

The Biosphere's Dynamics and Organization

Fundamental Rules and Grand Patterns

Cosmic energy determines the pressure of life that can be regarded as the transmission of solar energy to the Earth's surface.
Vladimir Ivanovich Vernadsky, *Biosfera*

It is interesting to contemplate a tangled bank, clothed with many plants of many kinds, with birds singing on the bushes, with various insects flitting about, and with worms crawling through the damp earth, and to reflect that these elaborately constructed forms, so different from each other and dependent upon each other in so complex a manner, have all been produced by laws acting around us.
Charles Darwin, *The Origin of Species*

The biochemical unity of life's structures, universally shared molecules and processes of genetic coding, and auto- and heterotrophic metabolism were described in some detail in second and third chapters. But life's underlying commonalities extend far beyond the molecular, cell, and tissue levels. Even casual observations reveal that the astonishing diversity of species is organized in a limited number of basic patterns ranging from small-scale plant communities to continent-spanning biomes, assemblages of life whose structure and function is largely determined by climate. And closer observations have shown that the lives of diverse species are governed by some reliably quantifiable, and hence highly predictable, regularities. This chapter will survey such grand patterns.

I will look first at perhaps the most remarkable set of laws of life at the organismic level, at the quarter-power scaling of animal and plant metabolism that applies across an entire range of body sizes and metabolic pathways. Then I will review the much less orderly process of energy flow through the biosphere, which involves transfers between successive trophic levels. In the second part of this chapter I will offer brief introductions to the biosphere's principal terrestrial and oceanic biomes, whose structure and dynamics have become better understood only during the second half of the twentieth century.

I will close the chapter by examining two sets of fundamental relationships that appear to have had an enormous influence on the evolution of life and on the persistence of the biosphere. The first deals with symbioses, various mutually beneficial arrangements whose effects are evident at levels ranging from intracellular organelles to ecosystems. The second topic of my concluding inquiry, the role played by life's complexity as a key determinant of biomass productivity and ecosystemic stability and resilience, is a contentious one, with experimental and empirical proofs offered to support either side of the dispute.

Energetics and Scaling

Compared to physics, which abounded in universally applicable laws long before the end of the nineteenth century, biology remained an overwhelmingly descriptive science—deficient in quantitative expressions that would generalize across broad spectra of organisms or ecosystems—well into the twentieth. In 1932 Max Kleiber changed that by publishing a simple double-logarithmic graph plotting body weights of various mammals against their basal metabolic rates (BMRs) and by deriving a generally valid equation for the relationship (Kleiber 1932). Subsequent studies using allometric equations, the principal tool of scaling studies, have provided a great deal of fascinating information regarding one of the biosphere's fundamental energetic imperatives.

And exactly a decade later came Lindeman's (1942) pathbreaking posthumous paper quantifying for the first time energy transfers between an ecosystem's successive trophic levels in terms of efficiencies (assimilation at level n/assimilation at level $n - 1$). This approach opened the way to understanding the performance of ecosystems in terms of large-scale efficiencies and ultimate limits. As so many other disciplines, ecological energetics matured rapidly after World War II, and by now it has quantified processes on levels ranging from molecules to ecosystems. The rapid extension of this knowledge to progressively smaller scales has been impressive, but the basic principles, laid down on organismic and ecosystemic levels more than half a century ago, have been neither fundamentally subverted nor greatly enriched by any new and universally valid findings.

Individual Metabolism

The most fundamental measure of the intensity of heterotrophic life is an organism's BMR: energy expenditure at complete rest, in a postabsorptive state (after the last meal was fully digested), and in a thermoneutral environment (digestion as well as shivering or sweating increases metabolism). An organism's area goes up as the square (l^2), and its mass (volume) as the cube (l^3), of its length, and so the body surface available for heat dissipation, and hence the BMR, could be expected to go up as the 0.667 power of its mass (l^2/l^3). But this is not the case: BMRs do not scale isometrically as a direct function of surface area (in which case the plotted function would be a straight line when plotted on linear axes); they instead progress allometrically, plotting as a curve on linear axes.

Kleiber, a Swiss German agricultural chemist trained at the Federal Institute of Technology in Zürich who joined the Animal Husbandry Department at the University of California at Davis to study the energy metabolism of animals, concluded that mammalian BMRs (expressed in watts) depend on total body weight (w in kg) as a function of $3.52w^{0.74}$. Although his initial data included organisms whose body weights ranged over three orders of magnitudes, it contained only a limited number of species. Gradually, the BMRs of many mammals were added to the original data set, extending the plot from Kleiber's rat-to-steer line to a mouse-to-elephant line without basically changing the exponent.

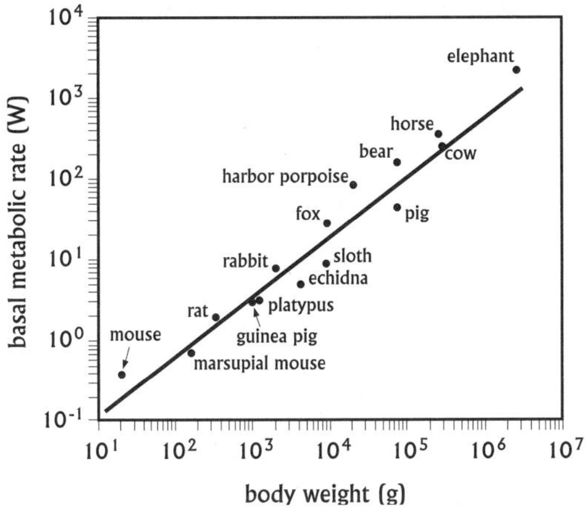

8.1 Kleiber's three-quarters law relates mass to basal metabolic rates and fits organisms ranging from prokaryotes to whales; this graph shows an excellent fit for mammals, the famous mouse-to-elephant line (Smil 1999d).

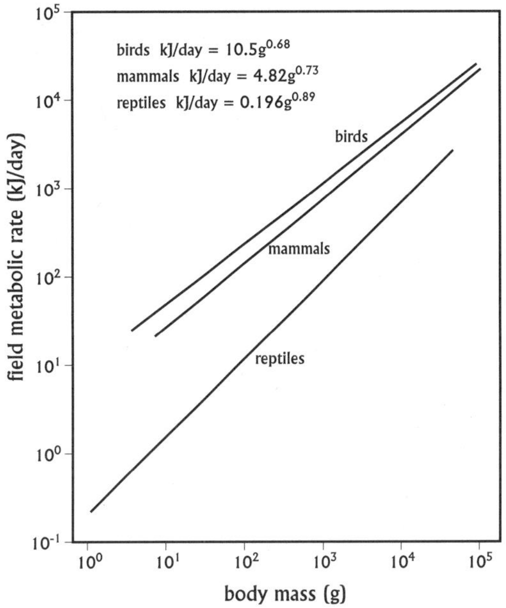

8.2 Departures from the three-quarters law among different taxa. The slopes are 0.48 for desert rodents, 0.59 for marsupials, 0.68 for birds, and 0.89 for reptiles. Based on Nagy et al. (1999).

Three decades after his original paper Kleiber opted for the three-quarters rule, recommending the expression of $70w^{0.75}$ for the BMRs in kilocalories per day, or $3.4w^{0.75}$ when the rate is given in watts (Kleiber 1961). The log-log mouse-to-elephant line and the three-quarters law became one of the most important and best-known generalizations in biology. Hundreds of BMRs have been measured subsequently, and the values are now available not just for mammals but for also for numerous invertebrates and microorganisms. Slopes for various invertebrate groups are close enough to the 0.75 line to conclude that the exponent is broadly representative for all ectotherms (Schmidt-Nielsen 1984). Consequently, the three-quarters law appears to apply to prokaryotes as well as to whales (fig. 8.1).

Not surprisingly, there are significant scaling differences between various taxa and dietary and habitat groups (Nagy et al. 1999; fig. 8.2). The most notable departures include greater allometric slopes for reptiles (0.89) and carnivorous mammals (0.87) and lower slopes for birds (0.68), marsupial mammals (0.59) and especially for desert rodents (0.48). As expected, there are also substantial differences between equally massive warm-blooded vertebrates (endotherms) and organisms that do not metabolically regulate their body temperature (ectotherms). Carrying a constant thermal environment allows endotherms to thrive in some thermally forbidding niches (polar regions, deep seas), but it exacts a high energy cost. Endothermic BMRs are twenty to forty times higher than are those of similarly massive ectotherms, who thermoregulate by selecting suitable thermal microenvironments and often resort to prolonged periods of cryptobiosis.

Field metabolic rates (FMRs) of wild terrestrial vertebrates measured by the doubly labeled water technique also reveal interesting differences in specific metabolisms. The slopes may be nearly identical, but the intercepts are relatively far apart (Nagy et al. 1999). The FMRs of marsupials are about 30% lower than those of the eutherian mammals, and those of Procellariiformes (storm-petrels, albatrosses, and shearwaters) are nearly 1.8 times as high as those of desert birds and about 4.3 times higher than those of desert mammals. The low FMRs of desert vertebrates reflect their adaptation to periodic food shortages and to recurrent or chronic scarcity of water. Some notable individual outliers, such as the very low FMRs of sloths, can be explained by obvious environmental adaptations, but there are also many unexplained specific departures from the norm.

Before I describe the latest major findings of scaling research I must make clear that I will not be referring to recently popular ecological scaling, which uses information on one spatial or temporal scale to infer attributes on another scale, be it from a leaf to the globe or from a cell to a landscape (Ehleringer and Field 1993; Van Gardingen et al. 1997). I will be concerned solely with allometric power functions that are used to express the change of features or attributes—be they structural, energetic, metabolic, or behavioral—in proportion to the scale of organisms or systems (McMahon and Bonner 1983; Peters 1983; Schmidt-Nielsen 1984; Brown and West 2000).

Scaling from Microstructures to Ecosystems

The enormous genetic variation among species is expressed in an astonishing diversity of specific sets of metabolisms, niches, behaviors, and life histories. But does life conform to any other rules as fundamental and as sweeping as those embedded in genetic codes? Kleiber's equation hinted at such a possibility, but it has taken several decades to appreciate both the ubiquity and the fundamental importance of scaling laws and to offer convincing explanations for their universality and simplicity.

The search for order across the huge size diversity on the organismic level takes the general form of

$$y = ax^b$$

where y is an attribute measured in relationship to the size of the organism, a is a specific constant (the multiplier needed to express the result in particular units), x is the variable of size (most often mass), and b is the scaling exponent that defines the relationship. Three properties enhance the appeal of this mathematically simple law: it tends to encompass the real world of organismic diversity with often a surprisingly small variance; when plotted on logarithmic axes the relationship is linear. And, as already noted in the case of Kleiber's law, it works across an enormous range of sizes.

The last attribute is the best confirmation of the universality of scaling laws: size is, of course, the most obvious attribute of any organism. Because the molecules of life as well as life's microstructures are shared by all organisms, biodiversity is to a large extent a matter of size. Molecules of life are, of course, identical throughout the biosphere. Basic organelles (mitochondria, chloroplasts) show typically no more than two- to fivefold differences in size, and cells span over nine orders of magnitude in mass. In contrast, the range from a *Mycoplasma* at about 0.1 pg or 10^{-13} g) to the blue whale (*Balaenoptera musculus*) at 100 Mg (100 tons), spans twenty-one orders of magnitude (appendix B). Comparisons limited to vascular plants and vertebrates span, respectively, about twelve (from aquatic weeds to sequoias) and seven orders of magnitude (from shrews to whales).

Organisms must adapt their structures and life strategies to compensate for important consequences of being small or large. Tiny organisms, whether airborne or locked in a droplet of water, live in a world of forbidding viscosities where gravity is unimportant; large bodies can propel themselves through air or water with little impediment because of the viscosity of those media, but gravity rules their lives. Tiny organisms acquire their nutrients by simple diffusion, large ones by targeted mass flows. Tiny organisms cannot ever maintain a constant inside temperature, whereas the large ones must resort to special structural or metabolic arrangements to lose their internal heat at an adequate rate. Yet Kleiber's three-quarters law appears to be valid across the entire size span, and its obvious corollary is that mass-specific metabolism (BMR divided by body mass) will follow a negative one-quarter law: with power in W and mass in kg, the equation is $3.4W^{-0.25}$.

A 10 g rodent will metabolize 10.8 mW/g, a 100 kg calf just 1.08 mW/g: a 10,000-fold jump in body mass reduces the specific metabolism by 90% or, conversely, supporting one gram of a tiny rodent will require ten times as much energy as maintaining a gram of a large ruminant (fig. 8.3). This reality limits the size of the smallest warm-blooded animals: creatures lighter than shrews and hummingbirds would have to feed incessantly to compensate for rapid heat losses. At the other extreme, the size of terrestrial mammals is not capped by the capacity to dissipate heat from massive bodies but rather by the structural limits of weight-bearing bones.

And because BMRs determine the intensity of life processes at levels ranging from cellular biosynthesis to an organism's growth, it is hardly surprising that so many other rates, starting with heartbeat and fast muscle contraction, scale as $w^{-1/4}$. An inevitable obverse of this relationship is that the times of these life processes scale as $w^{1/4}$: the heart-

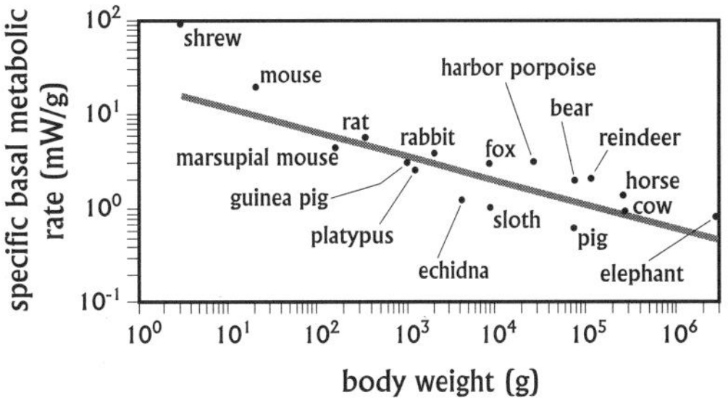

8.3 Declining specific basal metabolic rates (negative one-quarter law) for mammals whose body masses span almost six orders of magnitude (Smil 1999d).

beat of a 10-g rodent will be ten times faster than that of a 100-kg carnivore, and the maximum lifespan of a 100-kg herbivore will be ten times longer than that of a 10-g omnivore (fig. 8.4). And one-quarter scaling is not limited to heterotrophs. The total number of leaves, their surface area, the total number of branches, the cross-sectional area of trunks, and the overall rate of new tissue production are among the important attributes that scale as three-quarters power of total plant mass. The last relationships requires that relative growth rate decreases with increasing plant size as $w^{-1/4}$ (Enquist et al. 1999, 2000).

Although the interest in consequences of body sizes and in scaling is an old one, going back to Galilei's observations in the early seventeenth century and innovatively summed up by Thompson (1917) in his classic work on growth and form, allometric relationships were seen until fairly recently as purely empirical phenomena. McMahon (1973) offered the first explanation of the link based on mechanical requirements of animal bodies. Elastic criteria require a proportional relationship between the cube of the critical breaking length and the square of the diameter (d) of a loaded animal limb or spine. As the weight of these loaded

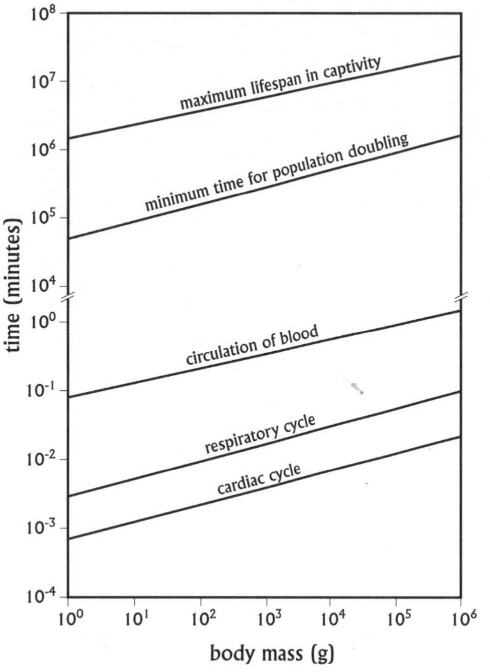

8.4 Examples of one-quarter scaling of processes among animals charting maximum life span (in captivity), minimum time for population doubling, circulation of blood, and respiratory and cardiac cycles. Based on Linstedt and Calder (1981).

members is a fraction of total body mass (w), their diameter will be proportional to $w^{3/8}$. The power output of muscles depends only on their cross-sectional area (proportional to d^2), and thus the maximum power output is related to $(w^{3/8})^2$, or $w^{0.75}$. If applicable to any particular muscle, the scaling should apply to the whole organism, and BMRs should be a function of $w^{0.75}$.

McMahon's (1973) mechanistic explanation of the three-quarters law remained unchallenged for more than twenty years. A more fundamental explanation, offered by West et al. (1997, 2000), is based on the geometry and physics of the network of tubes needed to distribute resources and remove wastes in organisms. They argue that rates and times of life processes are ultimately limited by the rates at which energy and material flows are distributed between the surfaces where they are exchanged and the tissues where they are used or produced. Three conclusions follow: distribution networks must be able to deliver these flows to every part of an organism; their terminal branches must have identical size, as they have to reach individual cells; and the delivery process must be optimized to minimize the total resistance and hence the overall energy needed for the distribution.

The first conclusion demands that many structures and functions of the delivery system—be they hearts or heartbeats—be scaled according to the size of organisms. Because the obviously tightly interdependent components cannot be optimized separately they must be balanced by an overall design which calls for a single underlying scaling. West et al. (1997, 2000) showed that the networks providing these fundamental services have a fractal architecture which requires that many structural and functional attributes scale as quarter powers of body mass, and that this requirements applies equally well to heterotrophs as well as to plants. The second conclusion regarding the invariant components of the delivery system is convincingly demonstrated by the identical radius of capillaries in mammals whose sizes span eight orders of magnitude (from shrews to whales), by the same velocity and blood pressure in their aortas and capillaries—and by the identical number of heartbeats per lifetime (Li 2000). The third conclusion rests on the economy of the evolutionary design: organisms develop structures and functions designed to meet but not to exceed the maximal demand.

Banavar et al. (1999) argued for a simpler explanation than the one offered by West et al. (1997), believing that the quarter-power law is an attribute of every optimally efficient network used to distribute nutrients. Hence there is no need to invoke fractality, merely to consider the need to feed roughly l^3 sites in a network distributing nutrients over the circulation length l. West et al. (1997, 2000) used their model based on the transport of essential materials through space-filling fractal networks of branching tubes to predict structural and functional properties of vertebrate cardiovascular and respiratory systems as well as those of insect tracheal tubes and plant vascular systems.

Finally, Gillooly et al. (2001) extended a general allometric model to all organisms. They concluded that once the resting metabolism is adjusted for mass and temperature, diverse species convert energy according to a predictable pattern and at a roughly similar rate, with the highest values, for endothermic vertebrates, being about twenty times above the lowest ones for unicellular organisms and plants. Enquist et al. (1998, 2000) also used this allometric relationship to develop a mechanistic model linking density and mass of resource-limited plants and found that the average plant size should scale as the negative three-quarters power of maximum population density (fig. 8.5).

Because the total amount of resources used per unit area is the product of population density (whose maximum scales as $w^{-3/4}$) and of the average resource use per individual (which scales as $w^{3/4}$), it then follows that the productivity of plants per unit area should not change with respect to their body size. This counterintuitive conclusion of invariant productivity in ecosystems with differently sized plants is confirmed by the fact that forests, grasslands, and crop fields sharing identical environmental conditions can be equally productive (Harper 1977).

Scaling phenomena are thus now well documented within organisms and among individuals, as well as among populations. Every complex organism combines invariant (or insignificantly varying) components (molecules, cells, capillaries) that are connected to optimally scaled structures that provide support, a metabolic framework, and neural links and that display the unmistakable self-similarity of hierarchically scaled fractal systems. Allometric relationships applied to species whose sizes span many orders of magnitude now include several thousand expressions scaling variables ranging from enzymatic functions to life spans and territory sizes. And scaling of populations helps uncover the limits to density and the frequency distribution of abundance.

But as ubiquitous as the quarter-power scaling (and its simple multiples) is in the biosphere, some relationships formerly claimed to exemplify it actually do not conform to it on closer examination. By far the most notable case of this mismatch involves the relationship between body mass and the average population density. This link appears to be nearly perfect for vascular plants: data for 251 populations of plants ranging from *Lemma* to *Sequoia* explain 96% of the $w^{-0.75}$ fit (Enquist et al. 1998). Enquist et al. noted that the fit is in agreement with comparable relationships in animals, a conclusion based primarily on Damuth's (1981) plots of mammalian densities and body masses (fig. 8.6).

Damuth's study showed a slope of –0.75, and as the BMR is proportional to the 0.75 power of the body mass, Damuth's rule implied that energy harvested daily per unit area is independent of the average body size of the feeding animals. No herbivorous species of mammals could thus outstrip another energetically only because of its larger size, and individual energy requirements would appear to be the ultimate determinant of maximum population densities. This hypothesis has been always controversial, because differences of up to three orders of magnitude can be found for identically massive animals with weights

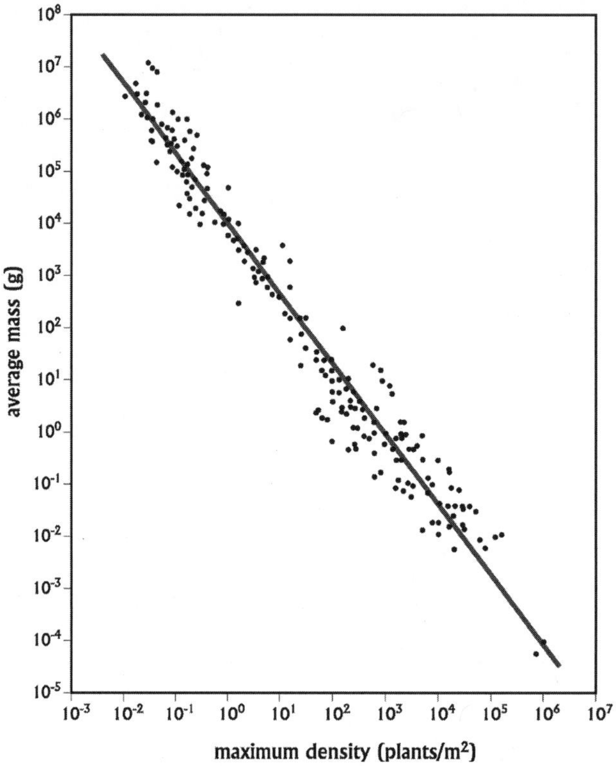

8.5 Relationship between average plant mass and maximum plant density for species ranging from *Lemma* to giant *Sequoia*. Based on Enquist et al. (1998).

between 0.1 and 100 kg, and as a result, size does not account for as much variance as in other scaling relationships (Calder 2000).

Cyr (2000) reexamined the link and found the exponents of global density–body size relationships steeper than predicted by the energetic equivalence hypothesis in both terrestrial ecosystems, with values between –0.92 and –0.98, and aquatic communities, where they ranged from –0.89 to –0.92, for populations of phytoplankton, zooplankton, and fish (differences explained largely by data sets used, e.g., area or volume densities of dominant species). Similarly, Schmid et al. (2000) analyzed the relation between population density and body size for more than 400 invertebrate species, ranging from testate amoebas to insects, in two European stream communities and found slopes practically indistinguishable from –1.0 when they used what they considered the best regression method.

Schmid et al.'s regressions also show substantial variations among functional and taxonomic groups (very shallow slopes for rotifers and crustaceans, steep slopes for

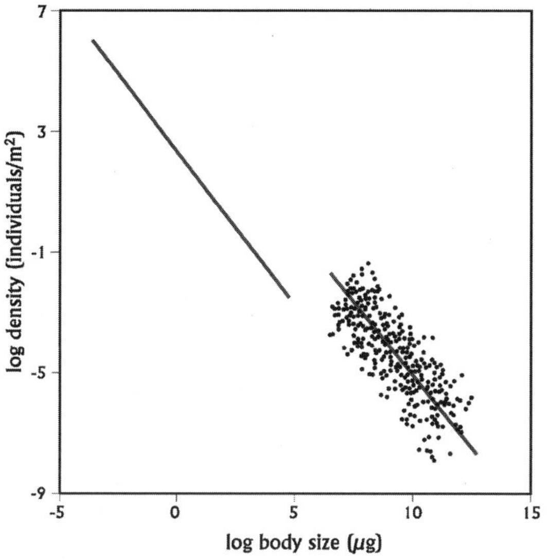

8.6 Inverse-scaling relationship between population density and body size. The upper line is the best fit for organisms ranging from amoebas to insects in two stream communities; the bottom line is the best fit for all mammals, often referred to as the Damuth line. Based on Schmid et al. (2000) and Damuth (1981).

hydracarina and insects), which means that the pattern cannot be explained only by the energy equivalence rule. Moreover, Marquet (2000) notes that when Damuth's (1981, 1991) mammalian data are fitted with the ordinary least squares bisector method, they too have a slope virtually identical with –1.0 rather than –0.75. Consequently, and not surprisingly, natural population densities are not simply a function of the energy needs of individuals.

Energy Flows through the Biosphere

Scaling laws make it possible to offer a number of fairly accurate quantitative conclusions relating properties of organisms to their body mass, but as far as the energy flows through the biosphere are concerned, we have not been able to come up with any such impressively orderly relationships, only with qualified approximations and broad qualitative conclusions. Elton (1927) was the first ecologist to recognize the declining numbers of heterotrophs in higher food web levels while pointing out the often-increasing size of those organisms. A decade later Hutchinson redefined the principle of the Eltonian pyramid of numbers in terms of productivity and opened the way for Lindeman's (1942) pioneering research on energy transfers among trophic levels.

Lindeman's (1942) study of trophic efficiencies in Minnesota's Lake Mendota showed the primary producers assimilating 0.4% of the incoming energy, while the primary consumers incorporated 8.7%, secondary consumers 5.5%, and tertiary consumers 13% of all energy that reached them from the previous trophic level. These results led to the formulation of the often-invoked 10% law for typical interlevel energy transfers, and Lindeman thought that the progressively higher efficiencies at higher trophic levels might represent a fundamental ecological principle.

Ten years after the publication of Lindeman's work, Odum began his detailed study of an aquatic ecosystem at Silver Springs in central Florida, which resulted in trophic efficiencies of about 16% for herbivores and, respectively, 4.5% and 9% for first- and second-level carnivores (Odum 1957). A decade later Teal (1962), in his study of a Georgia salt marsh, obtained trophic efficiencies of nearly 40% for bacteria, 27% for insects, and only about 2% for crabs and nematodes. These now classic studies, and a great deal of subsequent empirical evidence, make it clear that neither of Lindeman's original generalizations is defensible.

All we can do in the absence of any grand patterns is to describe the substantial variability in energy transfer rates and differentiate between various measures used to quantify the process. As detailed in chapter 4, photosynthetic efficiency, the share of radiation reaching the Earth's surface that is used in converting carbon in CO_2 to new phytomass, is at best about 4% when calculated for the PAR. After reducing that rate by carbon losses due to respiration, the efficiency of the NPP, that is, the phytomass actually available for heterotrophic consumption, averages only about 0.33% for terrestrial ecosystems, and because of the low rates of absorption of PAR by phytoplankton, less than 0.2% for the entire biosphere.

Energy losses between subsequent trophic levels are never that high but, on the average, they are definitely lower than 10%. Calculation of such losses must begin with determining exploitation efficiencies, shares of the production at one trophic level that are actually ingested by organisms at the level above it (the rest being accumulated and degraded by decomposers). Shares of the NPP actually consumed by all herbivores range from just 1–2% in many temperate forests and fields to 5–10% in most forest ecosystems. They peak at 25–40% in temperate meadows and wetlands, and, exceptionally, at 50–60% in rich tropical grasslands; in the ocean they can even surpass 95% for some patches of phytoplankton (Crawley 1983; Chapman and Reiss 1999). When the calculation includes only above-ground heterotrophs, the transfer is rarely above 10% in any temperate ecosystem, and when it is restricted to vertebrates, it is mostly around just 1%.

The next calculation step must consider a wide range of assimilation efficiencies, shares of ingested energy that are actually absorbed in the gut. Although there is no orderly scaling relating body mass to the efficiency of new zoomass production, there is an obvious link between the assimilation rate and the quality of the diet. Herbivores, subsisting on often digestion-resistant plant polymers, absorb their feed with low efficiency (often less than 30%), whereas carnivores absorb nutrients easily from their high-lipid, high-protein feed, and their assimilation efficiencies may surpass 90%. But these differences do not translate proportionately into less or more abundant zoomass whose biosynthesis depends on production (growth) efficiency, the share of assimilated energy that is neither respired nor spent on reproduction.

Here the generalization along the thermoregulatory divide is obvious. The theoretical maximum is about 95%, and actual top efficiencies are 50–65% for bacteria and 40–50% for fungi and protists. Ectotherms have high production efficiencies, with both herbivorous and carnivorous invertebrates channeling commonly more than 20% and nonsocial insects even more than 40 or 50% of all assimilated energy into growth. In contrast, the shares for social insects, with the much higher respiration requirements of their frenetic lives, are just around 10%, averages for large mammals are only about 3%, and the means for small mammals and birds are only 1–2% (Humphreys 1979).

When these shares are multiplied (exploitation × assimilation × production rates), the trophic, or ecological or Lindeman's, efficiency (the share of energy available at one level that is actually converted to new biomass at the level above it) may be only a small fraction of 1% or it can be well above 10%. Actual values of Lindeman's efficiency can thus depart quite significantly from the often cited typical 10% transfers, and it is also impossible to offer any representative large-scale means for the net energy transfer, as there is no evidence of predictable taxonomic, ecosystemic, or spatial variation for either grazers or predators.

The only permissible generalizations are that at the primary consumer level (excluding decomposers) the rates would fit overwhelmingly within the range of 1–10% and that the efficiencies of carnivores, whether invertebrate or vertebrate, are significantly higher than those of herbivores inhabiting the same environments. Nor should the high trophic efficiencies be seen as expressions of evolutionary success: they are significantly lower because of the maintenance of complex arrangements among social insects, and they are reduced even more by vertebrate endothermy. But the high energy price that is continuously paid for these arrangements has clearly been no obstacle to a massive penetration of temperate and tropical ecosystems by ants and termites, and it is obviously more than compensated for by competitive advantages enjoyed by endotherms of every size in every terrestrial and marine environment.

The higher metabolic needs of smaller endotherms result in some huge consumption-to-zoomass ratios, as populations of the smallest rodents can consume every year phytomass weighing more than 700 times than do their tiny bodies. In contrast, large and long-lived animals have a high capacity to accumulate zoomass, but they do so very slowly, and hence the energy flux through their populations is relatively low; for example, elephants' annual consumption-to-zoomass ratios are less than 10. These differences are also reflected in average production-to-zoomass ratios (turnover rates): as already explained, they are between 2 and 3 for small mammals and birds and a mere 0.05 for elephants (see fig. 7.8). Arthropods are the most prolific producers, with production-to-biomass ratios of up to 10 for termites, and their aggregate consumption rates may rival those of the largest herbivores, as many ant species harvest annually as much phytomass per unit area as do elephants.

Large differences in average lifespans and masses of producers and consumers in terrestrial ecosystems result in typically very broad-based trophic pyramids. Total phytomass is commonly twenty times more abundant per unit area than the mass of primary consumers, and zoomass in the highest (carnivorous) trophic level may be equal to a mere 0.001% of the ecosystem's phytomass. The short life spans of marine phytoplankton and high-energy and material throughputs among consumers usually reverse the layering in the ocean, producing inverted trophic pyramids, as the standing heterotrophic biomass is at least twice, and not uncommonly even three or four times, as large as the oceanic phytomass.

Biomes

Even somebody who had never heard the term or never seen it defined would understand the idea of a biome (bioclimatic zone) when, after traveling thousands of kilometers and crossing an ocean, an environment never seen before nonetheless felt completely familiar. Similar climates produce and maintain such distinct combinations of plant and animal life that their grand patterns may appear nearly identical, and only subtler clues will show that the site is on a different continent. Such familiarity may be due to identical tree species dominating near-identical forest communities, or it may come from impressive evolutionary convergence with near-identical environments molding the outward appearance of different species.

The first explanation is readily appreciated by comparing an undisturbed boreal forest in Europe and in Canada, whereas the plant communities of the Mediterranean biome offer an excellent illustration of the evolutionary convergence. This biome, found between 30° and 40° latitude in both hemispheres and molded by dry, hot summers and cool and rainy winters, is dominated mostly by evergreen (and often highly aromatic and fire-adapted) shrubs. Its disjointed patches are known by their specific regional names: the *chaparral* of California, the *maquis* of southern France, the *matorral* of Chile, the *fynbos* of South Africa's Cape region, and the *mallee* scrub of Australia. Although these floristic regions are highly endemic and there may be no species overlap between two of these floristic

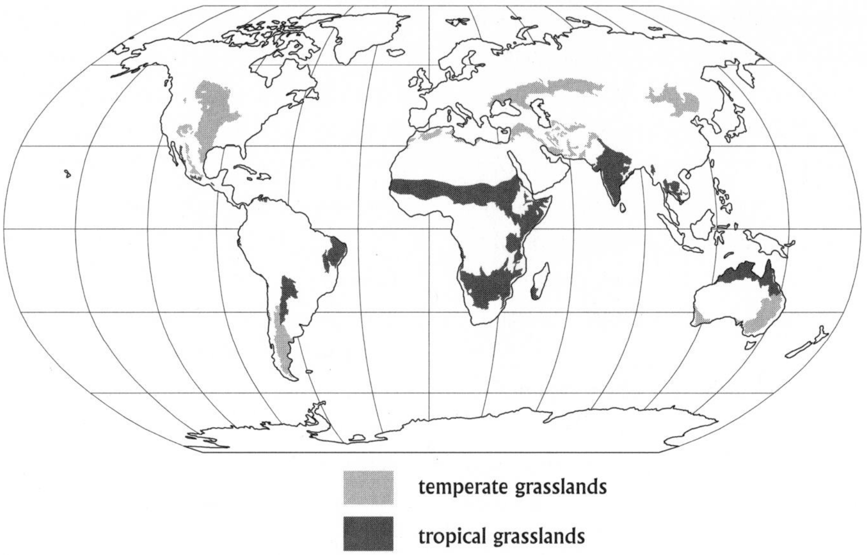

8.7 Approximate preagricultural extent of temperate and tropical grasslands. Derived from map available at <http://www.fao.org/forestry/fo/fra/gez2000.htm>

regions, the overall appearance conveys an unmistakable impression of a Mediterranean shrubland (Arroyo et al. 1995).

Because biomes are primarily the products of specific climates, their global distribution shows clear latitudinal regularities. The tundra, the least biodiverse biome of treeless but far from barren lands, skirts the northernmost coastlines of North America and Eurasia and envelopes the coastal regions of the lower portion of Greenland. Huge areas of the boreal forest (taiga) span the two northern continents from the Pacific coast to Nova Scotia and from Scandinavia to the Sea of Okhotsk. Temperate deciduous forests are naturally dominant in southeastern North America, throughout most of the Atlantic and Eastern Europe, and in parts of Northeast Asia.

Temperate grasslands are found in the arid centers of the two continents, but their extent is limited compared to that of tropical grasslands, which extend across Africa from Mali to Somalia and from Sudan to Mozambique, occupy large areas just south and north of the equator in Brazil and Venezuela, and dominate the environment in much of India and Australia (fig. 8.7). At their opposite ends the tropical grasslands blend into two very different biomes: subtropical deserts and tropical rain forests. Deserts are limited in both North and South America, but they span Africa from Mauretania to the Red Sea, continue eastward to northwestern India, and fill most of Asia between the Caspian Sea and north China. The tropical rainforest is the biosphere's most complex and diverse biome; its original extent encompassed not only all of Amazonia

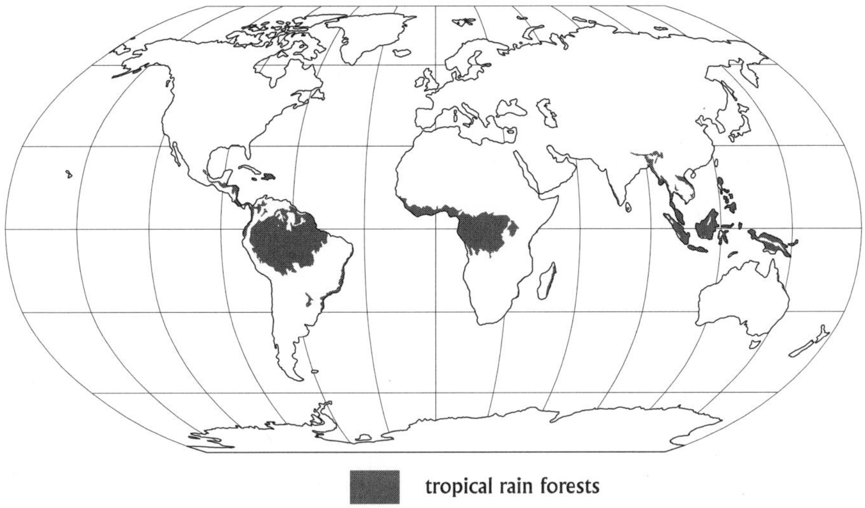

8.8 Maximum preagricultural extent of tropical rain forests. Derived from map available at <http://www.fao.org/forestry/fo/fra/gez2000.htm>

but also most of Central America and the Congo Basin, parts of West Africa, and large areas of continental Southeast Asia and the Indonesian archipelago (fig. 8.8).

The two most extensive marine biomes are the dark and cold abyssal waters that make up most of the ocean's enormous volume and the nutrient-poor surface waters of the open ocean. Both of these biomes have low densities of life compared either to nutrient-rich estuaries, where fresh water brings continental sediments and mixes with the salt water, or to the intertidal zone, repeatedly flooded and exposed by rising and ebbing waters. The ocean's two most diverse biomes, kelp forests and coral reefs, have a limited extent. Wetlands are the most complex freshwater biomes, containing much higher biodiversity than lakes and streams. What follows is not a systematic description of major biomes, but rather a selective survey stressing many of their unique features.

Extreme Environments

The most extreme of all biomes, unique assemblages of chemoautotrophs and heterotrophs congregating around hydrothermal vents on the sea bottom, has been described in some detail in chapter 6. Comprehensive surveys of this bizarre life domain make it clear that it is both more common and more productive than originally believed (Tunnicliffe et al. 1998; Van Dover 2000). Perhaps the best proof of the high productivity in this extreme environment comes from observations at a vent site along the crest of the East Pacific rise at a depth of 2,500 m that documented an extremely rapid colonization of the site by giant tube worms and their impressively rapid rate of growth, more than 85 cm a year (Lutz et al. 1994). Such rates are the fastest reported increments not only for any deep-sea organisms, but also for any known marine invertebrate!

Terrestrial environmental extremes seem mundane compared to this strange and disjointed hidden biome, but plants and animals living in tundra and in deserts must reckon with more than very low or very high temperatures. Life on the tundra involves coping with not just bitterly cold (minima often below –40 °C) but also largely dark winters that last for up to ten months. The growing season is usually over in less than ten weeks; no true soils lie above the permafrost, whose thawing and freezing often produces waterlogged patterned ground; and precipitation is limited to no more than 100 mm per year, with its distribution highly skewed.

Tundra vegetation may include dwarf trees (*Betula nana* and *Salix herbacea* may reach just 10 cm) and minishrubs, but it is dominated by tussock-forming grasses (*Poa, Puccinellia, Festuca*), mats or cushions of perennials, many of which produce ephemeral but colorful bloom carpets, and mosses and lichens (Wielgolaski 1997). Animal species are distinguished by inevitable adaptations to cold that include compact bodies, thick insulating fur and feather layers that turn white in winter, and summer accumulations of fat. Naturally, both biodiversity and NPP are very low in this biome, but the secondary productivity of small herbivores is often surprisingly high. The resulting cyclical fluctuations of rodent densities (lemmings, particularly genera *Lemmus* and *Dicrostonyx,* are the most notorious case) and predator populations are among the best-studied natural oscillations (Stenseth and Ims 1993). Arctic grazers include massive muskoxen (*Ovibos moschatus,* up to 400 kg) and large herds of migratory caribou and reindeer (*Rangifer tarandus*).

Deserts (associated with the subtropical latitudes, arid cores of continents, or rain shadows of major mountain ranges) have daily temperature maxima commonly above 30 °C and summer peaks above 50 °C. But it is the limited (less than 250 mm/year) and extremely variable and unpredictable precipitation that characterizes this biome (Evenari et al. 1985; Polis 1991; fig. 8.9). Small-leafed and often spiny or thorny shrubs with root mats extending beyond the canopies are the largest plants. Succulents, storing water in stems, leaves, roots or fruits, phreatophytes, plants with deep (commonly 2 to 3, and even up to 10 m) taproots, and perennials with bulky, water-storing tubers are also common. Cactaceae, ranging from tiny pincushions to majestic saguaro (*Carnegiea gigantea*) that can surpass 15 m, are the most diverse succulent family in the New World (Cullmann et al. 1986), whereas Euphorbiaceae and many species of Liliaceae fill the same niche in the old one.

Certainly the most remarkable phreatophyte is *Welwitschia mirabilis,* a gymnosperm growing in the Namib Desert of southwest Africa: it has just two succulent leaves, whipped into a frayed jumbled mass by desert winds, whose growth can support the plant for more than 1,500 years (von Willert 1994; fig. 8.10). At the other extreme of the longevity spectrum in deserts are the ephemeral plants that survive most of the time only as desiccation-proof seed and, once stimulated by a rare rain, can mature rapidly and produce new seeds in just a few weeks. Northern Arizona's Painted Desert is perhaps the most famous site of this ephemeral rain-dictated efflorescence.

Desert animals cope with extreme heat by being nocturnal or foraging or hunting during twilight hours, by hiding underground or seeking even the least available shade during the day, and by various body adaptations designed to maximize their heat loss and minimize their water loss.

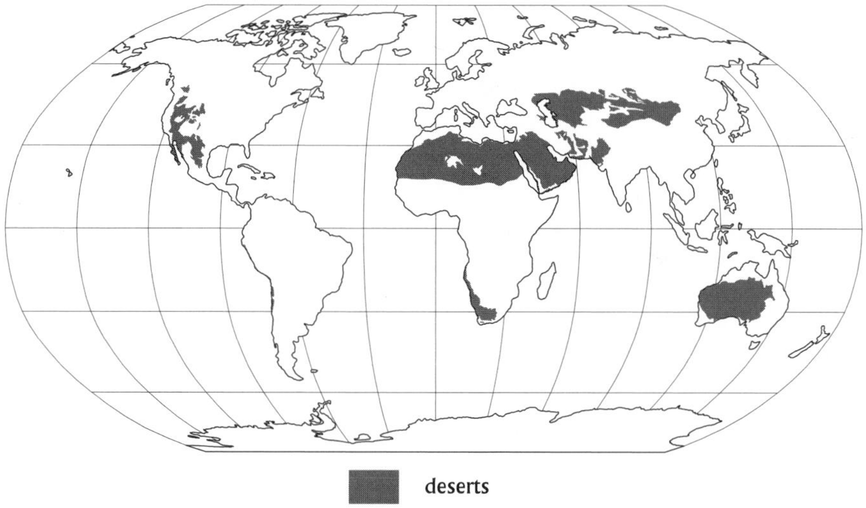

8.9 Current extent of subtropical and temperate deserts. Derived from map available at <http://www.fao.org/forestry/fo/fra/gez2000.htm>

The former type of adjustments are obvious in small bodies and in long ears and appendages, whereas the latter type include the use of countercurrent heat exchange in respiratory passages to cool exhaled air to temperatures much lower than body cores (both small rodents and camels can do this), the ability to live on air-dried food, production of concentrated excreta, and the tolerance of salt water (Schmidt-Nielsen 1964).

Forests

Biomes dominated by trees are both far-flung and diverse: in the Northern Hemisphere they range from the equatorial rain forests, the biosphere's richest terrestrial ecosystems, to marginal, species-poor growth of stunted trees above the Arctic Circle. No other ecosystem has elicited such an emotional response as forests. I will offer just a single quote, Darwin's summation after his 1831–1836 circumnavigation of the Earth:

Among the scenes which are deeply impressed on my mind, none exceed in sublimity the primeval forests undefaced by the hand of man; whether those of Brazil, where the powers of Life are predominant, or those of Tierra del Fuego, where Death and Decay prevail. Both are temples filled with the varied productions of the God of Nature. No one can stand in these solitudes unmoved, and not feel that there is more in man than the mere breath of his body. (Darwin 1897, p. 503)

As already noted (in chapter 6), forests also store the bulk of the Earth's biomass and account for most of the

8.10 *Welwitschia mirabilis* in its natural Namibian habitat: a mound formed by a very old specimen and a closer look at the split leaves of a younger plant. Photos courtesy of Dietrich J. von Willert, Institut für Ökologie der Pflanzen, Westfälische Wilhelms-Universität, Münster.

annual NPP. Not surprisingly, a great deal of research has been devoted to forest structure and function, and a number of summary volumes now also offer fascinating comparisons among the principal tree-dominated biomes (Persson 1980; Reichle 1981; Golley 1983; Landsberg 1986; Rohring and Ulrich 1991; Perry 1994; Waring and Running 1998; Barnes et al. 1998).

Conifers—often limited to just two of the four main genera of spruce (*Picea*), fir (*Abies*), pine (*Pinus*), and larch (*Larix*), or even to a single species—dominate the boreal forest (fig. 8.11). Most of the trees, able to survive minimum temperatures below –50 °C and up to half a year of average temperatures below freezing, have spindly trunks and narrow crowns to maximize snow shedding and limit branch loss under heavy snow loads. Many trees are malformed because of wind exposure or damaged because of waterlogging, as poorly drained bogs are common in glacial depressions. But it is a variant of the boreal forest, rather than a tropical growth, which has the world's highest accumulations of phytomass (Edmonds 1982).

Old-growth rain forests of the Pacific Northwest dominated by Douglas fir (*Pseudotsuga menziesii*) and noble fir (*Abies procera*) may store up to 1,700 t of biomass/ha and the maximum levels for coastal redwoods (*Sequoia sempervirens*) are at 3,500 t/ha of the above-ground phytomass! Even the richest tropical rain forests will store less than a quarter of this mass. These forests also have the largest living trees, indeed, the biosphere's most massive living creatures (although most of their phytomass is dead wood): giant sequoias (*Sequoiadendron giganteum*), growing to more than 90 m and able to live for over 3,200 years (fig. 8.12).

The boreal biome also includes such deciduous species as alder (*Alnus*), aspen (*Populus*), and birch (*Betula*) in its early successional growth. The temperate deciduous forest biome of the three northern continents, which shares various species of oak (*Quercus*), beech (*Fagus*), maple (*Acer*), chestnut (*Castanea*), and other genera of ancient Tertiary flora, is replaced on sandy soils by pines (*Pinus*), which are common throughout the U.S. Southeast and along the

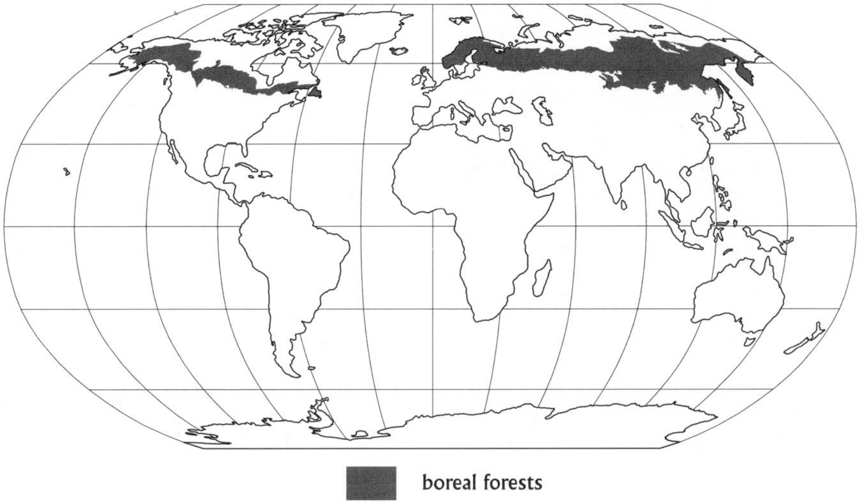

8.11 Maximum preagricultural extent of boreal forests. Derived from map available at <http://www.fao.org/forestry/fo/fra/gez2000.htm>

Atlantic coast both in the United States and in France. Regions of identical climate in the Southern Hemisphere are occupied by a mixed forest of an entirely different species composition.

The tropical rain forest can be monospecific, or dominated by a single family. Babassu palm (*Orbignya phalerata*) forms virtually pure stands on the Amazonia floodplain (Anderson et al. 1991), and forests in Southeast Asia and in the northeastern basin of the Congo River have most of their phytomass in trees belonging to the family Dipterocarpaceae (Connell and Lowman 1989). But a high level of biodiversity is the hallmark of this biome, with many large tree families, including Piperaceae, Moraceae, and Annonaceae, entirely or largely restricted to the biome. Each hectare in the central Amazonia may contain nearly 95,000 plants belonging to more than 600 species (Klinge et al. 1975), and the total above-ground phytomass ranges between 290 and 435 t/ha (Kauffman et al. 1995). Most of the tropical rain forest phytomass is stored in a few hundred trees belonging to about fifty species of the Leguminosae and Euphorbiaceae families, but this is still a much greater diversity than in temperate forests.

Tropical tree diversity acts as a passive defense against abundant seed predators and pathogens; in addition, tropical tree defenses include smooth barks, thick, coated leaves, hard woods, and symbioses with guardian ants. Some forests have a single tree stratum and a sparse ground cover of herbs, lichens, and mosses, but a layered structure is common. Five layers can be identified in the most diverse temperate deciduous forests, ranging from a ground layer of mosses and lichens through herb, shrub, and small tree layers to the surmounting tree stratum,

8.12 Giant sequoia (*Sequoiadendron giganteum*) in California's Yosemite Park. Photo from the author's collection.

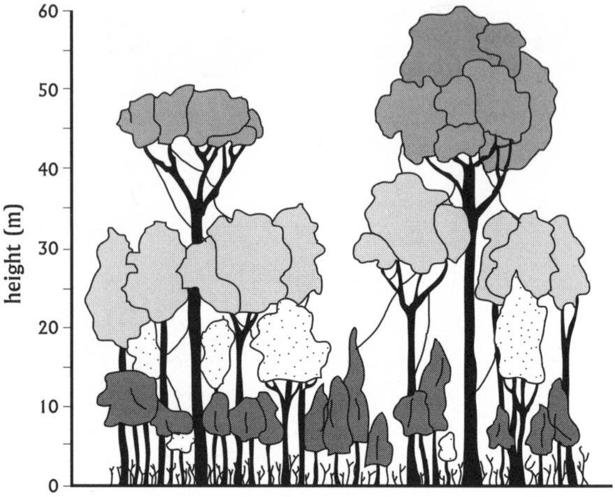

8.13 Layered structure of the tropical rain forest, crowned by protruding canopies of emergent trees (Smil 1999d).

whose canopies are mostly between 15 and 30 m above the ground. A sixth layer of exceptionally tall (30–40 m), widely dispersed emergent trees rising above the closed canopy level (at 20–25 m) is added in tropical rain forests (Brunig 1983; fig. 8.13).

Tropical forests are also distinguished by many species of buttressed trees, lianas (large woody vines), climbing vines (Vitaceae and Passifloraceae are prominent), and stranglers (some *Ficus* species are a particularly aggressive in that regard). Slow decomposition produces particularly thick litter layers in boreal forests, whereas the light-deprived floor in many tropical rain forests is comparatively bare. But tropical trees support an enormous variety of life in their canopies (Lowman and Nadkarni 1995). Vascular epiphytes growing on branches, particularly Orchidaceae (more than half of all epiphytic species), Araceae, and Bromeliaceae, are common, as are lichens, mosses, and free-living algae. Arthropod diversity in canopies is often stunning, with ants being the dominant family, often accounting for 25–40% of all biomass (Tobin 1995).

Large vertebrates, including herbivorous deer (*Cervus elaphus*) and moose (*Alces alces*), omnivorous bears (*Ursus americanus* and *Ursus arctos*), and carnivorous wolves (*Canis lupus*) and lynxes (*Felis lynx*), are relatively common in parts of the boreal biome. But large mammals are comparatively rare in tropical rain forests, where so many species are arboreal and hence have relatively small bodies: the jaguar (*Panthera onca*) and tapir (*Tapirus terrestris*) are the largest mammals in the Amazonian rain forest.

Grasslands

The deep and fertile soils that gave rise to temperate grasslands, exemplified by the famous Russian *chernozem,* have also been the main cause of their massive demise, as the natural extent of this biome has been drastically reduced by conversion of most of the U.S. Great Plains, Canadian prairies, Ukrainian, Russian, and Kazakh steppes, Argentinian pampas, and South African veld to cropland. In their undisturbed state these areas were under a continuous cover of perennial grasses (Graminae) and forbs dominated by species from two large plant families, Compositae and Leguminosae. Shortgrass formations consist of species growing no taller than 50 cm, whereas tallgrass prairies can exceed a height of 2 m. Temperate grasses, such as *Stipa* (spear grass) and tussocky *Lomandra* (iron grass), are admirably adapted to seasonal fire and drought, with most of their phytomass below ground. Root-to-shoot ratios in temperate grasslands can be as high as thirteen to one, and bulky root mats act as water storages and bind soils so tightly that they virtually eliminate water and wind erosion.

The mammalian diversity on preagricultural temperate grasslands was low, but migrating herds of *Bison bison* in North America and *Bison bonasus* in Europe represented enormous accumulations of zoomass. European bisons became extinct in their last two refugia in eastern Poland and northwestern Caucasus by, respectively, 1912 and 1927, with only fifty-four individuals surviving in zoos (Raczyński 1978). The total number of North American bisons was conservatively estimated at 30 million heads prior to 1830, but slaughters after that time left only some 300 free-ranging individuals by the early 1890s (Isenberg 2000).

In contrast to the great reduction it has brought about in the temperate grasslands, deforestation has been expanding the tropical grasslands. They now constitute the world's most extensive terrestrial biome, perhaps as much as 2.5 Gha, and the second-largest reservoir of phytomass after the tropical rain forest. The best-known regional expressions of this biome are East African savannas, Brazilian *cerrado,* and Venezuelan *llanos.* Tropical grasslands exist either as pure perennial grass formations or as open woodlands with varying densities of appropriately adapted (drought-, fire- and browse-resistant) shrubs and trees. Their presence clearly indicates that these are not climax communities: if not for fire and intensive grazing, a forest would occupy their place. Acacias and baobabs (*Adansonia digitata*) are the two signature trees of African savannas. Because the grassland root mats commonly account for 50–80% of all grassland NPP, many earlier studies that estimated or merely guessed at the below-ground production grossly underestimated the overall productivity of this biome. A revised total for tropical grasslands is up to 22 Gt C/year (Long et al. 1992), about twice as high as some of the earlier estimates of the biome's productivity.

Very few nonspecialists can name even a single grass species common in savannas: they include bluestem (*Andropogon*), annual or perennial *Themeda* (eighteen species in Africa, Asia and Australia, where it is called "kangaroo grass"), *Hyparrhenia* (jaragua or thatching grass), *Heteropogon* (tanglehead grass) and *Axonopus* (carpetgrass) (fig. 8.14). On the other hand, few images are as familiar as those of the largest terrestrial herbivores, above all elephants (*Loxodonta africana*) and massive herds of ungulates and their predators supported by grazing on African savannas. Wildebeest (*Connochaetes taurinus*), antelopes (eland, *Taurotragus oryx,* is the largest, weighing up to 900 kg, and Thomson's gazelle, *Gazella thomsoni,* the smallest, with

8.14 *Themeda* and *Hyparrhenia* are two common tropical grasses.

females weighing just 15–20 kg), and zebras (*Equus burchelli*) share the grassland without overgrazing it. Wildebeest, now numbering well over one million heads in the Serengeti, and other large species move around extensively after they have eaten down the vegetation to their maximum grazing depths, whereas the animals using the smallest depth can stay largely in the same place.

Studies in the Serengeti, where the mean amount of the NPP eaten by herbivores is about 80% (and the share is up to 95% in some areas), show that moderate grazing actually stimulates grassland productivity, raising it to twice the level of ungrazed plots (McNaughton and Banyikwa 1995). In contrast to African savannas, *cerrados* lack a distinct fauna, but tropical grasslands everywhere share high densities of termites, whose biospheric importance has already been stressed: the richest part of Serengeti supports more than 100 kg of herbivores per hectare, as well as 50 kg/ha of these social insects.

Ocean Biomes

The intertidal (littoral) zone, alternating between exposure and inundation of rocky, sandy, or muddy coastlines, is the most accessible of all ocean biomes. Only a few hardy seaweeds (*Fucus, Pelvetia*) live in the rocky zone, but aquatic animals well protected against dehydration (acorn and goose barnacles and crabs) may be fairly abundant. Littoral sands, nutrient-poor, ever-shifting, and abrasive, harbor a variety of tiny invertebrates and are visited by flocks of wading birds (plovers, curlews, sanderlings), as are mud flats, whose anaerobic layer, often filled with prokaryotic mats, starts just a few millimeters below their surface. Estuaries, mixing fresh and saline waters and receiving relatively large volumes of river-borne nutrients, are the only coastal biome able to support high levels of primary productivity.

Coral reefs, taking the forms of barriers, island fringes, and atolls, are the ocean's richest biome (Sorokin 1993). Their high level of productivity rests on the symbiosis of nocturnal polyps, cnidarians belonging to the same class of organisms as sea anemones and sea pens, with several species of microscopic protists of the genus *Symbiodinium* (Rowan and Powers 1991). All developmental stages of the swimming larval form of cnidarians already contain numerous spherical dinoflagellates (Glynn et al. 1994). These protists, 8–12 μm in diameter, are tinted brown by peridinin, a secondary photosynthetic pigment, and they translocate most of their photosynthate (as glucose and glycerol) to their hosts. Cnidarians have no centralized nervous system and only a simple blind gut, but they secrete a variety of protective limestone skeletons.

Most of the nearly 6,000 species of subtropical and tropical corals are colonial, and many of them tend to congregate in diverse communities, forming the marine counterparts of tropical rain forests. Coral reefs nourish and shelter an amazing variety of invertebrates and fish, including such improbably colorful and extremely poisonous species as *Amphiprion percula* (clownfish) and many bizarrely shaped organisms, some in constant motion, others perfectly camouflaged in prolonged immobility. Bellwood and Hughes (2001) found that despite steep longitudinal and latitudinal gradients in total species richness, the composition of reef corals and fishes in the Indian and Pacific Oceans is constrained within a rather narrow range, and that the regional diversity is best explained by large-scale patterns in the availability of shallow-water habitats.

In contrast to that of rich reefs, the biomass density of both the surface and deep waters of the open ocean is more akin to that of tundras and deserts than even to that of shortgrass savannas. The surface pelagic (open-sea) biome is limited by the depth of the euphotic zone: its maxima can extend below 200 m, but half of that extent is more common. The diameters of phytoplanktonic photosynthesizers range from 2 to 200 μm, and their communities are composed of various site- and region-specific shares of cyanobacteria, dinoflagellates, diatoms, and green algae (fig. 8.15). These autotrophs support an even greater variety of zooplankton. The biome's vertebrates range from tiny schooling fishes to the previously mentioned blue whales. The deep pelagic (abyssal) zone—the cold (about 3°C), permanently dark, and nutrient-poor region between the eutrophic layer and the ocean bottom—is the biosphere's most extensive, as well as most voluminous, biome. Throughfall of dead organic matter supports sparse populations of sightless and often also translucent invertebrates and strangely shaped fish, many of them possessing intriguing light-producing organs used for camouflage, detection, or signaling. As previously noted, the zone is repeatedly visited by a number of larger vertebrate divers, including seals and whales.

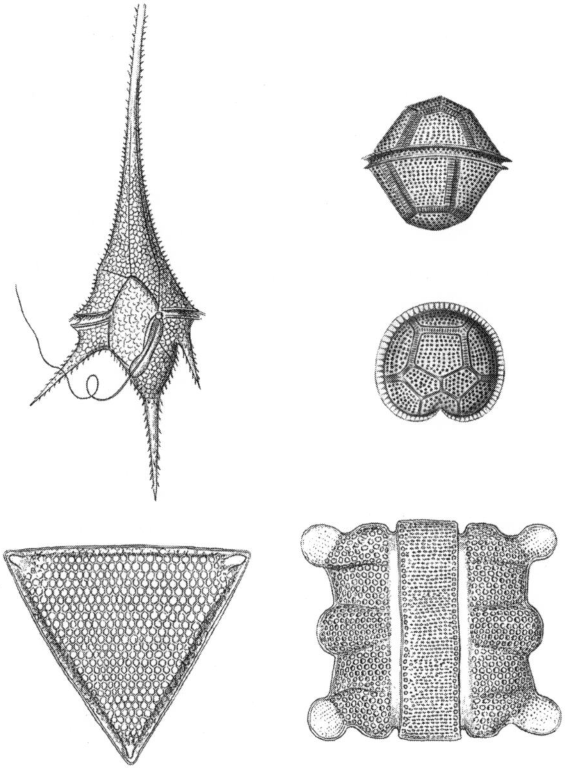

8.15 Four species of phytoplankton as portrayed in a classic volume by Oltmanns (1904). Clockwise starting with the top left image: *Ceratium macroceras,* a dinoflagellate from warm seas; two views (back and from below) of another dinoflagellate, *Goniodoma acuminata;* and two diatoms, *Biddulphia pulchella* and *Triceratium.*

The ocean bottom (benthic) biome is similarly divided into a shallow part confined to continental shelves (less than 200 m deep) and the ocean abyss. Shelf benthos is often rich in worms, mollusks, crustaceans, and fish, but by far the most spectacular shallow benthic communities are created by a giant brown algae (*Macrocystis pyrifera*) attached to the bottom and sprouting fronds up to 50 m long. These kelp forests provide an ideal shelter for many species of invertebrates and fish that feed on their detritus.

(The most remarkable deep benthic assemblages of life, hydrothermal vent communities, were described earlier in this chapter and in chapters 3 and 6.)

Symbioses and Complexity

The popular reduction of Darwin's complex thinking into a terse slogan ("survival of the fittest") continues to foster the idea of untrammeled competition as the driving force in the evolution of the biosphere and the emergence of new species. Competition's powerful effects are easily demonstrated by myriads of examples, but the consequences of cooperation and sharing have been even more profound. Competition is just one of the forms of symbiosis, the close and long-lasting association of two organisms of different species, or individuals of a single species. Some of its most remarkable demonstrations are reciprocal evolutionary changes as two populations are engaged in a race to avoid extinction. Ehrlich and Raven (1964) named this coevolution, and studies of plants and the organisms that attack them offer many excellent examples of this ubiquitous process.

In other symbioses, both species can remain unaffected (neutralism), one species can benefits and the other remain unaffected (commensalism), or one species can benefit while the other one is harmed (parasitism). In this chapter I will discuss only mutually beneficial relationships, whose frequency is much higher and whose impact is much larger than is commonly realized. Consequently, throughout the remainder of this book I will use the term "symbiosis" only in its mutualistic form. I will look first at the importance of symbioses in the evolution of complex organisms, and only then will I review the major categories of mutualism. Several key examples illustrate the biospheric indispensability of these symbiotic links between organisms. Perhaps the two most important symbiotic relationships on the organismic level are the sharing of resources between *Rhizobium* bacteria and their leguminous plant hosts and between plants and fungi attached to their roots. The first symbiosis relieves the shortages of the key nutrient whose supply limits terrestrial photosynthesis; the second enhances the uptake of water and of relatively immobile nutrients. Less ubiquitous, but no less remarkable symbioses abound. As noted in the preceding section, there are numerous symbioses between tropical marine invertebrates and dinoflagellates, above all those between polyps and *Symbiodinium,* creating ecologically and esthetically amazing coral reefs. And neither the digestion of cellulosic organic matter by large mammals nor the decomposition of phytomass in warm climates could proceed without symbiotic links between herbivores—ruminants in the first case, termites in the other—and their microbial endosymbionts.

Mutualism in the biosphere goes far beyond the myriads of ongoing interspecific actions, as ancient symbioses appear to be responsible for the very evolution of eukaryotic life. Mitochondria and chloroplasts, two key organelles in eukaryotic cells, have their origins in a symbiotic association with bacteria. This idea goes back to the work of two Russian biologists at the end of the nineteenth and the beginning of the twentieth century. After being determinedly promoted by Margulis (1970, 1993) the idea has received clear support from recent phylogenetic studies. In closing this varied chapter I will review one of the most interesting controversies in modern ecology, the strength of the relationship between higher levels of species diversity and higher levels of primary productivity and the greater resilience of ecosystems.

Mutualism in Ecosystems

The multitude of mutualistic relationships—they may be the most common links in the biosphere—falls logically into several functional categories (Janzen 1985). Pollination mutualisms between insects, birds, and bats and flowering plants may the best-known instances of rewarding symbioses. Harvest mutualisms embrace either the gathering or the processing of nutrients (or both). The most obvious demonstration of dispersal mutualisms is the diffusion of seeds by birds feeding on fleshy fruits and nuts. And the most common (though generally invisible) protective mutualisms involve fungi living inside plant leaves, stems, and barks. Of course, both agriculture and animal husbandry can be seen as special cases of a peculiar mutualism whereby human actions help crops and animals diffuse and thrive in return for rewarding food harvests.

Plant-pollinator mutualisms are usually diffuse, as foraging species visit a large variety of flowering plants. Bees are the world's best-known pollinators, and their importance for agriculture is fundamental, as they pollinate more than 70% of the world's top 100 crops (Nabhan and Buchmann 1997). Open flowers are accessible to any pollinator, whereas those with concealed nectar require the ability to crawl inside or to hover and probe with elongated tongues (as butterflies and moths do) or beaks (like hummingbirds or honeycreepers). Specific mutualisms can arise in the case of trees that flower year-round: various figs (*Ficus*) pollinated by tiny females of the gall-wasp *Blastophaga psenes* are perhaps the best-known example of pairwise plant-pollinator mutualism (Proctor et al. 1996).

Perhaps the most remarkable trait of endosymbioses is the combination of relatively few species of microorganisms with a great diversity, usually at least ten fold, of hosts (Douglas 1995). As previously explained (chapter 5), symbiotic fixation of nitrogen is carried out by bacteria belonging to just six small bacterial genera associated with more than 17,000 leguminous plants. Fewer than forty genera of algae and cyanobacteria form lichens, with nearly 1,600 genera of ascomycete fungi. And about 120 species of arbuscular mycorrhizal fungi are associated with an estimated total of 200,000 plants. Only the number of different species of ectomycorrhizal fungi nearly matches the number of their hosts.

Mycorrhizae are very common harvest mutualisms of plants and fungi, with some 90% of all plant species, and every conifer, being symbiotic with at least one or more kinds of fungi (Allen 1992). In woody perennials the fungal hyphae of one or several among thousands of species of *Basidiomycotina* and *Ascomycotina* wrap around the host's roots and penetrate the intercellular root spaces. Fruiting bodies of these fungi range from highly prized above-ground king bolete (*Boletus edulis*) to even more valuable underground black truffles (*Tuber melanosporum*). Vesicular-arbuscular mycorrhizae formed by *Zygomycotina* and common in the tropics produce knoblike swellings and bushlike hyphae that penetrate the host's root cells.

In either case the fungi receive sugars produced by photosynthesis and in return provide several highly valuable services: above all, they enhance the uptake of water and mineral nutrients, particularly of the relatively insoluble phosphates. They also protect roots against many pathogens, and their hyphal networks bind soil particles into large aggregates and link neighboring plants in nutrient exchange networks. Mycorrhizae were previously thought to be stable derivatives of initially antagonistic interactions between plants and parasitic fungi. The latest phylogenetic analysis of symbiotic and free-living (mushroom-forming) fungi confirm that symbioses with diverse hosts have evolved repeatedly from parasitic precursors, but they also show many reversals to free-living forms: mycorrhizae are thus unstable, evolutionarily dynamic symbioses (Hibbett et al. 2000).

Digestive symbioses are certainly not as widespread as mycorrhizae, but as noted above, without this type of harvest mutualism, termites could not become the principal decomposers of the enormous mass of detritus in warm climates. Lower termites (*Microtermitidae*) are able to digest cellulose only thanks to the flagellate protists living in their hindguts; bacteria provide the same service for the higher termites (*Macrotermitidae*). When Joseph Leidy looked at the content of subterranean termite (*Reticulitermes flavipes*) gut under a microscope, he was astonished to see that swarming "myriads of parasites . . . actually predominated over the food in quantity" (Leidy 1881). *Trichonympha, Pyrsonympha,* and *Dinenympha* are the most commonly present genera among scores of flagellate endosymbionts identified in the intestines of the lower termites, with some insects harboring more than ten different protist species. Without their enzymes, the biosphere's nearly 250 quadrillion termites could not process between one-third and two-fifths of all phytomass in their tropical and subtropical habitats, that is, perhaps as much as one-quarter of all terrestrial NPP (Zimmerman et al. 1982; fig. 8.16).

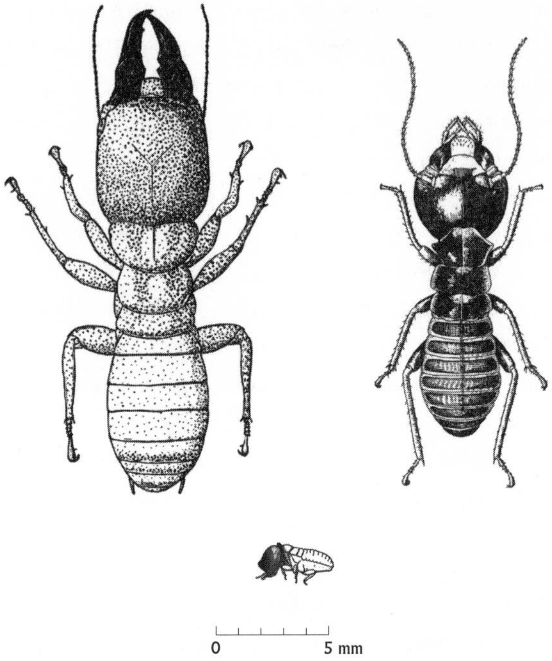

8.16 Three examples of termites. Clockwise starting with the top left image: a large soldier of the damp-wood termite (*Zootermopsis angusticollis*) from the Rocky Mountains; a smaller worker of *Hodotermes mossambicus* from East Africa; and a tiny soldier of *Cryptotermes havilandii,* a species native to Africa but now also found in the Caribbean and South America. Based on Harris (1971).

Termites are also myrmecophylous, with almost every termite nest harboring mutualistic ants. And some higher termites have evolved a more elaborate, extracorporeal scheme of phytomass breakdown by cutting leaves, dragging them into special chambers, and using them as a substrate for the cultivation of fungi. Tropical leaf-cutter ants (*Atta* and *Acromyrmex* are their two major genera) practice the same remarkable way of feeding (Hölldobler and Wilson 1990; fig. 8.17). And some ants (including *Formica* and *Lasius*) tend and protect aphids and scale insects that they "milk" for their carbohydrate- and amino acid-rich secretions (honeydew). I will mention just one more instance of a digestive symbiosis, remarkable for its high degree of specificity. Sap-sucking aphids are great destroyers of leaves, but their carbohydrate-rich diet contains only one amino acid, glutamine, in abundance; aphids convert it to glutamic acid, which is taken up by *Buchnera aphidicola,* and this intracellular symbiont, energized by carbohydrates provided by its host, uses the acid's nitrogen to synthesize all other essential amino acids needed for the host's growth and reproduction (Douglas 1998). This link is perhaps as much as 250 Ma old, and neither organism can now exist without the other.

Dispersal symbioses are particularly important for diffusing tree seeds, both in temperate forests, where over 60% of tree species are dispersed by animals, and in tropical ecosystems as well, where many canopy mammals join birds in eating nuts and fleshy fruits. In addition to dissemination through discarding and defecation, seeds are

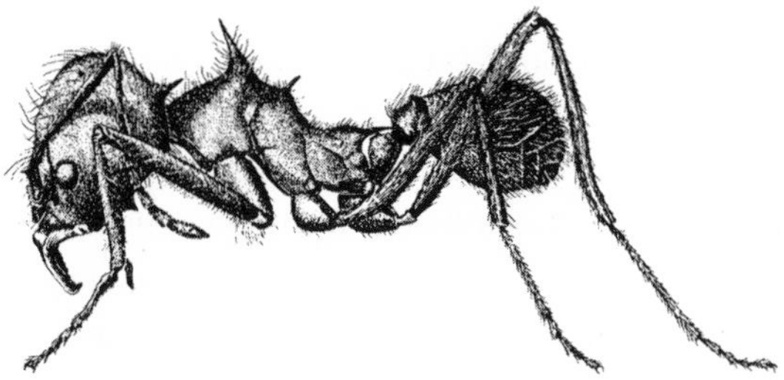

8.17 Tropical leaf-cutter ants belonging to genus *Atta* are the biosphere's most numerous cultivators of underground fungi. Based on a drawing reproduced in Hölldobler and Wilson (1990).

also dispersed by cache-building birds and rodents that either forget to retrieve their stores or waste the seeds in the process of storage or removal. Jays are particularly active horders, with individual caches containing thousands of seeds (Perry 1994).

Perhaps the most impressive protective mutualism is that of ants and swollen thorn acacias. An ant queen colonizes a young thorn acacia tree by laying eggs in swollen stipules, feeds on nectar secreted from the base of the tree's leaves, and gathers tiny nodules from the tips of the leaves that contain protein and lipids for larvae. Once a sizeable colony develops, swarming ants attack any insects that alight on the plant, as well as any climbing vines, and they often even clear a bare circle around the trunk (Janzen 1966). Removal of the ants result in the early death of the tree, which is otherwise defenseless. Invisibly, but no less effectively, fungal endophytes, protected inside trees, inhibit the growth of pathogens and infestations by bark beetles. Some acacias house several ant species that either coexist or occupy a tree in succession (Young et al. 1997).

In closing, I will provide just one more of many possible examples of near-magical symbioses. The bobtailed squid, *Euprymna scolopes,* spends its days buried in the sand and hunts among shallow reefs during the night—a potentially risky endeavor, as the moonlight would silhouette its body for carnivorous fish swimming below it. The squid effectively camouflages itself, however, by collecting luminescent bacteria of *Vibrio fischeri* into its light organ, where they lose their flagella and produce light that makes the squid blend better with moonlit waters (Visick and McFall-Ngai 2000).

Symbiotic Evolution

Decades of indefensible preoccupation with the modes and outcomes of interspecific competition, as well as the recent stress on random mutations as the only source of organismic novelty, have overshadowed the undeniable fact that mutualism, no less than selection through exclusion or the dictate of selfish genes, has been a key mechanism of evolutionary innovation. By far the most important expression of this ancient process is the very formation of eukaryotic cells through symbiogenesis, the emergence of new species via the genetic integration of endosymbionts.

The idea of symbiogenesis, indeed the term itself, arose from the work of two Russian scientists, both of them Vernadsky's contemporaries. Andrei Sergeevich Famintsyn (1835–1918)—botanist, university professor in St. Petersburg, and member of the Imperial Academy of Sciences—was the first biologist to notice the similarity between chloroplasts and cyanobacteria and concluded that these organelles had a symbiotic origin (Famintsyn 1891). Konstantin Sergeevich Merezhkovsky[1] (1855–1921)—a

1. As in Vernadsky's case, the proper transcription is "Merezhkovskii," and the author used "Mereschkowsky" in his German publications. Khakhina (1992) provides the best survey of the history of the Russian symbiogenetic research.

professor at the Kazan University who trained as a zoologist but was also active in anthropology and died in exile in Geneva—further developed the idea of endosymbiotic organelles derived from bacteria (Merezhkovskii 1909).

The work of these two biologists was well known in Soviet Russia, but even more so than Vernadsky's post-1917 writings, Merezhkovsky's and Famintsyn's ideas were largely ignored abroad. They received a wider exposure only with the 1959 reprint of Wilson's (1925) book, and especially after Lynn Margulis began enthusiastically promoting the symbiogenetic or, using Taylor's (1974) term, which she prefers, serial endosymbiosis hypothesis (Margulis 1970). After a period of doubt, this hypothesis received indisputable confirmation from the flood of phylogenetic studies whose advances were described in chapter 3.

In the early 1980s Christian de Duve (winner of the 1974 Nobel Prize in physiology and medicine for his work on structural and functional organization of the cell) still wrote cautiously about the endosymbiont hypothesis as a theory that "has much to commend it" but declined "to settle an argument that is still dividing experts" (de Duve 1984, pp. 149–150). A dozen years later he noted that "proofs of the bacterial origin of mitochondria and plastids are overwhelming" (de Duve 1996). After an anaerobic protist engulfed a respiring bacterium, the resulting chimera between symbiont (bacterium turned mitochondrion) and host led to the evolution of most, and perhaps even all, existing eukaryotes (Levings and Vasil 1995). Later a respiring eukaryote acquired a second endosymbiont, almost certainly a cyanobacterium, which gave rise to the chloroplast and endowed some eukaryotic cells with photosynthetic capability (Reith 1995).

All plastids (not just pigmented chloroplasts, the sites of photosynthetic conversions, but also nonpigmented organelles involved in fatty acid and amino acid biosynthesis) and mitochondria thus originated from endosymbioses between a bacterium and a eukaryotic host. Their conversion from independent organisms to organelles was accompanied by a tremendous reduction in the size and gene content of the original genome, and most of the organelle's proteins are now encoded by nuclear genes (Reith 1995; Moreira et al. 2000). Delwiche (1999) gives the following example: the genome of the cyanobacterium *Synechocystis* contains 3,200 genes, the plastid of the red alga *Porphyra purpurea* has just 250 genes, and that of the liverwort *Marchantia polymorpha* has less than half that many. Most of these gene transfers took place early in eukaryotic evolution, but Adams et al. (2000) demonstrated that whereas the functional transfer of mitochondrial genes has ceased in animals, there have been repeated recent and diverse transfers of mitochondrial genes to the nucleus in flowering plants.

The best current evidence also suggests that there are only three lineages of photosynthesizers equipped with primary plastids: Glaucocystophyta (a small group of uncommon unicellular freshwater algae); ancient Rhodophyta (red algae); and all Chlorophyta (green algae and plants) (Delwiche 1999). Several groups of algae acquired their plastids through secondary endosymbiosis when a plastid-containing eukaryote was incorporated into another eukaryotic cell. Heterokontophyta (kelps, diatoms, and chrysophytes) and Euglenophyta (unicellular aquatic protists) belong to this group, but there are even tertiary endosymbionts among diatoms and dinoflagellates. And there is also a new hypothesis about the origin of eukaryotes. After comparing the biochemistry of energy metabolism, Martin and Müller (1998) suggest that eukaryotes originated through the symbiotic association of an anaerobic autotrophic hydrogen-dependent archaeon acting as the host and a symbiotic bacterium that could respire but released H_2 as a waste product of its anaerobic heterotrophic metabolism.

The biosphere's evolution is unimaginable without symbioses. We see them in the very formation of eukaryotic cells, in the intricate coevolutionary patterns in coral reefs—where about fifty fish and shrimp species act as cleaners of ectoparasites, often entering even into the gill chambers and mouth of the host fish—and in flowering plants and their pollinators. Without endosymbioses there would be no cattle husbandry and beef empires, and termites, those miniature tropical cows, could not process a large share of the biosphere's litter fall.

Lichens—symbioses in which green algae, cyanobacteria or both photobionts enclosed inside a fungus provide it with carbohydrates—are among the oldest expressions of terrestrial mutualism (fig. 8.18). About one-fifth of all existing fungal species form these ancient symbioses, and today's major lineages of nonlichen-forming Ascomycota are actually derived from lichen-forming ancestors that eventually ceased to be symbiotic (Lutzoni, Pagel, and Reeb 2001). Moreover, lichens also commonly contain microbial communities (diatoms and chrysophytes) and hence may be thought of as miniature ecosystems (Schwartzman 1999). And given the fact that nearly all plants have some mycorrhizal association, it is not too far-fetched to call grasses and trees, as did Pirozynski and Malloch (1975), "inside-out lichens." Simply, as Margulis concludes, "if we care to, we can find symbiosis everywhere" (Margulis 1998, p. 5).

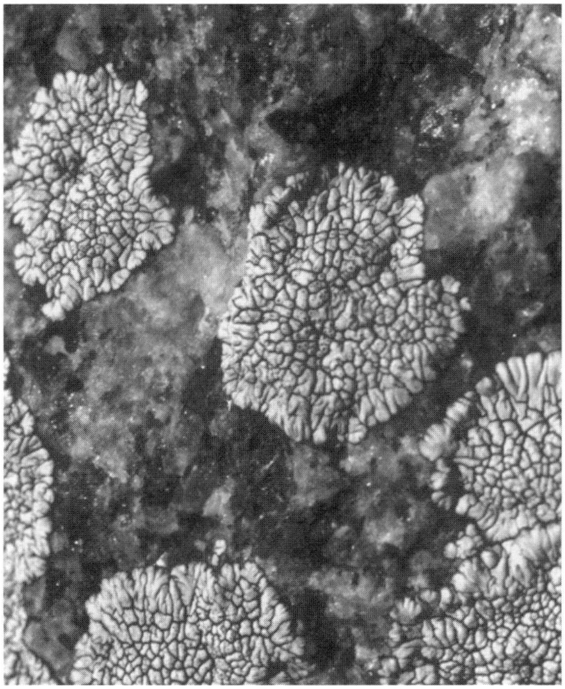

8.18 A crustose lichen (*Dimelaena oreina*) growing on granite in the Canadian Rockies. Macrophoto from the author's collection.

Diversity, Productivity, and Stability

Hypotheses that connect diversity, productivity, and stability are easily stated. A greater variety of species in an ecosystem makes it more resilient, buffering it against environmental disruptions. Conversely, a more diverse ecosystem makes better use of the site's production potential and hence yields more biomass than a less biodiverse community in the same environment. If these widely held opinions were invalid, it would much more difficult to make a persuasive case for the maintenance of the highest possible natural diversity and for its restoration in degraded ecosystems. But finding conclusive proofs for these two hypotheses has not been easy, and the debate concerning the links between biodiversity and ecosystem properties and functions has yet to find a broadly acceptable middle ground. The definitional problems are obvious.

Grimm and Wissel (1997) reviewed 163 definitions of seventy different stability concepts and concluded that the general term "stability" is so ambiguous as to be useless. They suggest using one of six better terms, including "constancy" (staying essentially unchanged), "resilience" (bouncing back to the reference state), "persistence" through time and "resistance" (staying basically the same in spite of disturbances). And because ecosystems may be described by many variables, all of these attributes must be evaluated in relation to a particular property. Stability of what? is thus the key question. Of the existing species variety, or entire populations? Of basic ecosystem structure, that is, of species richness, evenness of distribution, and food web complexity? Or of such key processes as photosynthesis or nutrient cycling? In any case, naturally changing and evolving environments, through more or less cyclical processes or sudden catastrophic shifts, make the use of even well-defined concepts of ecosystemic stability elusive. The demise of some key species or aggressive invasions may irretrievably change the structural aspects, but the integrity of essential ecosystemic functions may be maintained equally well by an entirely different set of species. Both empirical observations and theoretical predictions cut both ways.

Some of the world's most extensive (and even ancient) ecosystems, including boreal forests, bogs, and heathlands, are not notably species-rich. As noted earlier in this chapter, apparently stable natural mono- or oligocultures exist even in the tropical regions. Mayr's (1974) models predicted that species-poor ecosystems should be more stable than the species-rich ones, and evidence from different environments has also shown a unimodal relationship between diversity and productivity, with the former declining over the highest productivities (Rosenzweig and Abramsky 1993). In contrast, highly diverse tropical rain forests are seen as the prime example of stability underpinned by complexity, whereas the susceptibility of modern agroecosystems to pests or droughts is cited as the obvious consequence of their simplification.

A wave of new experiments conducted during the 1990s intensified these long-standing debates about the links between diversity, productivity, and resilience. Two studies opened up the controversy. Experiments with model plant communities grown in enclosed chambers with added insects and worms at the Ecotron in southern England measured the biomass yields of the communities in relation to the number of species (Naeem et al. 1994). And Tilman and Downing (1994) looked at the effects of diversity on key ecosystem properties (primary productivity, nitrogen mineralization, and litter decomposition) in synthesized plant assemblages at Cedar Creek in Minnesota.

The Ecotron studies found a link between biodiversity and total biomass yield, and species-rich grassland plots were found to be more drought-resistant than species-poor ones. Huston (1997) criticized both the inappropriate experimental design of these studies and the indefensible interpretation of resulting data. Soon afterward, similar studies showed different outcomes. Continued work with experimental grassland plots in Minnesota indicated that ecosystem properties are related to differences in the functional characteristics of the dominant plants rather than to greater species diversity (Tilman et al. 1997). And comparisons of fifty relatively untouched islands in an archipelago in the Swedish boreal zone showed that higher microbial biomass and more rapid rates of litter decomposition and nitrogen mineralization coincide with the lower levels of plant diversity and the earlier successional state of the vegetation on the larger islands (Wardle et al. 1997). The functional characteristics of the dominant plants, rather than their variety, would then be the principal controllers of ecosystem productivity and biogeochemical intensity, a conclusion that was first reached in the early 1980s (Lepš et al. 1982).

A very different approach to diversity-productivity studies relies on manipulating laboratory-scale microcosms, and it, too, has come up with more nuanced results. McGrady-Steed et al. (1997) manipulated microbial biodiversity in aquatic microcosms and concluded that ecosystem respiration became more predictable as biodiversity increased. Experiments by Naeem et al. (2000), who manipulated both algal and bacterial species in laboratory microcosms, did not support a simple relationship between diversity and productivity. Systems with moderate to high algal diversity had a higher combined productivity than systems containing a single algal species, but their bacterial biomass was relatively low regardless of bacterial diversity. Bacteria in highly productive systems used a greater variety of carbon sources, but this dependence was determined not solely by bacterial diversity, but also by greater algal variety. And Sankaran and McNaughton (1999) concluded on the basis of their controlled field experiments in savanna grasslands in southern India that low-diversity plant communities can have greater compositional stability than the high-diversity ones when subjected to experimental perturbations. They argue that those extrinsic influences, rather than diversity itself, may be the primary determinants of certain measures of community stability.

So far the most extensive diversity-productivity experiment followed synthesized grassland communities at eight European sites, ranging from Sweden to Portugal and from Ireland to Greece, for two years (Hector et al. 1999). The study concluded that in addition to a great deal of expected local variation, there was as well an overall log-linear reduction of average above-ground biomass with loss of species: each halving of diversity lead to a 10–20% reduction in productivity. Tilman (1999) hailed Hector et al.'s work as a landmark study, but its critics believed that sampling effect and the addition of a legume to some sites explain much of the reported link and argued that the experiments offer no evidence that having 200 species is any better than having just 50 (Huston 1997; Kaiser 2000).

Perhaps the best foundation for assessing this controversy is to approach the problem from a large-scale empirical point of view. Waide et al. (1999) reviewed about 200 cases of published relationships between productivity and species richness and found that the link was unimodal in 30% of all cases, positively linear in 26%, negatively linear in 12%, and insignificant in 32%. Their review of the literature concerning deserts, boreal and tropical forests, lakes, and wetlands led them to conclude either that the existing data are insufficient for any conclusive resolution of the link between diversity and productivity or that both patterns and underlying mechanisms are varied and complex. This may remain the best possible judgement for a long time to come.

Finally, I should mention the link between growth and biodiversity. After working in both tropical rain forests and coral reefs, Huston (1993) found that the slower the rate of growth in a particular community is, the higher is its level of diversity. Terrestrial growth rates are, of course, directly related to soil quality, and hence levels of plant diversity should be higher on unproductive soils, whereas fertilized soils produce more biomass of lower diversity: indeed, this pattern is found throughout the world under a wide variety of conditions. Controversially, Huston (1993) also interpreted this inverse relationship between species diversity and plant productivity as one of the key determinants of the wealth of nations, as many temperate-zone countries with the best soils are inherently better off in terms of agricultural productivity than species-rich, soil-poor tropical societies.

9

Civilization and the Biosphere

The Earth Transformed by Human Action

Man, alone, violates the established order.
Vladimir Ivanovich Vernadsky, *Biosfera*

Man has long forgotten that the earth was given to him for usufruct alone, not for consumption, still less for profligate waste.
George Perkins Marsh, *Destructiveness of Man*

The biosphere is a product of constant change. Of course, the rates and impacts of this change vary greatly as periods of surprisingly conservative evolution are followed by truly revolutionary shifts in the composition of species or abundance of biomass or as incremental and localized changes get suddenly swept away by new environments brought about by the impacts of large extraterrestrial bodies (see chapter 3). Comparisons of conditions on the Earth 4 Ga ago (just before the first prokaryotes began forming the biosphere) and 4 Ma ago (before the first individuals of our genus inhabited East Africa) show how profoundly the evolution of life had modified the third planet of the solar system before the appearance of our species.

Above all, the Earth's atmosphere was changed in the intervening time because of the accumulation of molecular oxygen generated by water-splitting bacterial and plant photosynthesis. In turn, the oxidizing atmosphere made it possible to shield the biosphere against UV radiation with stratospheric O_3 and to produce such abundant compounds as the (now economically indispensable) metallic oxides and biogeochemically important soluble nitrates and sulfates needed for biomass production. Plant and microbial metabolism also became a new source of CO_2, CH_4, and N_2O, greenhouse gases that were previously liberated only by inorganic processes. Decomposition of biomass introduced a number of other trace compounds, ranging from NH_3 to DMS, whose presence and atmospheric reactions affected the properties of the medium on a variety of spatial and temporal scales.

Life also molded the appearance, structure, and chemistry of the planet's surfaces and composition of its waters. The Archean and Proterozoic eons saw the rise of prokaryotes remarkably adapted to often extremely hot environments. Phanerozoic evolution clothed most of the land and filled the oceans with more than a quarter million plant species and an even greater diversity of invertebrates. Dense terrestrial vegetation changed substantially the albedos of many continental surfaces. The penetration of the crust's topmost layer by plant roots and associated symbiotic prokaryotes and fungi and the gradual accumulation of dead organic matter formed the Earth's most productive soils. The combination of physical and chemical effects of terrestrial vegetation accelerated the rates of weathering, withdrawing more atmospheric CO_2 and hence lowering the surface temperatures and changing the composition of fresh waters, but it also reduced the denudation rates, thanks to the ground cover and root mats, which provide very effective protection against water and wind erosion.

Life in the ocean, sustained primarily by myriads of tiny photosynthesizers and invertebrates, changed the planet's marine environments through its rapid processing and transfers of carbon. Sedimentation and burial of a small fraction of biomass formed huge deposits of hydrocarbons (whose extraction now supports our high-energy civilization) and led to the oxygenated atmosphere. And marine life changed the planet's appearance through its cumulatively astonishing biomineralization, which has amassed enormous volumes of carbonate sediments, and subsequent tectonic processes turned many of these sediment into the Earth's most formidable mountain ranges. Lovelock (1972, 1979) went even further, considering the Earth to be a superorganism capable of homeostasis on a planetary scale maintained by life on the surface. He labeled the notion of the biosphere as an active adaptive control system the "Gaia hypothesis" and suggested that the feedback between cloud-seeding DMS production by marine algae and higher cloudiness could have acted through time as an effective global thermostat (Lovelock et al. 1972).

As already noted, the very direction of this particular feedback appears questionable (Watson and Liss 1998; see chapter 5 and fig. 5.14), and Lovelock's entire Gaia hypothesis has met with reactions ranging from enthusiastic acceptance (Williams 1996; Volk 1998) to justifiable skepticism (Schneider and Boston 1991; Gillon 2000). Lovelock (1988) eventually refined his hypothesis by introducing a simple mathematical model, Daisy World, wherein the competitive growth of light- and dark-colored plants keeps the climate of an imaginary planet constant and comfortable. Perhaps the most obvious arguments against such a homeostatic control of the biosphere by life are the recurrent episodes of near-total extinctions and other, less drastic and more frequent biosphere-altering climate changes that have repeatedly shifted the Earth's conditions far from constant and comfortable. The evolution of the planet's oxygenating atmosphere is another key example of not just uncomfortable but distinctly lethal change visited upon the anoxic Archean biota by the evolution of cyanobacteria.

This example leads to another obvious question of what Gaia is supposed to accomplish. Is it to maximize the standing biomass, biodiversity, or primary productivity, or all of them? Lovelock (1988) suggested that "Gaia is best thought of as a superorganism . . . made up partly from living organisms and partly from nonliving structural material" (p. 15). He concluded that this superorganism has the capacity to regulate its temperature and that its health can be measured by the abundance of life. Given the paltry biomass and productivity rates over vast desert areas as well as in the ocean's vast nutrient-poor waters, Williams's (1992) caustic response to this is that "Gaia must be sick indeed over most of the Earth" (p. 483). And geneticists refuse to be converted to the idea that evolution can select for environment-altering traits that will benefit not just the individual but the entire biosphere. This is perhaps the most fundamental challenge to be answered by Gaian enthusiasts: how does natural selection at the individual level produce self-regulation at all higher levels up to the planetary one? But there is no doubt that the search for proofs and refutations of Gaian homeostasis has added to our understanding of the biosphere as scientists of many disciplines have looked for global-scale coupling and coevolution of biota and their environment (Volk 1998; Schwartzman 1999).

The biosphere is thus unlikely to be a superorganism, but there is no doubt that some of the changes brought by life have been obvious enough to be detectable even from space. Flybys of the Earth by the Galileo spacecraft in December 1990—it came as close as 960 km—provided a unique control experiment in remote detection of the Earth's life (Sagan et al. 1993). A widely distributed surface pigment with a sharp absorption edge in the red part of the visible spectrum and atmospheric methane levels far beyond thermodynamic equilibrium were the two main observations suggestive of life, whereas the detection of narrow-band, pulsed, amplitude-modulated radio transmission appeared as a unique sign of intelligence.

This remote-sensing experiment also illustrates the taxon-specific nature of the biosphere's evolution. Some organisms absorbing in the red part of the visible spectrum, most notably the hypobradytelic cyanobacteria, have retained their fundamental characteristics across billions of years, and others have remained largely unchanged for hundreds of millions of years. In contrast, the species responsible for the amplitude-modulated signals has undergone an extremely rapid encephalization that moved it from small groups of hesitant, vulnerable, naked bipeds on the plains of East Africa to more than six billion individuals inhabiting every terrestrial biome and dominating the biosphere through their technical prowess. This remarkable evolutionary achievement is also the most worrisome of all problems immediately facing the biosphere. As I will explain before closing this book, the biosphere's eventual fate, whether in the presence or absence of *Homo sapiens,* is sealed by the inexorability of cosmic forces, but it is its immediate future that appears to place our species at perhaps its most decisive evolutionary junction.

This conclusion is not an indefensible exaggeration based on catastrophists' fears but a simple description of realities based on the recent evolutionary history of our species. The diffusion and complexification of human societies have led to a large array of environmental changes that have transformed this planet during the past 5 ka, and particularly during the past 100 years, more rapidly than any other biogenic process in the planet's history. Even the early foraging societies could leave a profound, although relatively shallow, mark on their environment by widespread use of fire and by mass killings of herbivorous megafauna. Both the rate and the depth of the anthropogenic environmental change accelerated with sedentary agriculture, whose practice was often preceded by large-scale deforestation and maintained by intensifying inputs of nutrients and water.

The rise of urban civilizations expanded the environmental footprint of human needs, but whereas the local or regional-scale consequences of these changes were profound, their global impacts were either nonexistent or very limited until the advent of industrial civilization. At the beginning of the twenty-first century it is easy to argue that there is no place on this planet that is untouched by human action, that the entire Earth is now to a certain degree a human artifact, and that the extent and degree of this interference can only increase to accommodate additional billions of people and, even more importantly, their rising material consumption. The unprecedented combination of very rapid rates of change and truly global impacts is the source of justified concern (Turner et al. 1990; Smil 1993, 2000a; Moore et al. 1996; Beckel 1998; World Resources Institute 2000). I will merely highlight here the latest understanding of key transformations subsumed under larger categories of land cover and soil changes, loss of biodiversity and invasions of species, and anthropogenic interferences in the global cycling of water, doubly mobile elements, and minerals.

Genetic engineering is yet another process with potentially immense implications for the fate of the biosphere. Its techniques allow for much faster changes in the makeup of species than natural evolution or traditional plant and animal breeding, but its eventual global impact remains highly uncertain. Will its future resemble the fate of nuclear electricity generation (whose enormous promise and early successes were not enough to sustain the technique), or will designed species take over a large part of the biosphere by this century's end? All we can say now is that rapid and widespread introductions of numerous transgenic organisms have the potential of changing the biosphere in ways perhaps even more troubling than the concurrently unfolding global climate change.

Global Transformations

A brief definitional aside first. The adjective "global" has been used indiscriminately to refer to two fundamentally different kinds of environmental impacts. The first category includes many local or regional environmental changes whose widespread distribution makes them global (or nearly so) but whose increasingly negative effects or, conversely, improving trends in one location have little or no direct, near-term impact on the situation in other localities, regions, or continents. Such ubiquitous problems as excessive soil erosion in both traditional and modern cropping, depletion and pollution of aquifers, eutrophication of lakes and coastal waters, photochemical smog generated by transportation and industrial emissions, and bioinvasions of species belong to this category.

The adjective "global" is more fitting for those anthropogenic interferences whose eventual effects have planetwide impacts no matter where they originate and whose moderation and successful management has global benefits. The Earth's changing radiation balance, which may result in a rapid global warming, and a major loss of stratospheric ozone, which would expose life to higher doses of UV radiation, are the two most obvious examples in this category.

The following reviews of anthropogenic changes of the biosphere aim to cover only the most important and the most worrisome phenomena in both categories, stressing the latest findings and, where appropriate, major regional or continental differences and diverging trends. Uncertainties complicating any assessments of long-term environmental consequences of these changes will be stressed in the following section.

The Changing Atmosphere

There is, fortunately, little we can do to alter the atmosphere's N_2 or O_2 content. Atmospheric N_2 used as the feedstock for synthesis of NH_3 amounts to an insignificant share of the element's huge tropospheric stores, and nearly all of the nitrogen used in such synthesis is eventually returned by denitrification. Even complete combustion of all known fossil fuel resources would lower the tropospheric O_2 by less than 2%, a marginal reduction to which life could easily adapt (Smil 1991). Anthropogenic risks to atmospheric integrity are thus a matter of changing concentrations of trace compounds. These aerosols and gases are either directly emitted from combustion of fossil fuels and biomass and from various industrial processes or are formed subsequently by reactions in the atmosphere. Widespread atmospheric changes include the emissions of particulate matter and of acidifying compounds and the recurrent formation of photochemical smog resulting from complex atmospheric reactions of compounds emitted from combustion of hydrocarbons in large urban and industrial areas (Colbeck and MacKenzie 1994).

Most of the thousands of air pollutants either are present only in trace quantities (parts per billion or even trillion) or their distribution is spatially very limited. Particulate matter (PM, solid or liquid aerosols with diameter less than 500 μm) and carbon monoxide are the most abundant air pollutants in global terms. Anthropogenic aerosols, mostly from fossil fuel combustion and biomass burning, now account for about 10% of the total atmospheric loading (NASA 1999). Thanks to extensive controls (electrostatic precipitators), fly ash from coal burning has become a minor part of these emissions, which are now dominated by sulfates and nitrates (secondary PM generated by oxidation of sulfur and nitrogen oxides) whose small size (0.1–1 μm) is responsible for most of the light scattering. Their greatest impact has been in the northern polar latitudes, where they are the primary constituents of the Arctic haze, a thick (up to 3 km) brownish or orange pall covering a circumpolar area as large as North America (Nriagu et al. 1991).

Inefficient fossil fuel combustion is not the only anthropogenic source of carbon monoxide: low-temperature combustion of phytomass generates CO-rich plumes that can be detected by satellite observations above parts of Africa, Asia, and Latin America distant from any major urban or industrial centers (Newell et al. 1989). SO_2 originates in the oxidation of sulfur present in fossil fuels and in metal sulfides. Its annual global emissions rose from about 20 Mt S in 1900 to nearly 70 Mt by the late 1970s. Subsequent controls in Western Europe and North America and the collapse of the Communist economies reduced their emission growth rate, but Asian emissions have continued to rise (McDonald 1999; fig. 9.1). NO_x (NO and NO_2) are released during any high-temperature combustion that breaks the N_2 bond. Their annual emissions, mostly in densely populated industrialized regions of North

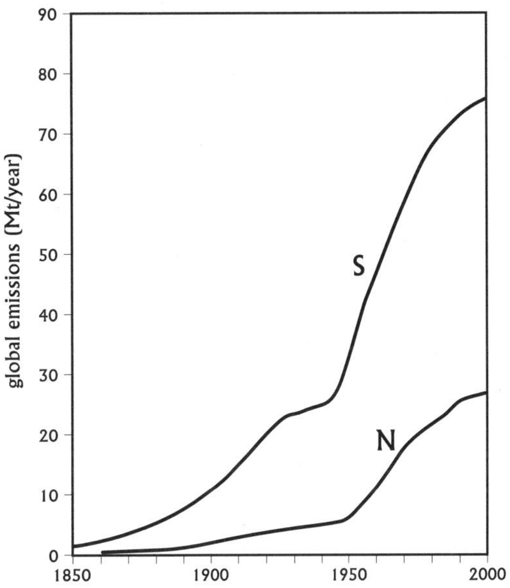

9.1 Global SO_2 and NO_x emissions expressed in sulfur (S) and nitrogen (N) equivalents, 1850–2000. Based on Smil (2000a).

America, Western Europe, and East Asia, are now about 25 Mt N (fig. 9.1). Volatile organic compounds (VOCs) result mostly from the incomplete combustion of fuels and from the processing, distribution, marketing, and combustion of petroleum products.

Acidifying deposition (wet and dry) arises from atmospheric oxidation of sulfur and nitrogen oxides. The resulting generation of sulfate and nitrate anions and hydrogen cations produces precipitation whose acidity is far below the normal pH of rain (about 5.6) acidified only by carbonic acid derived from the trace amount of CO_2 present in the atmosphere. Eastern North America, most of Europe north of the Alps, and most of the eastern third of China are the regions with the highest acidity (pH commonly below 4.5) of precipitation (Smil 2000a; fig. 9.2). In the presence of sunlight, NO_x, VOCs, and CO take part in complex chains of chemical reactions producing photochemical smog; O_3, its most aggressive constituent, is now repeatedly present in excessive concentrations in all megacities and their surrounding countryside (Colbeck and MacKenzie 1994; Mage et al. 1996).

Two global anthropogenic changes in the atmosphere's trace constituents have been much more worrisome. The first involves a relatively rapid alteration of the Earth's radiation balance brought about by rising concentrations of anthropogenic greenhouse gases; the second global risk has come about as a result of the unexpected destruction of stratospheric O_3 by chlorofluorocarbons (CFCs). The first problem will be with us for generations to come, whereas the second one may be, fortunately, a matter of a purely historic interest by the year 2050.

Fossil fuel–derived CO_2 is now the most important anthropogenic greenhouse gas (natural gas flaring and cement production are comparatively minor industrial sources). The annual rate of its emission rose from less than 0.5 Gt C in 1900 to 1.5 Gt C in 1950, and by the year 2000 it surpassed 6.5 Gt C, with about 35% originating from coal and 60% from hydrocarbons (Marland et al. 2000). After subtracting carbon incorporated by newly planted trees and by vegetation whose growth was accelerated by higher levels of atmospheric CO_2, there is also a net annual release of 1–2 Gt C from the conversion of natural ecosystems (mainly from tropical deforestation) to agricultural, industrial, and urban uses. Historical reconstructions show that until about 1910 more CO_2 was released annually from the biosphere than from fossil fuel combustion, with the cumulative input of plant- and soil-derived carbon amounting most likely to between 150 and 200 Gt between 1850 and 2000 (Warneck 2000). In contrast, CO_2 emissions from fossil fuels, gas flaring, and cement production totaled about 280 Gt C during the same period (Marland et al. 2000; fig. 9.3). We can now see this human interference in a very long perspective.

The first systematic measurements of background CO_2 levels (i.e., far away from major anthropogenic sources of the gas as well as from extensively vegetated areas) began in 1958 at Mauna Loa and at the South Pole, with stations in north Alaska and American Samoa added later (Keeling 1998). Mauna Loa's 1958 average CO_2 was 320 ppm, whereas in 2000 the mean observation surpassed 370 ppm (fig. 5.8). This increase of more than 30% in 150 years is of concern because CO_2 is a major greenhouse gas whose main absorption band coincides with the Earth's peak thermal emission (see chapter 4). Anthropogenic CO_2 has already increased the energy reradiated by the atmosphere by 1.5 W/m^2. Until the early 1980s global warming studies focused only on CO_2, but total forcing by other greenhouse gases now roughly equals that of CO_2 (Ramanathan

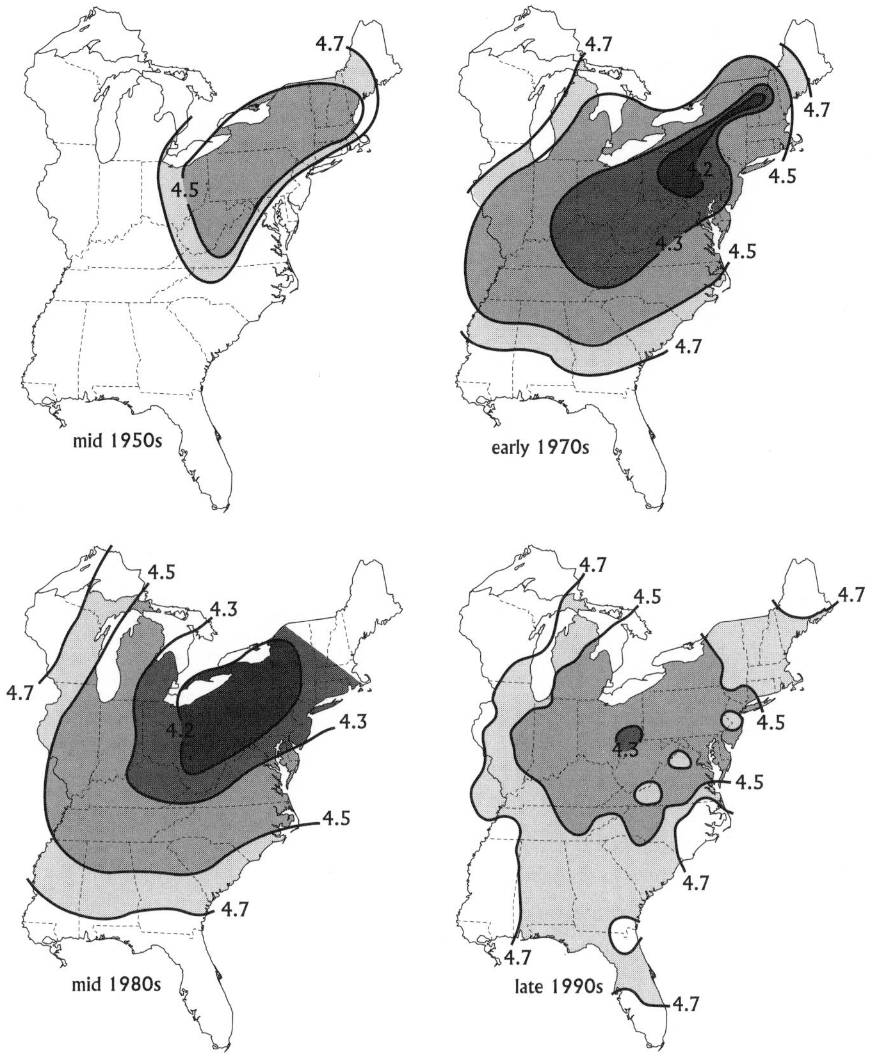

9.2 Changing acidity (pH) of precipitation in Eastern North America during the latter half of the twentieth century. Based on Smil (2000a) and National Atmospheric Deposition Program (NADP) maps available at <http://nadp.sws.uiuc.edu/isopleths/>

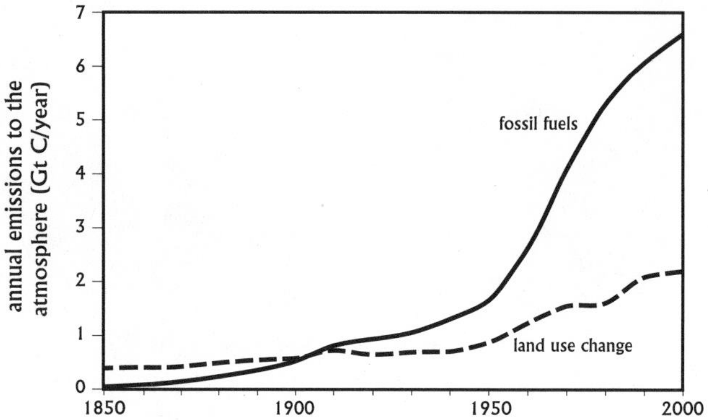

9.3 Long-term global trends of carbon released from land use changes and from the combustion of fossil fuels. Plotted from data in Marland et al. (2000) and Houghton (1999).

1998; Hansen et al. 2000). Atmospheric concentrations of these gases are much lower than those of CO_2, but they absorb (per mole or per unit mass) more of the outgoing IR radiation than does CO_2, and their atmospheric concentrations have been also increasing at a faster rate.

As already noted, methane levels have roughly doubled during the industrial era, to about 1.7 ppm by the late 1990s, and N_2O concentrations rose from about 280 ppt to around 310 ppt during the same period (Battle et al. 1996). Anthropogenic sources of CH_4 include anaerobic fermentation in landfills and garbage pits, methanogenesis in flooded farm soils, above all in the rice paddies of Asia, enteric generation of the gas by livestock, and direct emissions from coal mines and natural gas wells and pipelines. Global emissions add up to about 500 Mt CH_4 per year (Warneck 2000). The largest anthropogenic sources of N_2O are the denitrification of nitrogen fertilizers, industrial syntheses, and biomass and fossil fuel combustion. Their emissions total about 4 (1–9) Mt N (IPCC 1994). CFC emissions have been the principal cause of the stratospheric O_3 destruction (details later in this section), but the compounds are also greenhouse gases.

N_2O, CH_4, and CFCs are much more efficient absorbers of outgoing longwave radiation than CO_2: their global warming potentials relative to CO_2 are, respectively, about 60, 270, and 4,100–6,000 times higher during the first twenty years after their release (IPCC 1994). The combined direct and indirect effect of all greenhouse gases resulted in total anthropogenic forcing of about 2.8 W/m^2 by the late 1990s (Hansen et al. 2000; fig. 9.4). This is equivalent to a little more than 1% of the solar radiation reaching the ground, and a gradual increase of this forcing would eventually lead to the doubling of preindustrial greenhouse gas levels and to tropospheric temperatures 1–3.5°C higher than today's mean: the consequence of these changes will be detailed later in this chapter.

CFCs are synthetic compounds that were chosen during the 1930s to replace the previously common refrigerants (including NH_3 and CO_2) because they were stable, noncorrosive, nonflammable, nontoxic, thermodynamically

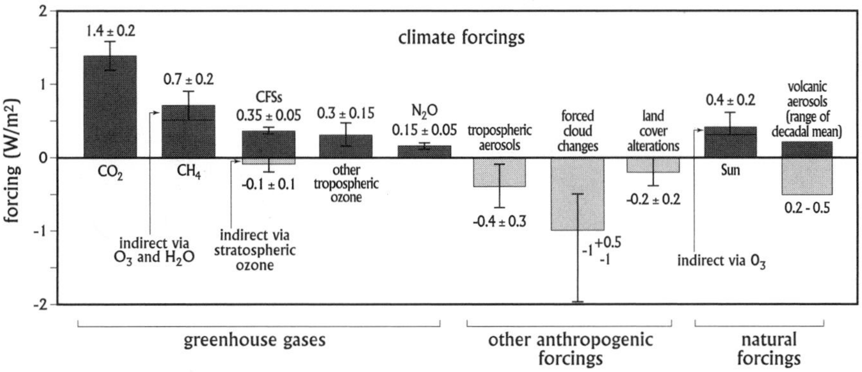

9.4 Estimates of cumulative radiation forcings by greenhouse gases and aerosols between 1850 and 2000, according to Hansen et al. (2000).

superior, and relatively inexpensive (Cumberland et al. 1982; Cagin and Dray 1993). They were later used also as aerosol propellants, as foam-blowing agents in producing polyurethane, polystyrene, and polyolefin, and for cleaning electronic circuits and extracting edible and aromatic oils. Three compounds—CCl_3F, CCl_2F_2, and $CHClF_2$—accounted for about 90% of the global output of freons (commercial name for CFCs). Halons, used for fire protection, contain bromine ($CBrClF_2$, $CBrF_3$, $C_2Br_2F_4$), as does the most common soil fumigant, methyl bromide (CH_3Br). In the stratosphere these gases can be photodissociated by wavelengths shorter than 220 nm that are entirely absent in the troposphere. This breakdown releases free chlorine atoms, which break down O_3 molecules and form chlorine oxide, which, in turn, reacts with O to produce O_2 and frees the chlorine atom to destroy yet another O_3 molecule. Before its eventual removal a single Cl atom can destroy on the order of 10^5 O_3 molecules.

The most alarming confirmation of this destructive activity, outlined first by Molina and Rowland (1974), came only in 1985, when measurements at the British Antarctic Survey base at Halley Bay showed that the lowest spring (October) concentrations of ozone in the Antarctic atmosphere had dropped from about 300 Dobson units (roughly equal to 1 ppbv) to about 200 Dobson units (Farman et al. 1985; Center for Atmospheric Science 2000; see fig. 1.12). The annual recurrence and the deepening of this decline (fig. 9.5) and concerns about its extension outside Antarctica led rapidly to international agreements (Montreal, 1987; London, 1990) to cut and eventually to eliminate all CFC emissions (UNEP 1995). Although the long atmospheric lifetime of CFCs means that the stratospheric effect will be felt for decades to come, atmospheric concentration of these compounds have been falling since 1994 (Hall et al. 2001), and the stratosphere may return to its pre-CFC composition before 2050.

Land Cover, Soils, and Biodiversity

Expansion of human populations has been responsible for the most obvious transformation of the biosphere, the

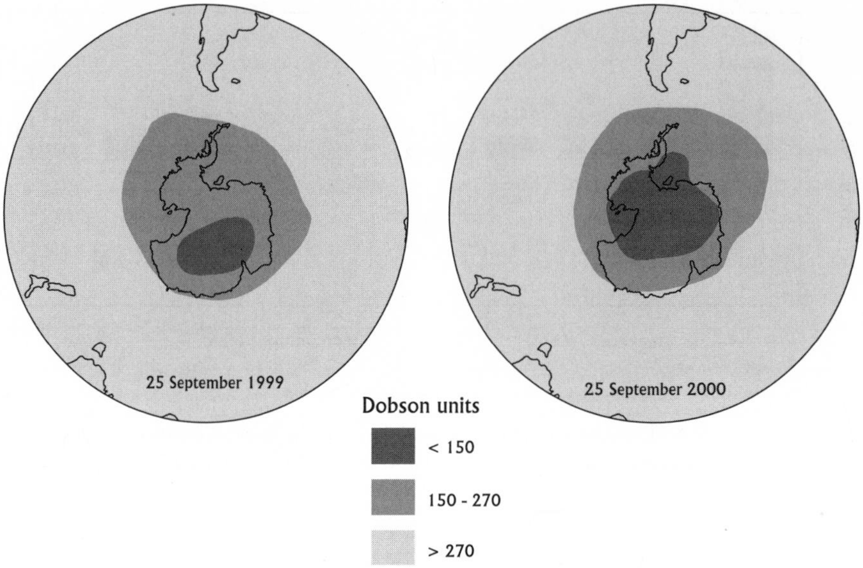

9.5 Extent of the Antarctic ozone hole on September 25, 1999, and on the same day in 2000. Archive of daily color images available at <http://www.cpc.ncep.noaa.gov/products/stratosphere/tovsto/archive/sp>

large-scale conversion of natural ecosystems to fields, pastures, settlements, and more recently also to a great variety of industrial and transportation uses. In the overwhelming majority of cases, these conversions are accompanied by soil loss or degradation of soil quality and by declining biodiversity. The richness and stability of ecosystems are also threatened both by deliberate introduction of new species and by uncontrolled bioinvasions.

These extensive changes can be assessed in a variety of ways: as comparisons of estimated areas of preagricultural biomes with today's land cover pattern; as historic reconstructions documenting major land use trends of the past 100–300 years, the period for which there is enough information to derive regionally disaggregated sequences of deforestation and extension of agricultural land; or as inventories of the amount of remaining wilderness, that is, of large, contiguous areas that are still shaped overwhelmingly by natural processes. Yet another approach is to account, as comprehensibly as possible, for materials outflows from industrial economies (Matthews et al. 2000a).

Because of the uncertainties inherent in reconstructing the potential preagricultural vegetation, classifying individual biomes, and defining "forest" (more than ninety definitions can be found in the literature; World Resources Institute 2000), the first comparison can be only a matter of rough approximations. Matthews (1983) put the global area of preagricultural closed forests and woodlands at 61.5 million km^2 and estimated that it was reduced to about 52 million km^2 by the 1980s, a decrease of some 15%. Ramankutty and Foley (1999) started with the potential global forest and woodland area of 55.3 million km^2 and estimated a 20% reduction (11.35 million km^2) by 1990, and Richards (1990) indicated that 12 million km^2 of forests were cleared between 1700 and 1980. He also concluded that the same period saw the loss of 5.6 million km^2 of grasslands and the gain of 12 million km^2 of croplands, while Ramankutty and Foley put the agricultural gain at 13.87 million km^2.

More restrictive definitions of "forest," limiting it to areas of closed or partially closed canopies, result in only 29–34 million km^2 of global coverage during the 1990s (Matthews et al. 2000b; FAO 2000), and more liberal estimates of original vegetation then imply a cumulative loss of nearly 50% of tree cover (Bryant et al. 1997). Two conclusions transcend any inevitable differences in such estimates: human actions have converted a significant share of all natural terrestrial biomes to other uses, and they have also modified, and often degraded, much of the remaining area of these ecosystems. Completely transformed surfaces include all arable land and areas under permanent crops, whose total the Food and Agriculture Organization (FAO 2000) puts at 15 million km^2, and the surface claimed by settlements, industries, transportation links, and water reservoirs. Urbanized areas, so vividly outlined by nighttime satellite sensing of lights, now amount to about 5 million km^2, and water reservoirs cover about 500,000 km^2. Human activities have thus entirely refashioned at least 20 million km^2, or 15% of all ice-free land surface.

Areas that still resemble, to a greater or lesser extent, natural ecosystems but have been modified by human actions are much larger. Fires set annually by pastoralists to keep shrubs and trees from encroaching on grazing land are the most obvious manifestation of a recurrent modification. Satellite monitoring has made it possible to follow this process through its annual progression and to estimate fairly accurately the extent of the burned area. No other continent is affected as much as Africa, where vegetation fires sweep annually over at least 3 million km^2, across the Sahel from Guinea to Somalia between November and April and across the southern semiarid belt from Angola to Mozambique between May and October (Barbosa et al. 1999; fig. 9.6). The median interval between burnings, based on satellite observation, is about four years, with ranges from up to twenty years in the

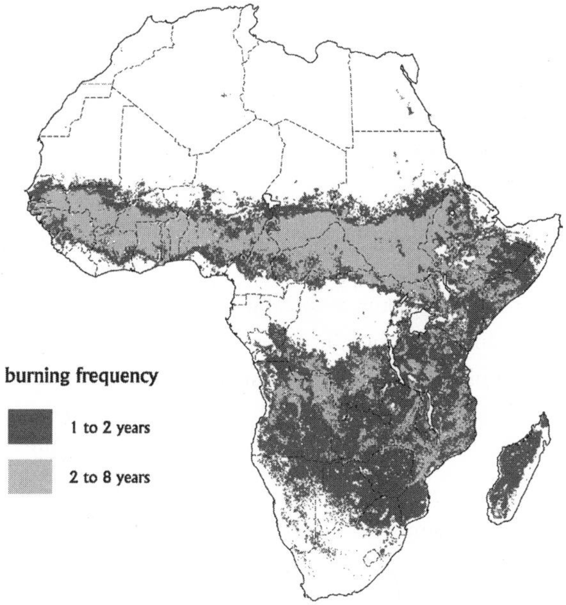

9.6 Satellite monitoring shows the extent of African grassland burning: shaded areas indicate grasslands burned every one to two and every four to eight years (the continental mean is about every four years) during the period 1981–83 and 1991–93. Based on Barbosa et al. (1999).

Sahel to every year in the Guinean zone. The total area of global annual savanna burning has been estimated at 7.5 million km^2, and the total carbon release from all biomass burning is 3–3.5 Gt annually (Levine 1991). This is equal to roughly half of the annual emissions from fossil fuels, but virtually all of the carbon released through burning in a given year will be reabsorbed by new grasses during the next growing season.

Permanent pastures, totaling about 34 million km^2 (FAO 2000), are thus the largest area that has been modified to different degrees by human actions. Tree plantations and forests actively managed for goods and services total about 6.5 million km^2 (Noble and Dirzo 1997). Road building, logging, and fires have degraded large areas of remaining forests. Nepstad et al. (1999) found that the present estimates of annual deforestation for Brazilian Amazonia capture less than half of the forest area that is impoverished each year and that the share captured is even smaller during the years of severe drought. A very conservative estimate of the global extent of degraded forests (including the effects of atmospheric deposition) is at least 5 million km^2, and the real extent may be twice as large.

Adding up these impacts reveals that the total area strongly or partially imprinted by human activities is about 70 million km^2, or no less than 55% of all nonglaciated land. For comparison, Vitousek et al. (1997a) concluded that 33–50% of the land surface has been transformed by human action. And this enormous transformation is far from finished. Although forests have been regaining some ground in affluent temperate countries and perhaps even in parts of China, tropical deforestation continues at an annual rate of anywhere between 50,000 and 170,000 km^2 (Tucker and Townshend 2000), and large areas of grasslands and wetlands continue to be lost through misuse, conversion to cropland, or aquaculture.

Inevitably, only fragmented pieces of terrestrial ecosystems that retain, more or less, their preagricultural form (although they may be subtly modified by long-distance atmospheric deposition) are now found in densely populated regions, forming relatively small islands protected in national parks or reserves amidst agricultural and urban-industrial sprawl. Large areas of largely unchanged ecosystems remain only in more extreme environments. A reconnaissance-level inventory of wilderness that identified only contiguous blocks over 400,000 ha concluded that during the 1980s about one-third of the global land surface was still wilderness, totaling about 48 million km^2 in just over 1,000 tracts (McCloskey and Spalding 1989).

Territorial shares of remaining wilderness range from 100% for Antarctica and 65% for Canada to less than 2% for Mexico and Nigeria and, except for Sweden (about 5%), to nothing for even the largest European countries. In terms of biomes and ecosystems, boreal forests (taiga) and circumpolar cold deserts remain the least affected (indeed, 40% of the remaining wilderness is in the Arctic or Antarctic), whereas there are no remaining undisturbed large tracts of such previously highly biodiverse ecosystems as the tropical rain forest of the Guinean Highlands, Madagascar, Java, and Sumatra, the temperate broadleaf forests of eastern North America and China, or the Mediterranean forests of Europe or California.

Yet another way to look at the extent and the intensity of the recent transformation of the biosphere is to estimate the share of GPP consumed or otherwise processed, managed, or destroyed by human actions. Vitousek et al. (1986) put this "appropriation" of the global terrestrial NPP at as much as 40%, but even their lower-bound estimate of about 25% illustrates the intensity of the biosphere's transformation. The implications of this impact on biodiversity and on future economic development are obvious, because even when one starts with the low share, two more consecutive doublings would do the impossible and appropriate the entire NPP for human use.

Removal of its natural vegetation almost always opens land to accelerated erosion. Excessive erosion has been a long-standing environmental concern, but we still lack reliable information needed to quantify the process on a global scale or, much more modestly, to discern clear national and regional trends (Pimentel 1993; Agassi 1995; Smil 2000b). Many fields around the world are losing their soils at rates twice or three times as high as their maximum (and highly site-specific) average annual losses compatible with sustainable cropping (mostly from 5–15 t/ha). A Global Assessment of Soil Degradation (GLASOD) study (Oldeman et al. 1990) found about 750 Mha of continental surfaces are moderately to excessively affected by water erosion and 280 Mha by wind erosion, with deforestation and overgrazing being the major causes. Mismanagement of arable land was judged to be responsible for excessive erosion on some 180 Mha of crop fields, or about one-fifth of all land affected by erosion.

Although it is reasonable to conclude that global soil loss increased during the past fifty years, it is very difficult to quantify the trend. The United States is the only major agricultural producer with regular nationwide soil erosion surveys. These surveys indicate that after a 39% reduction in the amount of topsoil lost to erosion between 1982 and 1997 there has been no further progress in recent years (Lee 1990; Soil and Water Conservation Society 1999). The most comprehensive study of basin-wide erosion, which measured the rate of alluvial sediment accretion in the agricultural Coon Creek Basin in Wisconsin, showed even greater long-term gains, with the losses for 1975–1993 being a mere 6% of the rate that prevailed during the 1930s (Trimble 1999). Given the large variability of sediment sources, sinks, and fluxes over time and space, results of these surveys cannot be readily extrapolated to other major Western crop producers or to other U.S. regions. Available evidence allows us only to argue that the African situation is particularly dire, whereas China's loss of soil organic matter may be not as severe as has been widely believed (Lal 1995; Lindert 2000).

Qualitative soil deterioration is of concern even in places with tolerable levels of erosion. Declining levels of soil organic matter, whose adequate presence is essential for the maintenance of both soil fertility and structure, is perhaps the most common concern in all places with inadequate crop residue recycling, low levels of (or no) manure application, and no planting of green manures or leguminous cover crops. Mining of macronutrients, micronutrient inadequacies, salinization caused by improper irrigation practices, compaction of soils by heavy agricultural machinery, and the presence of persistent agrochemicals are other common worries (Smil 2000b).

Selective hunting and introduction of aggressive predators were important causes of anthropogenic extinction of animals in preindustrial societies. Archeological finds demonstrate that these processes could be destructive even when the overall number of people involved and the changes they made to their environments were relatively limited. Polynesians colonizing Pacific islands that were inhabited by large numbers of endemic species exterminated more than 2,000 birds, about 15% of the world total (Pimm et al. 1995). Selective killing also brought several whale species to the brink of extinction. Moratoria on whale hunting are helping to restore some counts, but the global marine fish catch is approaching its limit, and overfishing has drastically reduced or nearly eliminated many previously abundant fish species (Botsford et al. 1997). Moreover, with the exception of early-maturing herring and related species, there has been very little evidence for any recovery of overfished stocks as much as fifteen years after 45–99% reductions in their reproductive biomass (Hutchings 2000).

Overfishing aside, it is the loss of habitat, particularly in the species-rich tropics, that is now by far the leading cause of anthropogenic extinction. Highly unequal distributions of biodiversity and regional differences in the tempo of environmental transformation mean that these anthropogenic eliminations have not been simply proportional to the areas of converted or modified ecosystems. Two contrasting examples illuminate this basic fact. Only a very small share of original forests east of the Mississippi survived the post-1500 European colonization of North America. But as this conversion progressed gradually and as forests reclaimed abandoned fields, a large part of the region was always under some tree cover (albeit a degraded secondary one in many places), and hence only 4 of the 160 bird species became extinct.

On the other hand, extinction rates have been particularly high on small, isolated islands. The highly endemic flora and fauna of such islands can be rapidly affected by conversion of a few key plant communities to cropping and by introduction of aggressive predators. Hawai'i (half of all bird species lost since 1778, another fifth endangered), the Caribbean islands (nineteen out of sixty mammalian extinctions worldwide), and the Indian Ocean islands are the textbook examples of such losses. Other environments whose species have been relatively prone to extinction include arid ecosystems (including the Mediterranean biome plants) and rivers, where bivalves have been especially affected.

But on the local and regional level, it is the loss of populations that matters. A species may thrive on one continent or in scores of other locations, but its demise in a particular area may seriously weaken the provision of a biospheric service (for more on this, see later in this chapter). Hughes et al. (1997) concluded that in tropical forests populations could be disappearing at a rate three to eight times higher than do species. In the long run, even the loss of a small fraction of species would definitely be detrimental if it were accompanied by deep, widespread losses of populations.

There is no doubt that the recent rates of extinctions have been far above the expected natural rate (at most some three species a year) that would maintain the existing biodiversity (Purvis and Hector 2000). Documented extinctions alone have equalled the expected rate since 1600, and the rate escalated during the second half of the twentieth century. Wilson's (1992) often-cited estimate, based on the relationship between biodiversity and areas of cleared tropical forest, put the recent annual extinction at 27,000 species, or one every 20 minutes. This and other similarly high estimates have been criticized, but there can be little doubt that the current rates of anthropogenic extinction are higher than those during many (possibly all?) past episodes uncovered in the fossil record.

In any case, there appears to be a remarkable opportunity to conserve a surprisingly large share of threatened species by concentrating on biodiversity hot spots, relatively small areas with an exceptionally high presence of endemic species. Their largest contiguous concentration is in Southeast Asia (Indo-Burma, Sundaland, Wallacea); the smallest hot spots cover the plant-rich areas of the Mediterranean, the Cape province in South Africa, coastal California, and Southwest Australia. As many as 44% of all species of vascular plants and 35% of all species in four vertebrate groups are confined to twenty-five hot spots whose areas add up to a mere 1.4% of land surface (Mittermeier et al. 2000; Myers et al. 2000).

Anthropogenic extinctions are also increasingly caused by introductions to particular environments of aggressively invasive nonnative species. The intentional diffusion of domesticated or otherwise valued plants and animals has been an integral part of human colonization of continents and islands, and this process has inevitably been accompanied by unintended dispersion of aggressively growing plants and weeds and troublesome pests and predators. The three most successful crops among these disseminated species—various cultivars of wheat (*Triticum aestivum,* originally from the Middle East), rice (*Oryza sativa,* from Southeast Asia) and corn (*Zea mays,* from Mesoamerica)—are now grown on every continent and occupy a combined area of about 5 million km^2, more than all the remaining tropical forests in Africa.

In contrast to unmistakably nonnative crop fields, noncrop species may blend into local vegetation, or they may even take over and create entirely new communities. Kudzu vine (*Pueraria lobata*) and cheatgrass (*Bromus tectorum*) are among the most dramatic American examples of such takeovers. Lush vegetation in many warm locations, including Florida and Hawai'i, contains particularly high shares of introduced species; even more surprisingly, 99% of all biomass in parts of San Francisco Bay is not native (Enserink 1999). Such realities raise the prospect of increasingly homogenized ecosystems rich in generalist species, be they pests or weeds.

Some fungal invasions have been no less destructive than plant introductions, and their usually unseen progress is even harder to control. Their most notable American victims have been the once majestic chestnut (*Castanea dentata*) trees destroyed by *Cryphonectria parasitica* accidentally introduced from Asia (Anagnostakis 1987) and the elms (*Ulmus americana*), whose stately shapes generations ago made them a preferred species for city planting and which are being destroyed by *Ophiostoma ulmi,* the fungus causing the Dutch elm disease (Rossman 2001).

Several generalizations characterize the process of bioinvasions in today's interdependent world, with its enormous foreign trade and massive personal travel. Bioinvasions progress in largely unpredictable ways, and once they are underway, they are extremely difficult to stop. Zebra mussels (*Dreissena polymorpha*) were carried by ships in European ballast water to North America for generations, but they took hold in Ohio and Michigan only in 1988, and by the end of 2000 they penetrated all of the Great Lakes and the Mississippi basin (USGS 2000; fig. 9.7).

Bioinvasions may have their origin in well-intentioned introductions that may look quite harmless for extended periods of time before turning into nuisances, as well as in obscure choices, as in the selection of *Caulerpa taxifolia.* This green marine alga grows in small, isolated patches in its native tropical (Caribbean, Indo-Pacific) habitat. During the late 1970s it was brought to Europe as a decorative aquarium species, and in 1984 its cultured clone was introduced into the Mediterranean from a public aquarium. The highly invasive alga is now widespread through most of the northwestern Mediterranean, and the identical aquarium-Mediterranean strain was discovered in 2000 on the Californian coast north of San Diego (Jousson et al. 2000).

Most importantly, once invaders get established in a new territory, it has proven almost impossible to eradicate them. Where the invaders are larger mammals, restorations of the original biodiversity in limited areas can rely on expensive total exclusions behind well-built and well-maintained fences, but this option is impractical for keeping away snakes or rodents. The rising volume of global trade and travel will only multiply the opportunities for unintended invasions. Two recent worrisome detections include Asian longhorned beetles (*Anoplophora glabripennis*)

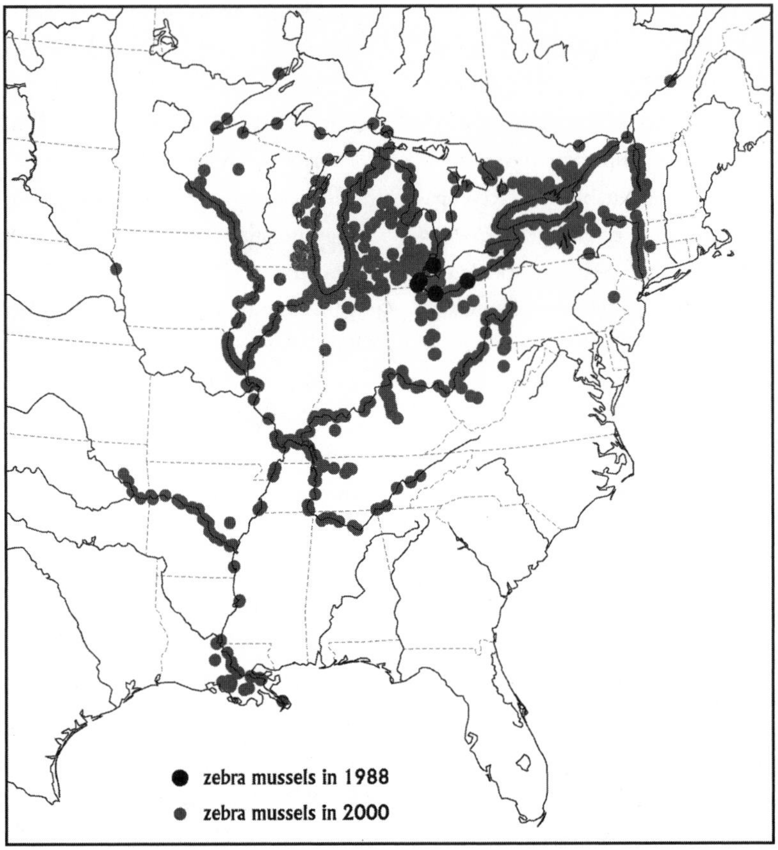

9.7 European zebra musssels (*Dreissena polymorpha*) were first seen at a few places in and around Lake Erie in 1988. By the end of 2000 they had invaded nearly twenty states and two Canadian provinces. Based on annual maps posted at <http://nas.er.usgs.gov/zebra.mussel/>

brought in Chinese wooden packing crates to the United States, where they could destroy many hardwood trees (Meyer 1998), and brown tree snakes (*Boiga irregularis*), who have already destroyed Guam's small vertebrates and were found in the wheel wells of jet planes landing in Honolulu. Finally, there are also obvious but unexplored consequences of the global spread of microbes, including those causing infectious diseases, in the ballast water of commercial ships (Ruiz et al. 2000).

On the other hand, not all of the bioinvasions that have been portrayed as having catastrophic consequences for native species have actually turned out to be so destructive. African honeybees are perhaps the best example of this reality. Continuous studies conducted for two decades in relatively intact forests of French Guiana and Panama have shown no declines in the abundance of native bees, and Roubik (2000) concluded, more generally, that there has been no detectable large impact on populations of native bees throughout tropical America from the presence of the African bees. And to give just one more example, invasive water lettuce (*Pistia stratiotes*), which forms dense floating layers (fig. 9.8), can be kept in check by adults and larvae of a South American weevil (*Neohydronomous affinis*) that feeds on its leaves.

9.8 Water lettuce (*Pistia stratiotes*) entirely covering a small pond on the outskirts of Big Cypress Swamp in Florida. Photo from the author's collection.

While land-use changes and introductions of troublesome species are long-standing problems, dramatic anthropogenic acceleration of microbial evolution began only during the 1940s with the introduction of antibiotics. Bacterial resistance to these compounds has been observed in as few as one to three years after their introduction, and after two generations of use penicillin is ineffective against most infections caused by such important bacteria as *Staphylococcus aureus;* resistant strains of bacteria have spread beyond hospitals to soils and water affected by effluents from farms that use prophylactic antibiotics in livestock production (Palumbi 2001). Analogical problems arise with crops and insects rapidly evolving in response to applications of herbicides and pesticides.

Biospheric Cycles

Human actions cannot fundamentally alter the global evaporation-precipitation cycle dominated by the ocean (see chapter 4). But changes in albedo and evapotranspiration (due to deforestation, overgrazing, agriculture, or settlements), excessive withdrawals from aquifers, diversions of stream and lake water for irrigation, rising water demand (above all for irrigation and for processing and municipal uses), pollution of water by agricultural, industrial, and household chemicals, and ubiquitous dumping of untreated waste water have radically transformed local and regional water balances by affecting both the quantity of available water and its quality. Construction of dams and regulation of streams has greatly disrupted the natural flows of all but a few major rivers and changed their age and composition. And the transfer of pollutants and nutrients to coastal waters has brought toxification and eutrophication to expanding areas of shallow seas. The combined effects of these changes add up to a profound disturbance of terrestrial water flows.

The replacement of tropical rain forests by grasslands and crops increases albedo and reduces evapotranspiration and precipitation, but surprisingly, the global impact would not be severe even in the most extreme case of Amazonian deforestation. Nobre et al. (1991) modeled the consequences of replacing the Amazonian forest with degraded pastures and concluded that a complete and rapid deforestation would be irreversible only in the southern part of the basin. Henderson-Sellers and Gornitz (1984) also forecast some locally significant changes but did not foresee any major impact on regional or global circulation patterns. The maximum total global albedo change over the last fifty years was less than 0.001, corresponding to a global temperature decrease of no more than about 0.1 K, far below the interannual variability. But Betts (2000) concluded that the potential carbon sink from boreal forestation could be offset by decreases in surface albedo brought by global warming.

Global water withdrawals have been significantly outpacing population growth: since 1800 they have expanded about fifteen-fold, compared to the sixfold increase in population over the same period (L'vovich et al. 1990). With most of the water used in irrigation (currently at least 60% of the total), irretrievable water losses increased thirteenfold during the same period. The estimated annual total withdrawal of no more than 5,000 km^3 is less than 15% of the global supply, but the share rises to more than a third when only the accessible runoff (both in spatial and temporal terms) is considered. Growing populations and rising per capita consumption may result in withdrawals of more than 40% of all accessible water before 2050 (Shiklomanov 1997). Several arid countries already exploit more than 100% of their sustainable water supplies by overdrawing ancient aquifers, and several rivers (the Colorado, the Nile, and the Huanghe) no longer ever, or only seasonally, discharge fresh water into the sea.

The ubiquitous construction of dams and creation of water reservoirs has had a surprisingly large cumulative impact on the world's rivers. There are now more than 42,000 large (above 30 m) dams and several hundred thousand smaller impoundments. In aggregate, reservoirs now hold about 10,000 km^3 of water (more than half of it behind large dams), or more than five times the water volume in all streams (Rosenberg et al. 2000). Naturally, these impoundments have dramatically increased the average age of river runoff. Whereas the mean residence time for continental runoff in free-flowing river channels is between sixteen and twenty-six days, Vörösmarty and Sahagian (2000) estimated that the discharge-weighted global mean is nearly sixty days for 236 drainage basins with large impoundments. Moreover, the mouths of several of the world's largest rivers show reservoir-induced aging of runoff exceeding six months (the Huanghe, Euphrates and Tigris, and Zambezi) or even one year (the Colorado, Rio Grande del Norte, Nile, and Volta; fig. 9.9).

Most dams present insurmountable obstacles to the movement of aquatic species; river channel fragmentation, now affecting more than ¾ of the world's largest rivers, changes their habitats; deep reservoirs lower the temperature of river water; and the aging of the continental runoff has brought a variety of biophysical changes, ranging from increased evaporative losses from large reservoirs in arid climates to reduced levels of dissolved oxygen and increased H_2S toxicity (ICOLD 1994). Interception of sediment and its deposition within the reservoirs cuts

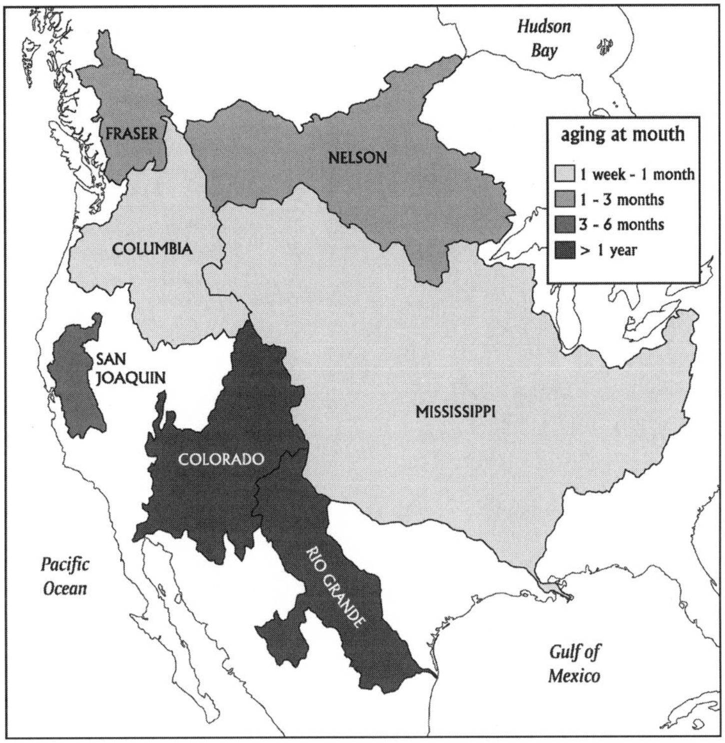

9.9 Aging of runoff due to damming of streams in several major watersheds of western North America. Based on Vörösmarty and Sahagian (2000).

the global sediment flow in rivers by more than 25% and reduces the amount of silt, organic matter, and nutrients available for downstream ecosystems, alluvial plains, and coastal wetlands. Retention of silicates is causing a shift from silicon-shelled diatoms to nonsiliceous phytoplankton in some estuaries and coastal regions (Ittekkot et al. 2000). Mining of aquifers, diversions of surface water, and deforestation contribute to a rise in sea level, accounting for about one-third of the twentieth-century change of 1.5–2 mm per year (Sahagian et al. 1994). Of course, reservoirs also retain part of the runoff that would have otherwise been stored in the ocean.

Human actions have intensified the biospheric flows of the three doubly mobile elements as well as the oceanward flows of minerals. The former intensifications have come largely from a combination of fossil fuel combustion, metal smelting, land use changes, and fertilizer application, the latter ones from accelerated erosion brought on by large-scale deforestation, improper agricultural practices, overgrazing, and urbanization. During the second half of the twentieth century, most of these interferences amounted to significant shares of the overall biospheric mobilization (release, fixation, oxidation, dissolution) of these elements, and the prospects are for even greater impacts before global plateaus are reached or these biospheric burdens are perhaps even reduced. Modifications of the global carbon cycle outlined earlier in this chapter have received most of the research attention, as well as most of the public policy attention, because of the possibility of a rapid climate change as a result. But in relative terms human interferences in the nitrogen and phosphorus cycles are even greater, and in some ways also more intractable.

Three processes account for all but a small fraction of the anthropogenic modification of the global nitrogen cycle: production of synthetic nitrogen fertilizers, combustion of fossil fuels, and plantings of leguminous crops. Both the traditional recycling of animal manures and human wastes and the first commercial nitrogen fertilizers, guano and Chilean $NaNO_3$, which were introduced before 1850, had only limited local environmental impacts. Only the Haber-Bosch synthesis of ammonia from its elements, commercialized by the Badische Anilin- und Soda-Fabrik (BASF) in 1913, opened the way for a large-scale inexpensive supply of fixed nitrogen (Smil 2001). The output of nitrogen fertilizers based on the synthesis expanded rapidly only after 1950, and it became particularly energy-efficient with the introduction of single-train plants equipped with centrifugal processors during the 1960s (fig. 9.10). The current rate of global NH_3 synthesis is surpassing 100 Mt N per year. About four-fifths of the NH_3 produced is used as fertilizers (most of it actually as a feedstock for producing urea and various nitrates, sulfates, and phosphates), the rest in industrial processes ranging from the production of explosives and animal feed to feedstocks for synthesis of dyes, plastics, and fibers (Ayres and Febre 1999).

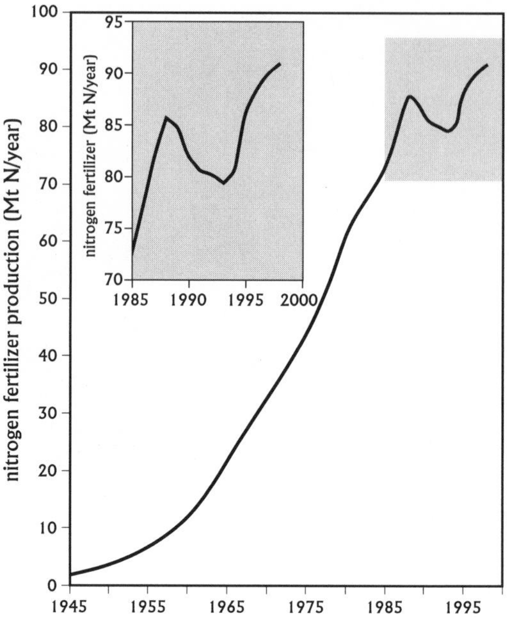

9.10 Global production of synthetic nitrogen fertilizers (Smil 2001).

As already noted, combustion of fossil fuels now emits almost 25 Mt N per year in NO_x. The planting of leguminous crops, practiced by every traditional agriculture, was the first major human intervention in the nitrogen cycle. The cultivation of pulses (dominated by soybeans) and leguminous cover crops (with alfalfa and various clovers being the most common choice) now fixes annually between 30 and 40 Mt N (the precise total is unknown, not only because of uncertain assumptions regarding the average fixation rates, but also because of fluctuations in cultivated area). The aggregate anthropogenic fixation of nitrogen thus amounts to 160–170 Mt N per year. In contrast, the best available estimate of the preagricultural terrestrial fixation is at least 150–190 Mt N per year (Cleveland et al. 1999). Human interventions have thus roughly doubled the global terrestrial fixation of the element.

The unwelcome biospheric consequences of this interference are due above all to numerous leakages during the process of production, distribution, and application of nitrogen fertilizers and to postapplication losses from leaching, erosion, volatilization, and denitrification (Smil 1999c). Uptakes are lowest (often less than 30%) in rice fields, highest in well-farmed temperate crops of North America and northwestern Europe. Because the primary productivity of many aquatic ecosystems is nitrogen-limited, the eutrophication of streams, ponds, lakes, and estuaries by runoff containing leached nitrogen from fertilizers promotes growth of algae and phytoplankton. Decomposition of this phytomass deoxygenates water and hence seriously affects or kills aquatic species, particularly benthic fauna. Algal blooms may also cause problems with water filtration or produce harmful toxins.

Nitrogen in eutrophied waters comes also from animal manures, human wastes, industrial processes, and atmospheric deposition, but there is a clear correlation between a watershed's average rate of nitrogen fertilization and the riverine transport of the nutrient. The worst affected offshore zone in North America is a large region of the Gulf of Mexico, where the nitrogen load delivered by the Mississippi and Atchafalya Rivers has doubled since 1965 and where eutrophication creates every spring a large hypoxic zone that kills many bottom-dwelling species and drives away fish. Other affected shallow waters include the lagoon of the Great Barrier Reef, the Baltic Sea, and the Black, Mediterranean, and North Seas.

NO_x from combustion are essential ingredients for the formation of photochemical smog, and their eventual oxidation to nitrates is a major component (together with sulfates) of acid deposition. Atmospheric nitrates, together with volatilized ammonia, also cause eutrophication of normally nitrogen-limited forests and grasslands: in parts of eastern North America, northwestern Europe, and East Asia their deposition (up to 60 kg N/ha per year) has become significant even by agricultural standards (Vitousek et al. 1997b; Smil 2001). The positive response of affected ecosystems to deposition of atmospheric nitrates is self-limiting, as nitrogen-saturation eventually leads to enhanced nitrogen losses. The intensifying anoxia of coastal bottom waters caused by the eutrophication of surface waters, observed in the Gulf of Mexico, in the Baltic Sea, and elsewhere, reduces the biodiversity of affected areas, but hypoxic conditions favor denitrification and hence the removal of excess nitrate. At the same time, higher denitrification produces more N_2O, and hence a future expansion of hypoxic zones around the world could be an appreciable source of that greenhouse gas (Naqvi et al. 2000).

Anthropogenic sulfur emissions come overwhelmingly (about 93%) from the combustion of fossil fuels, and the remainder of the total originates largely in smelting of color metals (Cu, Zn, Pb). Global sulfur emissions rose from about 5 Mt S per year in 1900 to about 80 Mt S/year in 2000 (Lefohn et al. 1999). Post-1980 declines in sulfur emissions in European countries (20–60% cuts) as well as in the United States and Canada (resulting from conversion to natural gas and desulfurization of flue gases) have been compensated for by rising levels of SO_2 generation in Asia, above all in China. SO_2 from fossil fuel combustion now accounts for at least 75% of the global emissions of the gas. Sulfates produced by atmospheric oxidation of SO_2 counteract global warming somewhat by cooling the troposphere: the combined global effect of natural and anthropogenic sulfates is now about $-0.6\ W/m^2$, and the

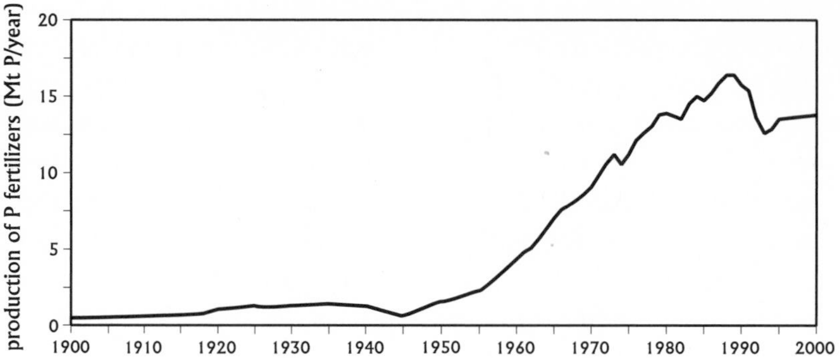

9.11 Global production of phosphorous fertilizers (Smil 2000c).

negative forcing is higher in eastern North America, Europe, and East Asia, the three regions with the highest sulfur emissions.

Deposited sulfates acidify waters and poorly buffered soils, and extensive research on acid deposition, including the impact of nitrates, has identified common effects of acidity in aquatic ecosystems devoid of any buffering capacity: leaching of alkaline elements and mobilization of toxic aluminum from soils and often profound changes in the biodiversity of lakes, including the disappearance of the most sensitive fishes, amphibians, and insects (Irving 1991; Driscoll et al. 2001). In contrast, the exact role of acid deposition in the reduced productivity and dieback of some forests remains uncertain (Godbold and Hutterman 1994).

As with nitrogen, the most important anthropogenic impact on the phosphorus cycle comes from inorganic fertilizers (Smil 2000c). The production of such fertilizers began during the 1840s with the treatment of phosphorus-containing rocks with dilute sulfuric acid. Discovery of large phosphate deposits in Florida (1870s), Morocco (1910s) and Russia (1930s) laid foundations for the rapid post-1950 expansion of the fertilizer industry. Consumption of phosphorus fertilizers peaked at more than 16 Mt P in 1988; after a 25% decline by 1993, global levels of use are rising once again (fig. 9.11). The top three producers (the United States, China, and Morocco) of phosphorus fertilizers now account for about two-thirds of the global output. The global food harvest now assimilates about 12 Mt P in crops and their residues, whereas no more than 4 Mt P are supplied by the weathering of phosphorus-bearing rocks and by atmospheric deposition. Fertilizer phosphorus is thus indispensable for producing today's harvest, but increased soil erosion from crop fields and degraded pastures now mobilize about twice as much phosphorus as did the natural denudation (because of the element's low solubility, the increase in dissolved flux is relatively small).

The past two centuries have also seen a roughly twelvefold expansion of the total mass of the nutrient voided in animal and human wastes. In 1800 anthropogenic mobilization of phosphorus through increased erosion was equal to about a third of the total continental flux of the nutrient. At the beginning of the twenty-first century, natural losses of the element from soils to air and waters amounted to about 10 Mt P per year, whereas intensified erosion introduces on the order of 30 Mt P into the global environment annually, mainly because human actions have roughly tripled the rate at which the nutrient reaches the streams. A variable part of this input is deposited in and along the streams before it enters the sea, but the total annual riverborne transfer of phosphorus into the ocean has at least doubled, and waterborne phosphorus, together with nitrates, is causing eutrophication of streams and water bodies to occur more commonly.

Long-Term Impacts and Uncertainties

How will these human actions affect the long-term integrity of the biosphere? Are they, seen from an evolutionary perspective, just ephemeral perturbations that will leave peculiar traces in the sedimentary record but will not fundamentally alter the natural course of the biosphere's development? Or will they amount, judged on a civilizational time scale, to a major derailment, if not to a catastrophic demise, of human aspirations as they undercut the very biophysical foundations of our existence? Plausible arguments can support either of these extreme positions. What we need is not more clever arguing, and what we cannot get, given the inherent complexities of biospheric transformations and major uncertainties concerning their outcomes, is a confident, albeit probabilistic, appraisal of our prospects.

What can be done is to outline the best evidence concerning the most likely impacts and then to qualify it by pointing out persistent ignorance and uncertainty. This combination does two necessary things: it alerts us to a multitude of nontrivial risks while tempering our verdicts and hence making it less likely that we will opt for hasty or extreme solutions. Appraising the human impact on the biosphere thus means above all asking fundamental questions about the prospect of our species and about its willingness to conform to biospheric limits. Although undoubtedly temporarily impoverished and in some ways permanently altered, the biosphere will survive even the most drastic human assaults, but modern civilization may be hastening its own demise by treating the biosphere in such predatory ways.

Anthropogenic climate change poses unprecedented risks to the biosphere, not only because it would be truly global, but also because it might proceed at a very rapid rate. Although we cannot forecast the actual degree of future warming with high confidence, we understand that once the troposphere becomes, on average, several degrees warmer, there would be inevitable and appreciable changes to a number of basic biospheric parameters, ranging from the length of the growing season to the amount and distribution of precipitation. As for the magnitude of change, we have to go back about 15,000 years to encounter the most recent rise of more than 1–2°C in the average annual temperatures of the Northern Hemisphere (Culver and Rawson 2000). And as it took about 5,000 years of postglacial warming to raise the mean temperature by 4°C, a global warming of more than 2°C during the twenty-first century would proceed at a rate unprecedented since the emergence of our species about 0.5 Ma ago (fig. 9.12).

But concerns about the effects of human intervention on the biosphere abound even in the absence of a relatively rapid global climate change. The anthropogenic interventions reviewed in this chapter have been, singly and in combination, compromising, weakening, undermining, and on local scales, even entirely eliminating biospheric services whose continued functioning is essential for the

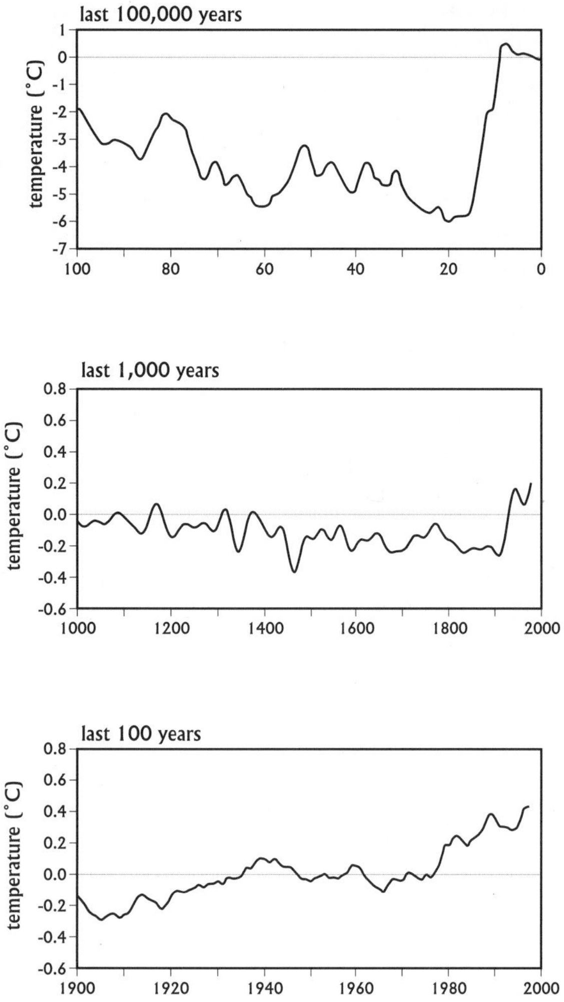

9.12 Reconstructed temperature trends during the past 100,000 years (from the Vostok ice core) and 1,000 years (for the Northern Hemisphere) and a five-year running mean from instrumental temperature measurements for the past 100 years. Based on graphs in Jouzel et al. (1987), Mann et al. (1999), and Hansen et al. (1999).

perpetuation of organisms. The appropriate valuation of biospheric services should help us make better choices, but I will show that recent efforts to do so have been fundamentally unsatisfactory. Significant uncertainties in our understanding of biospheric services make it extremely difficult to distinguish worrisome but transitory impacts from permanent damage and to formulate effective compromises between inevitable interferences and irreparable losses.

Climate Change

Recent multivariate calibration of widely distributed proxy climate indicators over the past six centuries shows that anthropogenic greenhouse gases emerged as the dominant temperature forcing during the twentieth century (Mann et al. 1998). A detailed reconstruction of Northern Hemisphere temperatures and climate forcing over the past 1,000 years confirms this conclusion. Preindustrial decadal-scale temperature changes were due largely to a combination of changes in solar irradiance and volcanism, whereas the removal of all forcing *except* greenhouse gases shows a large twentieth-century residual, as the greenhouse gas effect has already established itself above the level of natural variability in the climate system (Crowley 2000). As previously noted (chapter 2), variations in atmospheric CO_2 have also been widely accepted as the proximate driver of most, if not all, Phanerozoic climate changes. New findings support this hypothesis.

By far the most thorough examination of the heavily contaminated record of deep-sea sediment oxygen isotopic composition shows that the well-known 100,000-year periodicity of ice volume lags temperature, CO_2, and orbital eccentricity, and hence the cycle does not arise from ice sheet dynamics (Shackleton 2000). But there is no broadly accepted mechanism that would explain why atmospheric CO_2 levels were lower during the ice ages than they are today. Recent explanations have favored the role of marine photosynthesis, sequestering carbon into the abyss by the rain of organic matter. We are not sure if the process took place because the ocean's reservoir of nutrients was larger during glacial times (which would have intensified the sequestration at low latitudes, where nutrients are now growth-limiting) or because of more complete utilization of nutrients at high latitudes, particularly in the open ocean around Antarctica, where much of today's nutrient supply goes unused (Sigman and Boyle 2000).

Nor do these findings mean that we can confidently assume that variation in levels of atmospheric CO_2 is indeed the main driver of climate change on a geological time scale. After studying the oxygen isotopic composition of tropical calcite and aragonite shells, which broadly reflects the oscillations of sea surface temperature, Veizer et al. (2000) uncovered a mismatch between this record and proxy data for atmospheric CO_2. This finding casts doubt on the widely accepted conclusion that most, if not all, of the climate swings during the Phanerozoic can be linked to variations in atmospheric CO_2. As Kump (2000) notes, if valid (as always in such complex matters, there are uncertainties), such a conclusion would have huge policy implications: if large changes in CO_2 levels in the past did not produce the generally assumed response, then the case for reducing fossil fuel emissions would be greatly weakened.

And in spite of intensifying research into all conceivable aspects of global warming, we still do not have a solid understanding of numerous biospheric consequences that could result from further substantial increases in atmospheric concentrations of greenhouse gases. Palaeoclimatic analogies, laboratory and field tests, and complex quantitative models are all useful, but all fall short of providing the desired answers, as future developments are contingent on many unpredictable changes and will be determined by complex, often unprecedented interactions including variables that are poorly understood or even remain as yet unidentified. The eventual effect of much higher CO_2 levels on the ocean's heat circulation is a perfect illustration of these realities.

We do not know how sensitive the massive overturning of ocean waters may be to anthropogenic climatic change. If mechanical forcing by the Southern Hemisphere's winds has a significant role, then the ocean may be more resilient; if the overturning is largely buoyancy-driven, then even small increases in high-latitude precipitation (making surface water less salty and hence too light to sink below a few hundred meters) may be able to reduce it or even to shut it down (Broecker 1997). Any substantial reduction of water exchange resulting in a stagnating deep sea and decreased O_2 content could lead to drastic declines in oceanic biomass, leading in turn to additional releases of CO_2 into the atmosphere. And so we are reasonably confident that the expected emissions of greenhouse gases will induce global warming, but the magnitude of the change this increase may generate is poorly constrained.

Allen et al. (2000) concluded that, under the "business as usual" emission scenario, the projected temperature increase of 1–2.5°C in the decade 2036–2046 is relatively insensitive to the climate sensitivity of the models used, to oceanic heat uptake, or to effects of sulfate aerosols, as long as these errors are persistent over time. But changes in the current balance of greenhouse gases and aerosol-induced cooling could substantially displace the final equilibrium, and so the projected temperature increase has an uncertainty range of 5–95%. The cooling effect exerted by aerosols (sulfates, nitrates, organics, and fly ash) is now being accurately monitored over the tropical northern

Indian Ocean, the region affected by pollutants from the Indian subcontinent.

During the winters of 1998 and 1999 clear-sky radiative heating at the surface of the Indian Ocean was reduced by 12–30 W/m^2, similar to the cooling of 26 W/m^2 estimated for the western North Atlantic off the East Coast of the United States (Satheesh and Ramanathan 2000). On the other hand, Jacobson's (2001) simulations indicate that globally, the direct radiative warming from black carbon released from the burning of biomass and fossil fuels exceeds that due to CH_4 (0.55 vs. 0.47 W/m^2, being thus second only to that due to CO_2), and that this warming effect may nearly balance the net cooling effect of other anthropogenic aerosols. Continuing uncertainties regarding the future rate of average tropospheric temperature increases are perhaps best illustrated by the changing predictions of the IPCC panels: from 1–3.5 °C in the second (1995) report to 1.4–5.8 °C for the latest (2001) assessment.

Whatever the eventual rise will be, we are fairly certain that it will be more pronounced on land and during nights, with winter increases about two to three times the global mean in higher latitudes than in the tropics, and greater in the Arctic than in the Antarctic. The most worrisome changes arising from a relatively rapid global warming would include the following (Smith and Tirpak 1990; Gates 1993; Watson et al. 1996): intensification of the global water cycle, accompanied by unequally distributed shifts in precipitation and aridity; higher circumpolar runoffs, later snowfalls, and earlier snowmelts; more common and more intensive extreme weather events, such as cyclones and heat waves; thermal expansion of seawater and the gradual melting of mountain glaciers, leading to an appreciable (up to 1 m) rise in sea level; changes of photosynthetic productivity and shifts of ecosystemic boundaries; and poleward extension of some tropical diseases, including malaria (Rogers and Randolph 2000). Although such qualitative conclusions are fairly indisputable, almost every quantitative assessment, beginning with the rates of past and future global warming, extending to the biosphere's carbon budgets, and ending with particular health effects on humans, is questionable.

Absence of reliable long-term baselines is the first basic weakness in assessing both the degree of current and future problems and the rates of recent and anticipated change. Clever means for getting around this deficiency have been found in some cases. The use of air bubbles in ice cores to derive fairly reliable readings of atmospheric trace gas concentrations going back for more than 400,000 years is perhaps the most outstanding example of such a creative solution (Petit et al. 1999). Consequently, we know that both the magnitude and the rate of the recent atmospheric CO_2 (or CH_4) rise have been rather worrisome. Pinpointing actual changes in tropospheric temperature, even for the period of increasingly more widespread instrumental measurement after 1850, has not been easy. And it has been impossible to reconcile fully the discrepancy between temperature trends at the Earth's surface, where measurements show warming by 0.3–0.4°C since 1979, and in the midtroposphere, where models predict a stronger warming but where microwave sounding units on satellites find no change (NRC 2000; Santer et al. 2000).

A reliable assessment of historic changes in global-scale precipitation is even more difficult. Our understanding of long-term averages of global precipitation rests on a century of measurements from 8,300 gauges that were initially disproportionately concentrated in Europe and United States. Prior to the 1970s we had reliable precipitation data for only about 30% of the Earth's surface. Even now large parts of continents remain poorly sampled, few good measurements have been made for about two-thirds of the world's surface, and well-known problems in measuring precipitation (above all, efficiencies of snow catch in windy polar regions) affect many records (Hulme 1995). Only the availability of remotely sensed data by the microwave sounding unit mounted on satellites since 1979 has made it possible to get an accurate picture of global precipitation (Arkin and Xie 1994).

The fact that we do not have an accurate carbon budget for the biosphere and that we are not certain about long-term consequences of CO_2 enrichment for plants and soils greatly complicates any assessments of future interactions between the carbon cycle and other biogeochemical and climatic processes (Falkowski et al. 2000). Estimates of carbon sinks range widely, between 0.2 and 1.3 Gt C per year for the continental United States and between 0.1 and 1.3 Gt C per year for Siberian forests. Moreover, a modeling exercise has indicated that between 1980 and 1998, global terrestrial carbon fluxes were about twice as variable as ocean fluxes, and that whereas tropical ecosystems were responsible for most of these fluctuations during the 1980s, northern mid- and high-latitude forests dominated after 1990 (Bousquet et al. 2000). Time series measurements of atmospheric O_2 confirmed a significant interannual variability of carbon storage and led to the conclusion that the terrestrial biosphere and the oceans sequestered annually 1.4 ($\pm$0.8) and 2 ($\pm$0.6) Gt C between mid-1991 and mid-1997 and that this rapid storage of the element contrasts with the 1980s, when the land biota were neither a sink nor a source (Battle et al. 2000).

Additional photosynthetic sequestration of carbon due to the fertilizing effect of CO_2 has been estimated at 1.2–2.6 Gt C per year (Wigley and Schimel 2000), but we do not know if coming decades will see even higher rates of carbon sequestration or if the biosphere will become an appreciable source of CO_2. The immediate primary plant response to higher levels of CO_2 is to reduce stomatal conductance and transpiration (and hence to reduce water requirements) and to increase the rate of photosynthesis and nutrient use efficiency (Drake et al. 1996). The evolutionary response to higher atmospheric CO_2 levels, well documented by studies of herbaria specimens over the past 200 years, has been to decrease the number of plant stomata (Woodward 1987). How long will these trends continue? Gray et al. (2000) discovered that *Arabidopsis thaliana* has a gene that prevents changes in the number of stomata in response to any further increases of atmospheric CO_2 and that roughly doubling the current CO_2 level had no effect on the stomatal density in at least one strain of the plant. Can this finding be the base for concluding that any future vegetation response to higher levels of the gas will be marginal or nonexistent?

Feedbacks will further complicate the outcome, as the process of CO_2-induced global warming is subject to several potentially large twists that may work to accentuate it or to moderate it. A warmer climate may first result in an appreciable carbon sink in the terrestrial biosphere, but land ecosystems may become a net carbon source because of enhanced soil and plant respiration and declining vegetation cover as a result of regionally increased aridity (Cox et al. 2000). Warmer oceans may trigger substantial releases of CH_4, a potent greenhouse gas, from clathrates, which are methane hydrates found in oceanic sediment and in continental permafrost regions, in which a rigid lattice of water molecules forms a cage around the gas (Kvenvolden 1993).

In contrast, rising atmospheric CO_2 levels change the pH of the ocean surface and the overall carbonate chemistry, resulting in slower calcification rates not just in corals and algae, but also in coccolithophorids, such as the ubiquitous, bloom-forming *Emiliania huxleyi* and *Gephyrocapsa oceanica,* that dominate the transfer of $CaCO_3$ to the sediments (see fig. 5.16). Because calcification releases CO_2 into the surrounding water and hence into the atmosphere, a reduction in its rate may act as a negative feedback in the future high-CO_2 world (Riebesell et al. 2000; Gattuso and Buddemeier 2000). And yet other feedbacks have already produced oscillations. An excellent example is a study of the Alaskan tundra demonstrating the complexity of sink-source interactions: it found that a long-term sequestration of carbon in the ecosystem was reversed by a warming and drying of the Arctic climate in the early 1980s, but it also concluded that more recently, the continuation of the warming and drying trend has reduced the carbon loss and created again, in some cases, a summer carbon sink (Oechel et al. 2000).

Finally, there will be surprises, sometimes welcome ones. During the 1990s global CO_2 generation from fossil fuel combustion was substantially below the forecasts prepared during the 1980s because nobody envisaged the unraveling of the Soviet empire (which cut the region's energy use drastically) and market-driven modernization of China (which has sharply increased the country's previously dismal energy efficiency). Future anthropogenic forcing of climate change may also be less than has been commonly assumed because the growth rate of emissions of non-CO_2 greenhouse gases—not just of CFCs banned according to the Montreal Protocol of 1987, but also of CH_4 and tropospheric O_3—has declined during the 1990s, and aggressive attention to their future emissions could halt, and perhaps even reverse, the growth of their atmospheric concentrations and lead to a decline in the rate of global warming (Hansen et al. 2000).

But there will also be unwelcome surprises. Only recently Sturges et al. (2000) identified a new potent greenhouse gas, trifluoromethyl sulfur pentafluoride, whose radiative forcing is greater than that of any other greenhouse gas on a per molecule basis. The gas most likely originates as a breakdown product of sulfur hexafluoride (SF_6), which is widely used to insulate switch gear, transformers, and other high-voltage equipment. Its current atmospheric level is a mere 0.12 ppt, growing by 6% per year, but as it appears to be long-lived (on the order of 1,000 years), its generation requires monitoring. And although the current tendency is to focus on the evidence supporting the progress of anthropogenic global warming, the bigger picture may be different once we attempt to see the recent changes from very long-term perspectives. The present interglacial, on the evidence of ice, ocean, and lake sediment cores, is similar to the one 130,000–116,000 years ago: a warmer ocean may actually hasten the transition to a cold world by increasing the rate of water transport from the tropics to polar regions and the consequent ice buildup and surge of icebergs into midlatitude oceans (Kukla et al. 2000).

Valuing Biospheric Services

Even if there were no threat of global climate change, it is the second class of global anthropogenic impacts that poses increasing risks to the biosphere's integrity and to human civilization. These local changes, whose ubiquity adds up to global worries, have come to be seen as threats to biospheric services, processes that include a large and disparate group of benefits ranging from pollination of crops to soil formation and from pest control to the decomposition of dead organic matter and the recycling of nutrients. Westman (1977) published the first explicit definitions of environmental goods as benefits derived from the structural aspects of ecosystems, characterized by the properties of components, whereas nature's services are defined as benefits derived from the functional aspects of ecosystems, characterized by the way components interact. Two decades later Daily (1997) defined ecosystem services as "the actual life-support functions, such as cleansing, recycling, and renewal" (p. 3), which also confer many intangible aesthetic and cultural benefits.

The distinction between the structural (properties of components) and functional (interactions of components) benefits of ecosystems becomes clear in the contrast of wood (whose components, lignin and cellulose, make it a valuable commodity) with the biodiversity conserved in forests (which clearly results from interactions of many components making up that ecosystem). But forests, which provide many of the most often-cited environmental services (besides sheltering biodiversity, they also retain water, protect soil against erosion, reduce air pollution, and modify climate) can do so obviously only as long as their unique structural properties (stems built of lignin and cellulose, branches designed as cantilevered beams, roots acting as surface-maximizing mats holding soil) are intact. The functional benefits from interactions of these components would be impossible without their structural properties, and these structures could not be synthesized without the availability of appropriate inorganic building blocks obtained through numerous biogeochemical links and feedbacks.

Similarly, complex organic components making up the protoplasm of bacteria responsible for, literally, running the global biogeochemical cycles (from decomposition to denitrification) and hence delivering the bulk of irreplaceable environmental services could not do their work without appropriate structural confinement. They are encased in cleverly designed walls composed of peptidoglycan and one or two membranes whose properties make the cells both discrete and interactive. Consequently, many cell lives can oscillate between the extremes of intensive reactivity with their surroundings and astonishing cryptobiosis, a deathlike state in which they can remain for extended periods of time before reviving. Form is function, one could say, along with modern architects.

These examples show the arbitrariness of boundaries used to define individual services and demonstrate the fuzziness and overlaps of their questionable definitions. Undeniably, perceptions and interpretations, rather than environmental realities, determine this naming and classification game. But could all of this not be acknowledged while still maintaining that the practice of singling out individual services is a highly desirable one, as it may eventually permit their proper valuation? I think that this argument is misleading and counterproductive in the long run. This is not to say that I am content with the ways we price environmental goods today: by fundamentally ignoring concomitant and extensive destruction and degradation of other environmental goods and services. But I find the increasingly common quest for monetization of everything to be scientifically indefensible, ethically unacceptable, and practically confusing: in sum, to be inimical to the very quest for true valuation. Here are just a few of the most important reasons for rejecting this currently fashionable lure.

On the most general level, these valuation exercises are almost exclusively anthropocentric, yet at the same time they denigrate our capacities for holistic appraisal and for the pursuit of moral goals. The incessant flows of countless goods and services among millions of species and their interactions with abiotic environment maintain the biosphere. Valuing all those myriad links and feedbacks that have no direct benefits for humans is impossible both logistically and noetically. How could we, to give just a single example, monetize the irreplaceable service provided by microbes in the hindguts of termites that enables the insects to digest lignin, and hence be major decomposers of woody litter in tropical ecosystems?

Anthropocentric valuation exercises are thus very much in the Old Testament mode: the biosphere is seen as our dominion, where everything exists for our benefit. But this approach does a disservice to what we have become and what we have learned. Equating proper valuation with monetization is to believe that the grand process of human encephalization culminates in reducing complex realities, including deep-seated beliefs and convictions, to simplistic, unidimensional indicators. These conversions to a common denominator are caricatures of reality, akin to reducing human intelligence to a single and dubiously established IQ score or to measuring quality of life by the number of gadgets found in an average household.

Difficulties do not disappear even when we accept that these valuations should have only a limited function as surrogate (and rough) indicators. Given our inadequate level of understanding of complex biospheric processes and the consequences entailed by their degradation or disruption, any valuations of ecosystemic services are bound to be incomplete, influenced by reigning scientific paradigms, and hijacked by often irreconcilably opposed special-interest groups. Our incomplete understanding also means that today's consensus, no matter how broadly based or how narrowly held, may be dramatically overthrown tomorrow by new scientific paradigms and by new sociopolitical priorities.

All of these and many other problems well up in any attempt at actual large-scale quantification of environmental services, and two recent exercises on a global level are perfect examples of the futility of such estimates. The one prepared by Costanza et al. (1997) received the most attention. Costanza's group came up with a list of seventeen categories of goods and services provided by nature and partitioned the Earth into sixteen major biomes. Then, using a mixture of market prices, willingness-to-pay values, and costs of service replacement, they determined the lowest and the highest estimates for each item and ended up with the range of $16–54 trillion and an average annual value of $33 trillion for the environmental services the biosphere provides. In contrast, Pimentel et al. (1997) ended up with a global total just short of $3 trillion, an order of magnitude lower than the mean arrived at by Costanza's group.

This discrepancy alone speaks volume about the state of the art of these valuations and about any potential utility such studies might have. A closer look uncovers many deeper problems. Although they are not intended to be perceived that way, the sums are interpreted by many people as the amounts whose expenditure might provide equivalent services. This erroneous impression leads to a misleadingly relaxed assessment of the problem, particularly when one takes Pimentel et al.'s figure: their annual value of the biosphere's environmental services is less than 40% of the current U.S. gross domestic product (GDP) (about $8 trillion in 2000), and before too long we will have multinational business mergers in which the two companies combining will be worth that much.

Valuing all nitrogen fixed by free-living and symbiotic bacteria in the world's forests, grasslands, and wetlands simply by multiplying the estimated total nitrogen fixed by the average market price of nitrogen in synthetic fertilizers, as done by Pimentel et al., is particularly inappropriate. Matching the natural biofixation would require at least doubling the current ammonia synthesis, provided this fertilizer were used as efficiently as is the nitrogen fixed by bacteria. To make up for much larger losses of nitrogen from fertilizers, we would have to quadruple, rather than double, the current synthesis. And how would we ever distribute those synthetic fertilizers to all biomes? Instead of fertilizing about one billion hectares of farmland, we would have to fertilize all of the Earth's ice-free land, and the attendant logistics boggles the mind. And even the $33 trillion figure preferred by Costanza et al. is ridiculously low, being roughly equal to the current gross world economic product expressed in purchasing power parities. This becomes obvious as soon as we imagine that we would have to reconstruct all essential biospheric services on an Earth-like planet. Even if we knew how to do this—and so far our rather primitive, and botched, attempts at building small-scale closed life-support systems show that we are nowhere near that point (Cohen and Tilman 1996)—we could not replicate the biosphere's annual work by spending all of the gross world product!

Specific problems with these estimates are equally glaring. Applying uniform and ridiculously precise values for major vegetation types—in Costanza et al.'s account, every hectare of wetland renders annually $14,765 of services, compared to a mere $232 for grasslands—ignores the essence of biodiversity and hence undercuts the very reason for valuing this key environmental service: most definitely, all wetlands (or grasslands or forests) are not equal! Following this approach would lead to overvaluation, and hence in practice to overprotection, of some ecosystems, and to undervaluation, and underprotection, of others. One could also cut down a large area of tropical rain forest and replace it with a wetland (by building a low dam and causing shallow flooding), thus raising the price put on an average hectare from a mere $2,000 to nearly $15,000. Similar logic would lead us to welcome rapid global warming and the ensuing rise of the sea level due to the thermal expansion of water: the additional 0.5 m would create more extensive coastal seas, whose value Costanza et al. put at $4,000 per hectare, compared to a mere $252 per hectare of open ocean.

Planetary Management

Is such a task as planetary management not only realistically possible but, given civilization's increasing impact on the global environment, basically unavoidable? Or does the term amount to, at best, a wishful aspiration or to a ridiculous hubris? As with so many contentious matters, it is not difficult to gather arguments supporting either of these positions. Allenby (2000) argues that because "the biosphere itself, at levels from genetic to the landscape, is increasingly a human product" (p. 15) and because the actions of industrial civilization have transformed the Earth into a human artifact, we need to engage consciously in Earth systems engineering and planetary management. Assorted geoengineering solutions proposed recently to deal particularly with the human interference in the carbon cycle range from pumping liquid CO_2 into the deep ocean to supplying enough nitrogen to the ocean's euphotic layer to sequester millions of tons of CO_2 per year in a sinking, dead phytomass (Jones and Otaegui 1997; Schneider 2001).

Criticizing such extravaganzas is easy. Keith (2000) retorts that "we would be wise to learn to walk before we try to run, to learn to mitigate before we try to manage" (p. 28). My feeling is that although we may already (or soon) have some requisite technical capacities, effective planetary management is far beyond our intellectual and social capabilities—but that we are doing it anyway. And as in so many other instances where management has to reckon with inadequate understanding of many underlying processes, with irrational aspirations and with doctrinaire stances, we are doing much worse compared not just to what we should be doing, but to what we could have already done.

Truly effective management must be based on an understanding that would not just indicate the direction of change but preferably offer a relatively constrained estimate of its magnitude. As I have just argued, neither goal is yet achievable when dealing with the principal challenges of global change, and, in any case, facts are not the only decisive factors. For example, conservation of biodiversity faces not just the fundamental problems of unidentified redundancy, but also those of emotional attachment. What fraction of the astounding invertebrate biodiversity is functionally essential? Clearly not all of it. And to what extent should our sentiments dictate heroic efforts to save a few individuals of so-called charismatic species (condors, leopards) whose regrettable demise would not affect any fundamental functions of the ecosystems in which they lived (Ehrenfeld 1986)? More fundamentally, who will arbitrate between the apparently irreconcilable perceptions and appraisals of national needs and socioeconomic aspirations that reflect the deep divide between the affluent countries and the rest of the world?

And how are we going to structure concerted global actions requiring multigenerational commitment to goals that will not remain fixed but will have to be adjusted to take into account inevitable surprises? The second half of the twentieth century saw several notable examples of these by definition utterly unexpected and rapid changes, some of them having truly global, and potentially most worrisome, effects. The discovery in Antarctic penguins of long-lived pesticides applied mostly in rich countries of the northern temperate zone (Sladen et al. 1966) and the sudden thinning of the stratospheric ozone layer above the continent (Farman, Gardiner, and Shanklin 1985) must be the two best-known examples of global importance.

But even in the unlikely absence of any major surprises, we will be forced to confront the fundamental moral dimensions of biospheric salvation. Valuing the biosphere, even if it is to be across just the civilizational (merely 10^3 years) rather than the evolutionary (10^9 years) span, cannot be accomplished through largely arbitrary processes of subdividing nature's realm into discrete services and then assigning to them globally averaged, functionally irrelevant, and questionably derived values. As I have argued above, the biosphere can be valued effectively only as a matter of a binding planetary compact uniting humanity in the acceptance of natural precedences and the need for limits on material consumption. This may initially sound like an excessively idealistic and impractically vague declaration of grand principles, but on reflection, it is merely a concise description of reality. In practice this means striving for the greatest possible preservation of environmental integrity. This goal is achievable by leaving large chunks of some of the most valuable ecosystems entirely or largely alone and by continuously trying to limit the environmental impacts associated with population growth and economic development.

The first objective could be achieved with relatively low expenditures. Bruner et al. (2001) found that parks and other protected areas are an effective means to safeguard tropical biodiversity and that even modest increases in their funding would be very helpful. James et al. (1999) calculated that the protection of the world's biodiversity—by securing adequate budgets for maintaining the protected areas (national parks and various nature reserves, now covering nearly 5% of the world's area) and by buying additional land to extend their coverage to a minimum standard of 10% of the area in every major region—would cost annually about $16.6 billion on top of the inadequate $6 billion spent currently. And Pimm et al. (2001) estimate that securing another 2 million km² and adequately managing the same amount of land that is already protected for biodiversity would need a one-time investment of about $4 billion.

These sums are small in comparison to monies the rich world spends on many frivolous, or outright harmful, endeavors. After all, even a number as high as $23 billion is equal to only about 0.1% of the combined GDP of the world's most affluent economies, and is a small fraction of environmentally harmful subsidies that governments of affluent countries extend to their agricultural and industrial production.

Pursuing the second objective requires a complex combination of approaches, including some well-established fixes as well as a much greater emphasis on new solutions. The first category includes autonomous technical improvements as well as increasingly common, but often too expensive and difficult-to-enforce, technical and monetary fixes ranging from limits on pollutant releases to penalties and taxes on environmentally harmful processes. By substantially lowering energy and material intensities, continuing technical advances have greatly reduced the environmental burdens of modern consumption. Some control measures have also worked well: particulate and sulfurous emissions, causing acid precipitation and widespread damage to sensitive aquatic and forest ecosystems, declined remarkably throughout the Western world during the second half of the twentieth century.

However, even a combination of these approaches will not suffice, as even admirable technical progress and aggressive controls may not be enough to keep in check the overwhelming tide of increased personal consumption in both affluent and modernizing countries. No deep analysis is required to see that human needs compatible with decent life are fairly limited; human wants, generated by the powerful combination of acquisitiveness, consumerist ethos, and relentless advertisement, are practically endless.

I have demonstrated that a decent quality of life, in both a physical and an intellectual sense (ranging from longevity to good access to postsecondary education) can be achieved with efficient annual energy use of about 80 GJ per capita (Smil 1991, 2000d). In contrast, the rich world's mean is now over 150 GJ per capita, and the North American average approaches 350 GJ per capita (BP 2000; fig. 9.13). If most of the modernizing world, led by China and India, were to follow this course, it is highly unlikely that the combination of technical innovation and mandated controls and taxes would suffice to preserve a desirable level of biospheric integrity.

Arguing for limits on consumption does not mean any Orwellian impositions of drastic restrictions, nor is the universal acceptance of such limits a realistic goal to be implemented globally within a set period of time. Gradual but sustained diffusion of such measures, practices, and approaches must begin in affluent countries, where populations acting in the spirit of moderation, regardless of whether they are motivated by higher scientific literacy or by tax incentives, could generate dramatic gains within relatively short periods of time. Recent claims of widespread opportunities for "factor ten" improvements in the efficiency of our resource use in a single generation must be rejected as utterly unrealistic (Hawken, Lovins and Lovins 1999). But there are no fundamental obstacles preventing affluent countries from cutting their use of energy, water, and raw materials to no more than half of today's rates within the next twenty-five to thirty years, and the attendant reductions in environmental impacts could be even more significant.

Concerns about population and environmental services can thus be best addressed by assuring at least a modicum of scientific literacy, by preferentially allocating both public and private investment toward the reduction of energy-intensive activities and toward the support of environmentally benign consumption. Even then, an essential ingredient would be missing: no lasting equilibrium between human desires and nature's capacity to sustain them

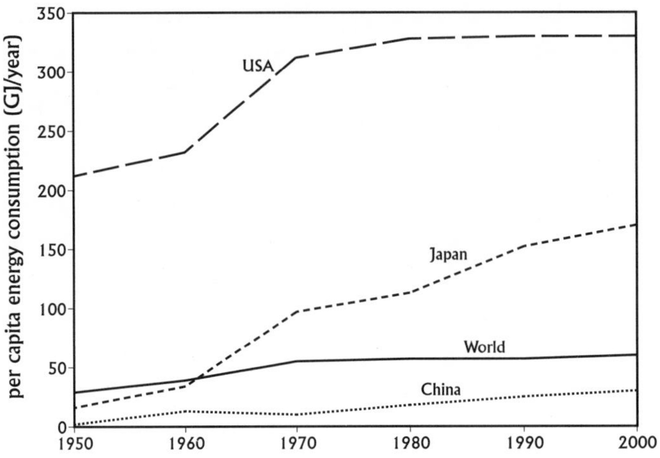

9.13 Comparisons of annual per capita consumption of primary commercial energy, 1950–2000: the North American rate remains an order of magnitude higher than the Chinese mean, and it is about four times as high as the minimum compatible with healthy and dignified life in a society striving to maximize the efficiency of its energy use. Data from UNO (1976) and BP (2000).

can be achieved without the willingness to accept, individually and collectively, limits on our acquisitiveness and to cap our expectations whenever they pose an irreparable threat to the biosphere. Technical fixes or tax incentives can ease the burden, but only the ethos of moderation derived from the sense of sharing and intergenerational responsibility (both being applicable to other life forms as well as to future populations) can guarantee the long-term integrity of the biosphere.

Guaranteeing such long-term biospheric integrity requires making its preservation an important purpose of human endeavor, a difficult task in a civilization in which biological orthodoxy worships random mutation and does not believe in purpose, and in which economists, using personal preferences, reduce value to taste (Daly 1999). But we should and can do better: life with dignity for billions of people cannot be sustained without a considerable toll on the biosphere, but we know that life with dignity does not have to be predicated on a massive destruction of the biosphere's irreplaceable functions. One of Horace's most famous sentences puts it better than tomes of computerized analyses:

Est modus in rebus, sunt certi denique fines, quos ultra citraque nequit consistere rectum.

(There are fixed limits beyond which and short of which right cannot find resting place.)

10

Epilogue

The parts of the whole, everything, I mean, which is naturally comprehended in the universe, must of necessity perish; but let this be understood in this sense, that they must undergo change.
Marcus Aurelius, *Meditations*

The growth rates of population and consumption that have marked the twentieth century cannot continue unabated during the twenty-first century without resulting in a massive degradation of the biosphere. If the Earth's population were to double yet again before the year 2100 and if the modernizing countries were to consume, on the average, at least a third of energy used per capita by affluent nations, the global demand for fuel and electricity could quadruple. If fossil fuels remain the dominant, albeit declining, source of that energy, then the cumulative carbon releases during the twenty-first century could easily be twice as large as during the twentieth century, and pronounced global warming could become evident in just a couple of decades. Such a climate change would weaken agricultural capacity just as the food requirements of the doubled global population and its higher demand for animal foods would eliminate all but a few token areas of tropical forests and as the use of irrigation water and applications of nitrogen and phosphorus fertilizers would more than double to sustain high yields.

The combination of climate change, biodiversity loss, water shortages, and nutrient overload could lead to severe disruptions of ecosystem functions and to the elimination of environmental services on local and regional scales, and some limits might be approached or reached even on the biospheric scale. Rapid warming could also disrupt the thermohaline circulation that transports huge amounts of heat from the Gulf of Mexico via the Gulf Stream to the North Atlantic and Europe and bring sudden, severe cooling to the continent (Stocker et al. 1997). Other civilizations have been set back, or even obliterated, by anthropogenic environmental change (Taintner 1988), but this would be the first instance of this process acting on the global scale. All of this could be made much worse by global epidemics of new infectious diseases, by the unintended creation and release of extremely invasive bioengineered organisms, or by large-scale military conflicts, if not a new global war (most of the nuclear warheads developed to date remain in place).

If biodiversity were reduced to a fraction of its current richness, if large parts of continents were to lose their capacity to perform essential environmental services, and if the ocean were also to experience mass anthropogenic extinction of species, then the biosphere's recovery could be a very prolonged affair. Kirchner and Weil (2000) showed that peaks in recoveries from mass marine extinctions seem to lag the peak of extinctions by about 10 Ma, a time span three orders of magnitude longer than the duration of human civilization. The only certain conclusion is that even a suicidal civilization could not pull down the curtain on the biosphere, though it could profoundly alter its course. But it would be absurd to pretend that we have even the merest inkling of what specific forms a recovered biosphere would follow in 10 or 50 Ma. Fortunately, there is nothing inevitable about this dismal scenario; indeed, it may have a rather low probability.

The biosphere's immediate future, in the first half of the twenty-first century, will definitely be marked by a continuing intensification of the human impacts upon it. Some encouraging local, regional, and national efficiency and conservation gains will be outweighed by environmental degradation elsewhere, and the net global effect of these changes will be a further deterioration of environmental quality and additional compromising of biospheric services. But there is a fairly high probability that the global population will not double yet again and that it can in fact stabilize at well below 10 billion people, perhaps only at 7.5–8.5 billion, during the second half of the twenty-first century (Lutz, Sanderson, and Scherbov 2001; UNO 2001; fig. 10.1). Such a stabilization, followed possibly by a gradual decline of human numbers, offers more hope for the second half of the century than for the first. Still, excessive per capita consumption could more than negate such a desirable shift, but technical innovations may go a long way toward accommodating the increasing material needs of the world's population.

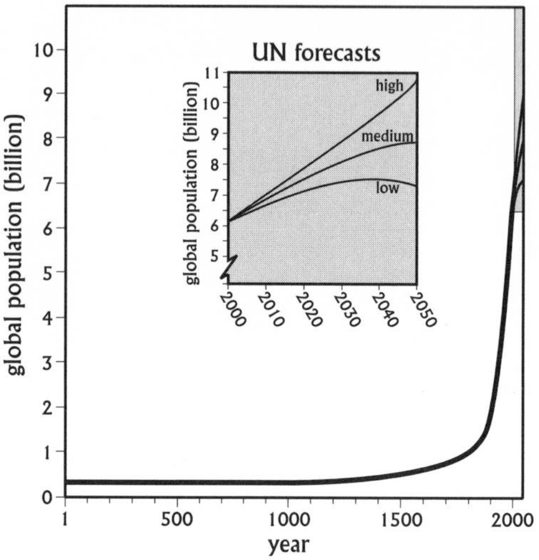

10.1 According to the latest United Nations forecasts, the global population may remain below 7.5 billion people by the year 2050, and it is highly unlikely that it will double yet again. Based on data in UNO (2001).

None of these technical advances would be more important than notable progress toward the long-awaited (and necessarily gradual) transition from fossil fuels to solar radiation as the dominant source of human energy needs. This transition should be accelerated, not because of concerns about imminent exhaustion of coal and hydrocarbon resources, but because the extraction and conversion of fossil fuels is the leading source of greenhouse gases as well as one of the most important causes of land degradation and water pollution. Consequently, the second half of this century may see the beginning of a new kind of civilization whose material needs and primary energy sources will be more closely matched with the biosphere's capacity to support them over the long term.

Other energy outcomes are imaginable. Low power densities and other difficulties of effectively harnessing immense flows of renewable energies (Smil 1991) may present such a challenge to modern societies dependent on high power fluxes that eventually the path of nuclear power could be aggressively pursued in addition to, or perhaps even instead of, the solar route. Fusion, rather than fission, would be the preferred conversion technique, as deuterium in the ocean, and later perhaps ^{3}He from the Moon and from outer plantes, would cover any imaginable energy needs for at least 10^3–10^4 years (Fowler 1997). But brilliant technical advances offering benign solutions to civilization's energy needs would not be enough, and they could actually worsen the human propensity for material consumption. Moral choices will be essential. Limiting and eventually reducing the human impact on the biosphere would require an unprecedented degree of international, indeed global, cooperation and, inevitably, collective willingness to forgo certain kinds of consumption.

Some scientists would dismiss this hope with a superior, DNA-based smirk. As Dawkins (1995) wrote, "time and again, cooperative restraint is thwarted by its own internal instability. . . . God's Utility Function betrays its origins in an uncoordinated scramble for selfish gain." To him, group welfare may be only a fortuitous consequence, but never a primary drive, in a world where there is "no evil and no good, nothing but a pitiless indifference" (p. 85). The combination of these genetic imperatives and the human propensity for irrational behavior and mass violence would then make any prospects for staving off the biospheric decline less, rather than more, likely as biospheric services weaken during the twenty-first century. Arthur C. Clarke (1999) imagined just such an eventual future when he described the musings of a superior civilization regarding the cataclysmic end of the Earth:

It is impossible to feel much regret in this particular case. The history of these huge creatures contains countless episodes of violence. Whether they would have made the necessary transition — as we did, ages ago — from carbon- to germanium-based consciousness, has been the subject of much debate. (p. 19)

But there are alternatives to Dawkins's merciless determinism and to Clarke's dark vision. Human societies have been violence-prone and have often shown an almost inexplicable resistance to change, but they are also capable of admirable cooperation and can embrace a new course swiftly if the necessity dictates. The very challenge of the unprecedented global change may actually foster both of these processes. Paul Ehrlich, who has for decades been very pessimistic about the human future, noted recently that "our challenge is to learn to deal sensibly with both nature and human natures," and although he is still pessimistic about whether we *will* do it, he now tends "to be optimistic in thinking that we *can* do it" (Ehrlich 2000, p. 330).

Undeniably, human societies have become progressively more complex, and as Teilhard de Chardin (1966) noted, their higher complexity begets greater awareness, and the growth of human numbers, the proliferation of connections, and the feeling of a common anxiety will catalyze the rise of a unified human consciousness. He argued that this process is inevitable if humanity is to survive and pointed out the undeniable progress already made in the formation of a new world:

[I]t is absolutely impossible for us to escape the forces that draw us together: in the pre-industrial periods of history their uncontrollable pressure increased almost unnoticed; today we see it brought suddenly into the open in all its strength. . . . Mankind is now caught up, as though in a train of gears, at the heart of a continually accelerating vortex of self-totalisation. (p. 99)

Nearly half a century after these words were written, we find ourselves amidst an accelerating progress toward the formation of what Teilhard de Chardin called "the noosphere (or thinking sphere) superimposed upon, and coextensive with (but in so many ways more close-knit and homogeneous) the biosphere" (Teilhard de Chardin 1966, p. 80).

The key ingredients of this new evolutionary step include not just those now so easily affordable means of instant global communication and the emergence of a readily accessible repository of civilizational knowledge within the World Wide Web, but also our incipient attempts at managing the global commons, thinking about sustainable economies, and slowly putting in place the basic structures of worldwide governance. I side with Teilhard de Chardin, rejecting Dawkins's DNA-foreordained destiny. I believe that our collective future is not predetermined, that cooperative endeavors can eventually correct most of our past blunders, and that human actions need not keep inflicting irreparable damage to the biosphere. These were also Vernadsky's feelings. In his writings he accurately described and correctly anticipated key universal concerns now facing the civilization: all is needed is to translate the labels.

Vernadsky's identification of humanity as a major transformer of the biosphere is now referred to as "global environmental change." His stress on the importance of electronic communications has become the cliché of the "global village," whose connections are now yet again refashioned by the Internet. And his insistence on a unity of humanity that cannot be opposed with impunity is evident in growing universal preoccupation with safeguarding human rights (Smil 2000a). Looking far ahead, Vernadsky considered the emergence of the noosphere as a critical evolutionary step needed for preserving and reconstructing the biosphere in the interest of humanity as a single entity. Consequently, his judgment about the long-term prospects for the biosphere was full of hope: "I look forward with great optimism. I think that we are experiencing not only an historical change, but a planetary one as well. We live in a transition to the noosphere" (Vernadsky 1945).

Because the future depends to such a large extent on choices we will make, we cannot assign any meaningful probabilities either to the transition to a cooperative biosphere-sparing world or to any scenario of a not too distant cataclysmic demise. In any case, forecasting any specifics for more than half a century ahead is a test of the imagination that must be accompanied by the realization that the probability of reality's conforming to such scenarios is basically zero. At that distance science fiction produced by paperback writers merges with arguments offered by physicists. For example, Dyson (1999b) predicts that the colonization of space will begin in about 2085 and that cloned and genetically manipulated *Homo* will look for suitable bodies most likely in the cometary Kuiper Belt, a ring-shaped region outside Neptune's orbit, or even further away in the Oort Cloud. He is convinced that "in the end we must travel the high road into space. . . . To give us room to explore the varieties of mind and body into which our genome can evolve, one planet is not enough" (Dyson 1999b, p. 113).

Still, some long-term conclusions about the biosphere's fate are possible. If effective global cooperation, drastically reduced violence, a stabilized population size, moderated consumption, and widespread biodiversity preservation were to become new planetary norms, then the next most likely jolt for the biosphere in general, and for circumpolar humans in particular, would be the new ice age. During the past 2.5 Ma the Earth has been considerably cooler than during the past 10 ka, with ice age conditions prevailing for about four-fifths of that time (see fig. 9.12). For the past 1 Ma the timing of ice ages has been driven primarily by inexorable cyclical changes in the eccentricity of the Earth's orbit, in its more or less elliptical course, with frequencies of 100 and 413 ka. This explanation, first detailed by Milanković (1941), has been confirmed by recent studies of oxygen isotopes (Rial 1999). Consequently, there is a very high probability that this remarkably persistent rhythm of glacial-interglacial eras will continue in the near future. A new ice age will not pose any threat to the survival of the biosphere, but its potential effects on civilization are incalculable.

The transition from foraging to settled agriculture and the subsequent complexification of human societies leading to the rise of the great extratropical civilizations took place during the unusually equable and moderate climates of the current interglacial, which has already lasted some 11,000 years. The last glacial episode, whose peak was about 18 ka ago, buried everything as far south as Chicago and Boston in North America and Wales and northern Poland in Europe under 10^2 m of ice and converted southern France and Kansas into cold grasslands (Wilson et al. 2000; fig. 10.2). But the populations affected at that time amounted to just 10^4 to 10^5 hunters and gatherers who simply slowly retreated before the advancing ice. A similar glaciation would now directly displace at least 150 million people and drastically alter agricultural productivity in areas that not only are today inhabited by about 250 million largely urban residents but also produce more than 3/4 of the world's grain exports.

Other natural disasters that could derail the recent course of evolution much more suddenly and in more dramatic ways were described in chapter 3. Supereruptions akin to those that created giant calderas of Toba in Sumatra (73.5 ka years ago) and in the Yellowstone (the last event, 630 ka ago) and were followed by years of decreased temperatures, accelerated glaciation, and drastic reductions in the size of the human population (Rampino and Self 1992; Ambrose 1998) are just a matter of time. Their impact will be both nearly instantaneous, as a result of the deposition of 10^9–10^{12} t of volcanic ash, which may cover everything within 10^3 km downwind under 10^1 cm of dust, and long-term, because of the persistence of volcanic aerosols in the stratosphere and the resulting volcanic winter and years of pronounced cooling.

And depending on how massive and how sustained a new flood-basalt volcanism would be, only a small part of the biosphere might disappear under layers of new lava (if the event were, for example, ten or fifty times the size of 1783 Laki event in Iceland), or the Earth could experience the loss of half or more of its species, after centuries of volcanism only one-tenth as extensive as the episode that created India's Deccan Traps 65 Ma ago.

Asteroid impact also remains always a worry. The odds are, of course, in favor of a smaller body affecting the open ocean, with limited damage to the biosphere. With larger asteroids, however, the point of impact is, of course, irrelevant, as the consequences would be global. The probability

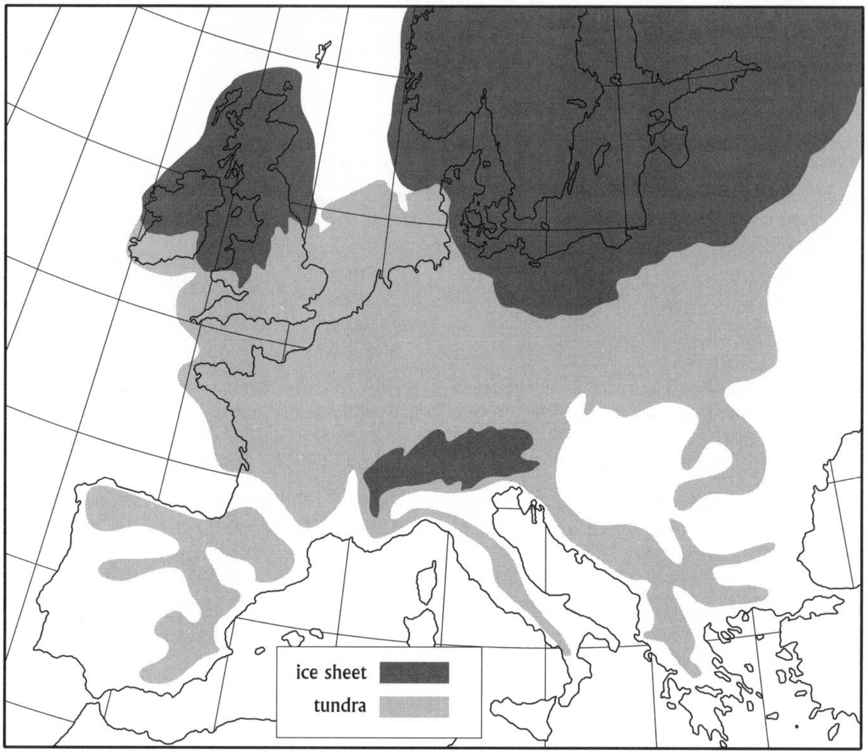

10.2 The greatest extent of glaciers and cold periglacial grasslands during the last ice age. Based on maps available at <http://www.esd.ornl.gov/projects/qen>.

of any of these global events taking place ranges from negligible during the next 5,000 years to virtually certain during the next 100 Ma (see fig. 3.15).

And there will be other events endangering human evolution, if not the very existence of the biosphere, of which we are still completely ignorant today. But there may be also welcome surprises. The biosphere may turn out to be much more resilient to human degradations than we now believe, or it may at least recover much more rapidly once we are not here to interfere anymore. In any case, the biosphere's inevitable end is much easier to envisage, because its ultimate fate remains tied to that of the Sun.

Long before that star will transform itself into a red giant, its increasing brightness will make the Earth uninhabitable. As the Sun's core contracts, its diameter will begin expanding, and increasing radiation from the red subgiant will heat the oceans. Huge volumes of the evaporated water will migrate to the stratosphere and be lost to space as fierce hot winds sweep through the Earth's atmosphere. After 4 Ga of evolution, the biosphere has perhaps no more than 500 Ma left: only about one-eighth of its entire existence lies ahead. Once the oceans are gone, even the sturdiest extremophiles will succumb to the soaring temperatures, and one billion years from now, only the lifeless, hot Earth will continue orbiting the brightening Sun. Then, some 5 Ga from now, the yellow G-2 star will leave the main sequence and will transform itself into a red giant, about 1,000 times more luminous than it is today. As its radius approaches the Earth's orbit, the planet will spiral inward and disappear in the red giant's core. Eventually the red giant will fizzle into a white dwarf, and this defunct degenerate core will cool slowly, a mere "clinker in the celestial ash heap" (Kaler 1992, p. 166).

But perhaps all of this, even the death of the solar system, will not be the end of the Earth's biosphere. Instead of any specific (and inherently worthless) forecasts, I will close the book by briefly outlining just one of many possible scenarios based on the undeniable rise of machines that could conceivably lead to the emergence of non-carbon-based consciousness. Signs of the steadily rising importance of machines in the biosphere have been obvious for a long time (Wesley 1974). Machine infrastructures—industrial plants, electricity-generating stations, roads, railroads, parking lots, switchyards, and warehouses—claim ever larger areas as they displace natural ecosystems. Although machines are generally much better converters of energy than humans, their metabolism produces a large array of airborne liquid and solid pollutants whose disposal, capture, and treatment have claimed increasing amounts of land for landfills, toxic dumps, and waste-cleaning devices and facilities and whose uncontrolled releases have destroyed or adversely affected biodiversity.

The aggregate mass of machines already greatly exceeds that of humans. The dry-matter anthropomass is about 100 Mt (see chapter 6), whereas the mass of motor vehicles (cars, buses, and trucks) alone is now an order of magnitude larger, in excess of 1 Gt. And machines now need more carbon every year than humans do. The global food harvest now amounts to about 1.3 Gt C per year (Smil 1999c), whereas almost 1 Gt of fossil carbon (mostly metallurgical coke and hydrocarbon feedstocks) is used annually to produce metals and plastics from which machines are assembled, and about 4 Gt C are used each year to power them, either directly with coal, oil, and natural gas, or indirectly with electricity generated in thermal stations.

Our ability to conceive, design, build, and perfect machines has been truly astounding. Even those machines whose design has been evolving rather conservatively (such as internal combustion engines or steam turbogenerators) have seen a growth of capacities and efficiencies that has been far faster than the agonizingly slow process of the biospheric evolution. And the new twentieth-century machines, distinguished by their information-processing capacities and their ability to make connections rather than by their mass, have been evolving much (10^6–10^7 times) faster than has any other organism or set of functional attributes in the long history of the biosphere. Moore's law of doubling the number of components placed on silicon-based integrated circuits (microchips) is certainly the best-known instance of this unprecedented evolutionary speed. Starting with a single planar transistor in 1959, the number of components per microchip doubled every year until 1972, and since then it has continued doubling every 18 months.

But the speed and density of computation had been increasing exponentially long before the advent of

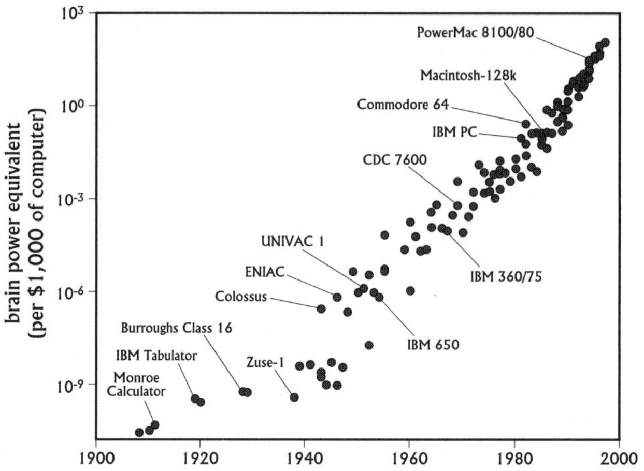

10.3 Continuous advances in mechanical and electromechanical calculators increased the speed of calculation 1000-fold compared to manual methods between 1900 and 1940, and electronic machines had brought a more than billion-fold increase above the 1940 level by 2000. Based on Moravec (1998).

microchips: this trend began with mechanical calculators and continued first with electromechanical machines and then with electronic computers that used first vacuum tubes and then discrete transistors before they incorporated integrated circuits (Kurzweil 1990; Moravec 1988, 1998; fig. 10.3). Consequently, there is a high probability that this trend will be sustained by other means once the current microchip-making techniques reach their physical limits. Bandwidth, the rapidity with which information can be transmitted through a network, has grown even faster, with the performance of optical fibers (measured in bits carried per second) doubling every nine months, and before too long the expected advances in fiber optics will supply network capacity approaching the infinite. Still, when compared on the basis of instructions per second and when judged as general-purpose performers, even the most powerful computers in 2000 had brain power equivalent to that of invertebrates.[1]

A considerable amount of technical evolution lies ahead before machines begin approaching human abilities and bridging the gap between nonliving and living matter (Brooks 2001). Even the synthesis of the simplest living cell with its hundreds of genes and thousands of proteins and other molecules is not yet a foreseeable reality, merely an imaginable goal (Szostak et al. 2001). In spite of these formidable obstacles, some machine enthusiasts have been

1. Special tasks are a different matter: in 1997 IBM's Deep Blue defeated Gary Kasparov, perhaps the most talented chess player in human history.

eager to predict the time when robotic intelligence rivals human capability and, controversially, look forward to the subordination of human intellect to machines (Moravec 1998; Kurzweil 1999). Without assessing the perspicacity or madness of such forecasts, I wish only to point out that if such new forms of consciousness were to arise, they would be incomparably less limited in choosing their new abodes, as they would not need the Earth's biosphere for their functioning.

Whenever it comes, if it comes, such a new form of machine-based consciousness would not need the Earth's biosphere for its functioning. Machines would not have to protect any molecules of life in order to effect intergenerational transfers of information and to produce new bodies. They would need just starlight or nuclear energy and access to raw structural materials for their replication and metabolism, as they can already perform in conditions far surpassing those tolerated by even the most extremophilic extremophiles (in a vacuum, near 0 K, in a radioactive environment, exposed to direct solar flux).

And so it is conceivable that the many truly intelligent machines that will have left the Earth long before the heat death of its biosphere would then have no need to terraform other planets as they diffuse through space. But other conscious machines may be so imbued with human values that they might settle down once they encounter planets resembling the Earth, and then use their exhaustive libraries of genomes, proteomes, and other molecules of life to direct synthetic processes that would recreate replicas of the Earth's biosphere on suitable celestial bodies. We have no way of appraising how fantastic or how timid these scenarios are, how ridiculous or how remotely realistic are the possibilities of this or any other tale of the very far future that we can think of.

The half billion years that are still available for the biosphere's evolution will surely bring developments no less fascinating, and obviously no less unpredictable, than those that have taken place during the equally long Phanerozoic eon. The natural and anthropogenic contingencies of that long journey will determine if we are among the last few generations to marvel at *Cataglyphis* ants and *Sequoiadendron* trees—or if these or similar organisms will survive even as the brightening Sun evaporates the Earth's biosphere.

Appendixes

Appendix A

Evolutionary Time Scale

Eon	Era	Period	Began (Ma ago)	Duration (Ma)
Phanerozoic	Cenozoic	Quaternary	1.64	1.64
		Tertiary	65	63.4
	Mesozoic	Cretaceous	144.2	79.2
		Jurassic	205.7	61.5
		Triassic	248.2	42.5
	Palaeozoic	Permian	290	45
		Carboniferous	362.5	72.5
		Devonian	408.5	46
		Silurian	439	30.5
		Ordovician	510	71
		Cambrian	570	60
Precambrian	Proterozoic			
	Sinian	Vendian	610	40
		Sturtian	800	190
	Riphean		1,600	800
	Animikean		2,200	400
	Huronian		2,450	150
	Archean			
	Randian		2,800	350
	Swazian		3,500	700
	Izuan		3,800	300
	Hadean		4,560	760

Sources: Harland et al. (1989) and Berggren et al. (1995).

Appendix B

Sizes and Masses of Organisms

Organisms	Length	Mass
	nm	ag
Viruses	50	60
	nm	fg
Minimum bacterial size	250	8
Oligotrophic bacteria	400	30
	µm	pg
Mycoplasma	2	< 1
Escherichia coli	3	2
Emiliania huxleyi	4	30
	µm	ng
Myxobacteria	10	50
	µm	µg
Dinoflagellates	100	50
	mm	mg
Amoebas	2	0.1
Coelenterates	10	50

Organism	Length	Mass
Bees	20	100
	cm	g
Grey voles	5	4
Mice	9	20
Rats	25	350
	cm	kg
Foxes	60	8
Chimpanzees	90	35
Humans (adult)	170	70
Gorillas	180	100
Camels	300	500
	m	t
Giraffes	4	1
Asian elephants	6	2.5
Blue whales	24	95
Giant sequoia	90	1,000

Sources: Prokaryotic sizes mostly from Margulis and Schwartz (1982); mammalian sizes and masses from Eisenberg (1981).

Appendix C

Milestones in the Evolution of the Earth and Its Biosphere

Event	Years before present
	Ga
Sun and accretionary disk formed	4.57
Earth accretion begins	4.55
Moon formed	4.51
Earth's accretion essentially complete	4.47
Earliest known zircon fragment	4.404
Earliest surviving continental crust	4.03
End of intense impacts	3.9
First hyperthermophiles	3.8–3.5
Cyanobacteria dominant	3.5–2.5
Oldest microfossils	3.465
Spurt of continental growth	3–2.6
First indications of eukaryotes	2.7
Eukaryotic/prokaryotic mats	2.3–1.5
Grypania	2.1
Diversification of marine algae	1.8–1.1
Rodinia formed	1.32–1.0
Algal mats, primitive lichens	1.5–0.85
Radiation of sulfide-oxidizing bacteria	1.06–0.64

Event	Years before present
	Ma
Divergence of protostomes	670
Freshwater algae	600
Appearance of chordates	600
Microscopic sponges	570
Gondwana formed	650–550
Cambrian explosion	533
Pangea formed	450–250
Land plants	450
Transition between fish and amphibians	365
Vascular plants dominant on land	360
Permian mass extinction	250
Radiation of modern insects	245
Gymnosperms dominant	200
First mammals	195
Archaefructus	140
Placental mammals	85
K/T boundary impact	65
Prosimians	54
Hominids begin to evolve	20
Hominid bipedalism	4
Homo sapiens	0.05

Appendix D

Biogeochemistry of Biospheric Cycles

Carbon Cycle

Weathering

Silicate weathering ($CaSiO_3$, wollastonite, representing all silicates)

$$CaSiO_3 + 2CO_2 + 3H_2O \rightarrow Ca^{2+} + 2HCO_3^- + H_4SiO_4$$

$$MgSiO_3 + 2CO_2 + 3H_2O \rightarrow Mg^{2+} + 2HCO_3^- + H_4SiO_4$$

Aluminosilicate weathering

$$CaAl_2Si_2O_8 + 3H_2O + 2CO_2 \rightarrow Al_2Si_2O_5(OH)_4 + Ca^{2+} + 2HCO^{-3}$$

$$2KAlSi_3O_8 + 3H_2O + 2CO_2 \rightarrow Al_2Si_2O_5(OH)_4 + 4SiO_2 + 2K^+ + 2HCO_3^-$$

Carbonate weathering (calcite and dolomite)

$$CaCO_3 + H_2CO_3 \rightarrow Ca^{2+} + 2HCO_3^-$$

$$CaMg(CO_3)_2 + 2H_2CO_3 \rightarrow Ca^{2+} + Mg^{2+} + 4HCO_3^-$$

Metamorphosis of carbonate sediments

$$CaCO_3 + SiO_2 \rightarrow CaSiO_3 + CO_2$$

Conversion of silicates to limestone

$$CaSiO_3 + CO_2 \rightarrow CaCO_3 + SiO_2$$

Chemoautotrophy

Aerobic chemoautotrophy (oxidation of H_2S by sulfur bacteria)

$$CO_2 + 4H_2S + O_2 \rightarrow CH_2O + 4S + 3H_2O$$

Anaerobic chemoautotrophy (oxidation of S or H_2S by green or purple sulfur bacteria)

$$3CO_2 + 2S + 5H_2O \rightarrow 3CH_2O + 4H^+ + 2SO_4^{2-}$$

$$CO_2 + 2H_2S \rightarrow CH_2O + H_2O + 2S$$

Anaerobic chemoautotrophy

$$CO_2 + 2H_2 \rightarrow CH_2O + H_2O$$

Photosynthesis

Anaerobic photosynthesis

$$3CO_2 + 2S + 5H_2O \rightarrow 3CH_2O + 2SO_4^{2-} + 4H^+$$

$$CO_2 + 2H_2 \rightarrow CH_2O + H_2O$$

Aerobic photosynthesis (cyanobacteria and plants)

$$CO_2 + H_2O \rightarrow CH_2O + O_2$$

Autotrophic respiration and decomposition (and weathering) of organic matter

$$CH_2O + O_2 \rightarrow CO_2 + H_2O$$

Appendix D

(continued)

Anaerobic decomposition

$$CH_2O + H_2O \rightarrow CO_2 + 2H_2$$

$$3CH_2O + H_2O \rightarrow CH_3COO^- + CO_2 + 2H_2 + H^+$$

Methanogenesis

$$4CO + 2H_2O \rightarrow CH_4 + 3CO_2$$

$$CO_2 + 4H_2 \rightarrow CH_4 + 2H_2O$$

$$2CH_3COOH \rightarrow 2\,CH_4 + 2\,CO_2$$

Methane oxidation

$$3CH_4 + 6O_2 \rightarrow 3CO_2 + 6H_2O$$

$$CH_4 + OH \rightarrow CH_3 + H_2O$$

Biomass combustion

$$CH_2O + O_2 \rightarrow CO_2 + H_2O$$

$$CH_2O + \tfrac{1}{2}O_2 \rightarrow CO + H_2O$$

Fossil fuel combustion

$$C + O_2 \rightarrow CO_2 \text{ (coal)}$$

$$CH_4 + 2O_2 \rightarrow CO_2 + 2H_2O \text{ (methane)}$$

Carbonate equilibria in seawater

$$CO_2 + H_2O \leftrightarrow H_2CO_3$$

$$H_2CO_3 \leftrightarrow H^+ + HCO_3^-$$

$$HCO_3^- \leftrightarrow H^+ + CO_3^{2-}$$

$$CO_2 + CO_3^{2-} + H_2O \leftrightarrow 2HCO_3^-$$

Precipitation and dissolution of carbonates

$$Ca^{2+} + 2HCO_3^- \leftrightarrow CaCO_3 + CO_2 + H_2O$$

Nitrogen Cycle

Nitrogen fixation by lighting

$$N_2 + O_2 \rightarrow 2NO$$

$$N_2 + O \leftrightarrow NO + N$$

$$N + O_2 \leftrightarrow NO + O$$

Appendix D

(continued)

Atmospheric oxidation of NO_x

$NO + O_3 \rightarrow NO_2 + O_2$

$2NO_2 + O_3 + H_2O \rightarrow 2H+ + 2NO_3^- + O_2$

$NO_2 + OH\cdot \rightarrow H^+ + NO_3^-$

NO_2 dissociation by sunlight

$NO_2 + O \rightarrow NO + O_2$

$NO_2 + O_2 \rightarrow NO + O_3$

Bacterial N fixation (*Rhizobium*)

$N_2 + 2H^+ + 3H_2 \rightarrow 2NH_4^+$

$N_2 + 3H_2 + 2H_2O \rightarrow 2NH_4{}^+ + 2OH^-$

Nitrification

$NH_4^+ + H_2O \rightarrow NO_2^- + 3H_2 + 2H^+$ (*Nitrosomonas*)

$NO_2^- + H_2O \rightarrow NO_3^- + H_2$ (*Nitrobacter*)

Nitrite reduction

$NO_2^- + 3H_2 + 2H^+ \rightarrow NH_4^+ + 2H_2O$

Nitrate reduction

$NO_3^- + 4H_2 + 2H^+ \rightarrow NH_4^+ + 3H_2O$

Denitrification

$2NO_3^- + 2CH_2O \rightarrow N_2O + H_2O + 2CO_2 + 2OH^-$

$N_2O + H_2 \rightarrow N_2 + H_2O$

$5S_2O_3^{2-} + 3NO_3^- + H_2O \rightarrow 10SO_4^{2-} + 4N_2 + 2H^+$ (*Thiobacillus denitrificans*)

NH_4^+ volatilization

$NH_4{}^+ + OH^- \rightarrow NH_3 + H_2O$

NH_4^+ oxidation

$2NH_4^+ + 3O_2 \rightarrow 2NO_2^- + 2H_2O + 4H^+$

Urea hydrolysis

$CO(NH_2)_2 + H^+ + 2H_2O \rightarrow 2NH_4^- + HCO_3^-$

Sulfur Cycle

Bacterial reduction of elemental sulfur (*Beggiatoa, Desulfovibrio*)

$S + H_2 \rightarrow H_2S$

Appendix D

(continued)

Oxidation of H_2S by sulfur bacteria

$$CO_2 + 4H_2S + O_2 \rightarrow CH_2O + 4S + 3H_2O$$

Oxidation of S or H_2S by green or purple sulfur bacteria

$$3CO_2 + 2S + 5H_2O \rightarrow 3CH_2O + 4H^+ + 2SO_4^{2-}$$

$$CO_2 + 2H_2S \rightarrow CH_2O + H_2O + 2S$$

Bacterial reduction of sulfate and related compounds (*Desulfutomaculum, Desulfobacter*)

$$SO_4^{2-} + 4H_2 + 2H^+ \rightarrow H_2S + 4H_2O$$

$$S_2O_3^{2-} + 4H_2 \rightarrow 2HS^- + 3H_2O$$

$$C_6H_{12}O_6 + 3H_2SO_4 \rightarrow 6CO_2 + 3H_2S + 6H_2O \text{ (\textit{Desulfovibrio})}$$

Oxidation of S or H_2S by aerobic chemoautotrophs (*Beggiatoa, Desulfurolobus, Sulfolobus, Thiobacillus*)

$$2S + 3O_2 + 2H_2O \rightarrow 2SO_4^{2-} + 4H^+$$

$$2H_2S + O_2 \rightarrow 2S + 2H_2O$$

Precipitation of sulfides

$$H_2S + Fe^{2+} \rightarrow FeS + 2H^+$$

$$2H_2S + 2FeOOH + 2H^+ \rightarrow FeS_2 + Fe^{2+} + 4H_2O$$

$$Fe^{2+} + H_2S \rightarrow FeS + 2H^+$$

$$FeS + S \rightarrow FeS_2$$

Oxidation of pyrite (*Thiobacillus ferrooxidans*)

$$FeS_2 + 3\tfrac{1}{2}O_2 + H_2O \rightarrow Fe^{2+} + 2SO_4^{2-} + 2H^+$$

Weathering of pyrite

$$2FeS_2 + 7\tfrac{1}{2}O_2 + 4H_2O \rightarrow Fe_2O_3 + 4SO_4^{2-} + 8H^+$$

Oxidation of S during the combustion of biomass and fossil fuels

$$S + O_2 \rightarrow SO_2$$

Atmospheric oxidations of SO_2

$$SO_2 + OH\cdot \rightarrow HOSO_2$$

$$SO_3 + H_2O \rightarrow SO_4^{2-} + 2H^+$$

$$SO_2 + H_2O_2 \rightarrow SO_4^{2-} + 2H^+$$

$$SO_2 + H_2O \leftrightarrow HSO_3^- + H^+ \leftrightarrow SO_3^{2-} + 2H^+$$

$$HSO_3^- + H_2O_2 \rightarrow SO_4^{2-} + H^+ + H_2O$$

Dissolution of sulfates

$$CaSO_4 \rightarrow Ca^{2+} + SO_4^{2-}$$

Appendix D

(continued)

Deposition of gypsum

$$8Ca^{2+} + 16HCO_3^- + 8SO_4^{2-} + 16H^+ \rightarrow 8CaSO_4{\cdot}2H_2O + 16CO_2$$

Neutralization of sulfate

$$2NH_3 + 2H^+ + SO_4^{2-} \rightarrow (NH_4)_2SO_4$$

$$2CaCO_3 + 2H^+ + SO_4^{2-} \rightarrow 2Ca^{2+} + 2HCO_3^- + SO_4^{2-}$$

Note: Phytomass is designated by the formula CH_2O, whose elemental ratios are typical of many common carbohydrates making up the bulk of living matter.

Appendix E

Estimates of the Biosphere's Phytomass

Year	Authors	Phytomass (Gt C) Land	Ocean	Total
1926	Vernadskii		10^6	
1940	Vernadskii		10^3	
1937	Noddack		270	
1966	Bowen	507	2.0	509
1970	Bolin	450		
1971	Kovda	1395		
1971	Bazilevich et al.	1080		
1975	Whittaker and Likens	827	1.75	829
1976	Baes et al.	680		
1979	Ajtay et al.	560		
1979	Bolin et al.	590		
1982	Brown and Lugo	500		
1983	Olson et al.	559	3	562
1984	Matthews	734		
1984	Goudriaan and Ketner	594		
1987	Esser	657		
1992	Holmén	560	3	563
1992	T. M. Smith et al.	737		
1993	Hall and Scurlock	560	3	563
1993	Siegenthaler and Sarmiento	550	3	553
1994	Foley	801		
1997	Post et al.	780		
1998	Amthor et al.	486		
1998	Field et al.	500	1	501
1999	Potter	651		

Appendix F

Estimates of the Biosphere's Heterotrophic Biomass

Organisms	Biomass estimates (Mt C)	Organisms	Biomass estimates (Mt C)
Prokaryotes		Land (continued)	
Soils	15,000–26,000	Elephants	0.1
Waters	1,500–13,700	Domesticated vertebrates	100–120
Subterranean	22,000–215,000	Humans	40
Subsea	?–303,000	Ocean	
Land		Invertebrates	300–500
Fungi	3,000–6,000	Fish	< 40
Invertebrates	400–1,000	Whales	5–15
Wild vertebrates	< 5		

Sources: Bowen (1966); Bogorov (1969); Whittaker and Likens (1973); Hinga (1979); Romankevich (1988); Smil (1991); Whitman et al. (1998); Wilhelm and Suttle (1999); and my new calculations based on the latest population estimates for elephants and whales collated, respectively, by the IUCN (2001) and by the IWC (2001).

Appendix G

Selected Estimates of the Biosphere's NPP

NPP (Gt C/year)				
Year	Authors	Land	Ocean	Total
1862	Liebig	63		
1882	Ebermayer	25		
1908	Arrhenius	27		
1919	Schroeder	16		
1937	Noddack	15	29	44
1944	Riley	20	126	146
1958	Fogg	21	27	48
1960	Deevey			82
1960	Müller	10	25	35
1964	Lieth	32–44		
1966	Bowen	106	29	135
1968	Koblents-Mishke et al.		27–32	
1969	Ryther		20	
1970	Bazilevich et al.	78	27	105
1972	Golley	40	25	65
1973	Lieth	45		
1975	Whittaker and Likens	53	25	78
1976	Baes et al.	56		
1979	Ajtay et al.	60		
1979	De Vooys		46	
1983	Olson et al.	60		
1985	Smil	51		
1988	Box	68		
1993	Hall and Scurlock	60	46	106
1993	Melillo et al.	53		
1993	Potter et al.	48		
1994	Foley	62		
1994	Warnant et al.	65		

Appendix G

(continued)

1996	Antoine et al.		37–46	
1996	Ruimy et al.	59		
1997	Behrenfeld and Falkowski		48	
1998	Field et al.	56	49	105
1998	Amthor et al.	59		
1999	Potter	56		

Appendix H

Selected Web Sites

AVHRR (global land cover)

<http://orbit-net.nesdis.noaa.gov/arad/fpdt/nwasat.html>

Biodiversity and Biological Collections Web Server

<http://www.biodiversity.uno.edu>

Biomass burning

<http://weather.engin.umich.edu/geia/projects/bio.html>

Carbon Dioxide Information and Analysis Centre

<http://cdiac.esd.ornl.gov>

CERES (Clouds and the Earth's Radiant Energy System)

<http://asd-www.larc.nasa.gov/ceres/ceres_bottom.html>

Climatology

<http://daac.gsfc.nasa.gov>

Deep Green Project

<http://ucjeps.berkeley.edu/DeepGreen>

ERBE missions

<http://asd-www.larc.nasa.gov/erbe/ASDerbe.html>

El Niño-La Niña

<http://topex-www.jpl.nasa.gov/science/elnino.html>

Extrasolar planet encyclopedia

<http://www.obspm.fr/encycl/encycl.html>

Global change master directory

<http://gcmd.nasa.gov>

Global Observing System Information Center

<http://www.gos.udel.edu>

International Geosphere-Biosphere Program

<http://www.igbp.kva.se/cgi-bin/php/first_page.php>

JASON (global ocean circulation)

<http://topex-www.jpl.nasa.gov>

Leaf area index (global)

<http://Earthobservatory.nasa.gov:81/Newsroom/Stories/LAI_FPAR_data.html>

Lightning (global)

<http://thunder.msfc.nasa.gov>

National Atmospheric Deposition Program

<http://nadp.sws.uiuc.edu/isopleths>

Oceans

<http://podaac.jpl.nasa.gov/info/ftp.html>

Precipitation (global)

<http://orbit-net.nesdis.noaa.gov/arad/gpcp>

Solar and Heliospheric Observatory

<http://sohowww.nascom.nasa.gov>

Stratospheric ozone

<http://jwocky.gsfc.nasa.gov>

<http://wwww.epa.gov/ozone/science/hole/size.html>

TERRA

<http://terra.nasa.gov>

Terrestrial biosphere

<http://daac.gsfc.nasa.gov/CAMPAIGN_DOCS/LAND_BIO/GLBDST_main.html>

TOPEX

<http://topex-www.jpl.nasa.gov/mission/jason-1.html>

Tree of Life

<http://tolweb.org/tree/phylogeny.html>

Vegetation (global)

<http://www2.ncdc.noaa.gov/docs/gviug/index.htm>

Visible Earth

<http://visibleearth.nasa.gov>

Water vapor (global)

<http://www.ghcc.msfc.nasa.gov/GOES/globalwv.html>

References

Abbott, A. 1999. Battle lines drawn between "nanobacteria" researchers. *Nature* 401:105.

Aczel, A. 1998. *Probability 1: Why There Must Be Intelligent Life in the Universe.* New York: Harcourt Brace.

Adams, K. L., et al. 2000. Repeated, recent and diverse transfers of a mitochondrial gene to the nucleus in flowering plants. *Nature* 408:354–357.

Agassi, M., ed. 1995. *Soil Erosion, Conservation and Rehabilitation.* New York: Marcel Dekker.

Ager, D. 1993. *The New Catastrophism.* Cambridge: Cambridge University Press.

Agren, G. I., and E. Bosatta. 1998. *Theoretical Ecosystem Ecology: Understanding Element Cycles.* New York: Cambridge University Press.

Ajtay, G. L., P. Ketner, and P. Duvigneaud. 1979. Terrestrial primary production and phytomass. In *The Global Carbon Cycle,* ed. B. Bolin et al., 129–181. New York: John Wiley.

Aksenov, G. P., ed. 1993. *Vladimir Vernadskii: Zhizneopisanie, izbrannye trudy, vospominaniia sovremennikov, suzhdeniia potomkov* (Vladimir Vernadsky: Biography, selected works, reminiscences of contemporaries, opinions of descendants). Moscow: Sovremennik.

Alberts, B. 1998. The cell as a collection of protein machines: Preparing the next generation of molecular biologists. *Cell* 92:291–294.

Albritton, C. C. 1975. *Philosophy of Geohistory, 1785–1970.* Stroudsburg, PA: Dowden, Hutchinson & Ross.

Albritton, C. C. 1989. *Catastrophic Episodes in Earth History.* London: Chapman & Hall.

Allen, M. F., ed. 1992. *Mycorrhizal Functioning: An Integrative Plant-Fungal Process.* New York: Chapman & Hall.

Allen, M. R., et al. 2000. Quantifying the uncertainty in forecasts of anthropogenic climate change. *Nature* 407:617–620.

Allenby, B. 2000. Earth systems engineering and management. *IEEE Technology and Society Magazine* 19(4):10–24.

Alley, R. B. 2000. Ice-core evidence of abrupt climate changes. *Proceedings of the National Academy of Sciences USA* 97:1331–1334.

Alphey, L. 1997. *DNA Sequencing*. New York: Springer-Verlag.

Alroy, J. 2001. Effects of sampling standardization on estimates of Phanerozoic marine diversification. *Proceedings of the National Academy of Sciences USA* 98:6261–6266.

Alroy, J. P., P. L. Koch, and J. C. Zachos. 2000. Global climate change and North American mammalian evolution. *Paleobiology* (Supplement to No. 4).

Altmeyer, P., K. Hoffman, and M. Stücker, eds. 1997. *Skin Cancer and UV Radiation*. Berlin: Springer-Verlag.

Alvarez, L. W., et al. 1980. Extraterrestrial cause for the Cretaceous/Tertiary extinction. *Science* 208:1095–1098.

Alvarez, W. 1997. *T. rex and the Crater of Doom*. Princeton, NJ: Princeton University Press.

Ambroggi, R. P. 1977. Underground reservoirs to control water cycle. *Scientific American* 236(5):21–27.

Ambrose, S. H. 1998. Late Pleistocene population bottlenecks, volcanic winter, and differentiation of modern humans. *Journal of Human Evolution* 34:623–651.

Amthor, J. S., and members of the Ecosystem Working Group. 1998. *Terrestrial Responses to Global Change*. Oak Ridge, TN: Oak Ridge National Laboratory.

Anagnostakis, S. L., 1987. Chestnut blight: The classical problem of an introduced pathogen. *Mycologia* 79:23–37.

Anderson, A. B., P. H. May, and M. J. Balick. 1991. *The Subsidy from Nature: Palm Forests, Peasantry, and Development on an Amazon Frontier.* New York: Columbia University Press.

Anderson, J. W. 1986. *Bioenergetics of Autotrophs and Heterotrophs*. London: E. Arnold.

Antoine, D., J. M. André, and A. Morel. 1996. Oceanic primary production. 2. Estimation at global scale from satellite (coastal zone color scanner) chlorophyll. *Global Biogeochemical Cycles* 10:57–69.

Arabidopsis Genome Initiative. 2000. Analysis of the genome sequence of the flowering plant *Arabidopsis thaliana*. *Nature* 408:796–815.

Archibald, J. D. 1996. *Dinosaur Extinction and the End of an Era: What the Fossils Say*. New York: Columbia University Press.

Arkin, P., and P. Xie. 1994. The global precipitation climatology project: First algorithm intercomparison project. *Bulletin of the American Meteorological Society* 75:401–419.

Arking, A. 1996. Absorption of solar energy in the atmosphere: Discrepancy between model and observations. *Science* 273:779–782.

Arrhenius, S. 1896. On the influence of carbonic acid in the air upon the temperature of the ground. *Philosophical Magazine* 41:237–276.

Arrhenius, S. 1908. *Worlds in the Making: The Evolution of the Universe*. New York: Harper & Brothers.

Arroyo, M. T. K., P. H. Zedler, and M. D. Fox, eds. 1995. *Ecology and Biogeography of Mediterranean Ecosystems in Chile, California, and Australia*. Berlin: Springer-Verlag.

Ashcroft, F. 2000. *Life at the Extremes*. Berkeley: University of California Press.

Aunan, K., et al. 2000. Surface ozone in China and its possible impact on agricultural crop yields. *Ambio* 29:294–301.

Ayala, F. J. 1998. Is sex better? Parasites say "no." *Proceedings of the National Academy of Sciences USA* 95:3346–3348.

Ayala, F. J., A. Rzhetsky, and F. J. Ayala. 1998. Origin of the metazoan phyla: Molecular clocks confirm paleontological estimates. *Proceedings of the National Academy of Sciences USA* 95:606–611.

Ayres, R. U., and L. A. Febre. 1999. *Nitrogen consumption in the United States*. Fontainebleau, France: INSEAD.

Baes, C. F., et al. 1976. *The Global Carbon Dioxide Problem*. Oak Ridge, TN: Oak Ridge National Laboratory.

Bailes, K. E. 1990. *Science and Russian Culture in an Age of Revolutions: V. I. Vernadsky and His Scientific School*. Bloomington: Indiana University Press.

Baker, D. 2000. A surprising simplicity to protein folding. *Nature* 405:39–42.

Baker, N., ed. 1996. *Photosynthesis and the Environment.* Dordrecht, the Netherlands: Kluwer Academic.

Baker, R., ed. 1981. *The Mystery of Migration.* New York: Viking.

Bakker, R. T. 1975. Experimental and fossil evidence for the evolution of tetrapod bioenergetics. In *Perspectives of Biophysical Ecology,* ed. D. M. Gates and R. G. Schmerl, 365–399. New York: Springer-Verlag.

Balandin, R. K. 1982. *Vladimir Vernadskii.* Moscow: Mir Publishers.

Banavar, J., A. Maritan, and A. Rinaldo. 1999. Size and form in efficient transportation networks. *Nature* 399:130–132.

Barbosa, P. M., et al. 1999. An assessment of vegetation fire in Africa (1981–1991): Burned areas, burned biomass, and atmospheric emissions. *Global Biogeochemical Cycles* 13:933–950.

Barnes, B. V., et al. 1998. *Forest Ecology.* New York: John Wiley.

Barns, S. M., et al. 1996. Remarkable Archaeal diversity in Yellowstone National Park hot spring environment. *Proceedings of the National Academy of Sciences USA* 91:1609–1613.

Barton, N. H., and B. Charlesworth. 1998. Why sex and recombination. *Science* 282:1986–1990.

Bateman, R. M., et al. 1998. Early evolution of land plants: Phylogeny, physiology, and ecology of the primary terrestrial radiation. *Annual Review of Ecology and Systematics* 29:263–292.

Bates, T. S., et al. 1992. Sulfur emissions to the atmosphere from natural sources. *Journal of Atmospheric Chemistry* 14:315–337.

Battle, M., et al. 1996. Atmospheric gas concentrations over the past century measured in air from firn at the South Pole. *Nature* 383:231–235.

Battle, M., et al. 2000. Global carbon sinks and their variability inferred from atmospheric O_2 and $\delta^{13}C$. *Science* 287:2467–2470.

Bazilevich, N. I., et al. 1971. Geographical aspects of biological productivity. *Soviet Geography* 12:293–317.

Beavan, C. 1997. Underwater daredevils. *Atlantic Monthly* 279(5):106–110.

Beckel, L. 1998. *The Atlas of Global Change.* New York: Macmillan.

Beerling, D. J., and R. A. Berner. 2000. Impact of a Permo-Carboniferous high O_2 event on the terrestrial carbon cycle. *Proceedings of the National Academy of Sciences USA* 97:12428–12432.

Beerling, D. J., C. P. Osborne, and W. G. Chaloner. 2001. Evolution of leaf-form in land plants liked to atmospheric CO_2 decline in the late Palaeozoic era. *Nature* 410:352–354.

Behrenfeld, M. J., and P. G. Falkowski. 1997. Photosynthetic rates derived from satellite-based chlorophyll concentration. *Limnology and Oceanography* 42:1–20.

Bell, G. 1997. *Selection: The Mechanism of Evolution.* London: Chapman & Hall.

Bellwood, D. R., and T. P. Hughes. 2001. Regional-scale assembly rules and biodiversity of coral reefs. *Science* 292:1532–1534.

Bengtson, S., ed. 1994. *Early Life on Earth.* New York: Columbia University Press.

Benton, M. J., M. A. Willis, and R. Hitchin. 2000. Quality of the fossil record through time. *Nature* 403:534–537.

Berger, A. 1991. Long-term history of climate ice ages and Milankovitch periodicity. In *The Sun in Time,* ed. C. P. Sonett, M. S. Giampapa, and M. S. Matthews, pp. 498–510. Tucson: University of Arizona Press.

Berggren, W. A., and J. A. Van Couvering, eds. 1984. *Catastrophes and Earth History: The New Uniformitarianism.* Princeton, NJ: Princeton University Press.

Berggren, W. A., et al., eds. 1995. *Geochronology, Time Scales, and Global Stratigraphic Correlation.* Tulsa, OK: Society for Sedimentary Geology.

Berner, R. A. 1998. The carbon cycle and CO_2 over Phanerozoic time: The role of land plants. *Philosophical Transactions of the Royal Society London B* 353:75–82.

Berner, R. A., A. C. Lasaga, and R. M. Garrels. 1983. The carbonate-silicate geochemical cycle and its effect on atmospheric carbon dioxide over the past 100 million years. *American Journal of Science* 283:641–683.

Bernstein, I. S., and E. O. Smith, eds. 1979. *Primate Ecology and Human Origins: Ecological Influences on Social Organization.* New York: Garland STPM Press.

Bernstein, M. P., C. A. Sanford, and L. J. Allamandola. 1999. Life's far-flung raw materials. *Scientific American* 281(1):42–49.

Berthold, P., and S. B. Terrill. 1991. Recent advances in studies of bird migration. *Annual Review of Ecology and Systematics* 22:357–378.

Bethe, H. 1996. *Selected Works of Hans A. Bethe.* River Edge, NJ: World Scientific.

Betts, R. A. 2000. Offset of the potential carbon sink from boreal forestation by decreases in surface albedo. *Nature* 408:187–190.

Bintrim, S. B., et al. 1997. Molecular phylogeny of Archaea from soil. *Proceedings of the National Academy of Sciences USA* 94:277–282.

Blackmore, S. 1996. Knowing the Earth's biodiversity: Challenges for the infrastructure of systematic biology. *Science* 274:63–64.

Blackwell, M. 2000. Terrestrial life—Fungal from the start? *Science* 289:1884–1885.

Blagden, C. 1775. Experiments and observations in a heated room. *Philosophical Transactions of the Royal Society London* 65:111–123.

Blankenship, R., et al. 1995. *Anoxygenic Photosynthetic Bacteria.* Dordrecht, the Netherlands: Kluwer Academic.

Blöchl, E., et al. 1997. *Pyrolobus fumarii,* gen. and sp. Nov., represents a novel group of archaea, extending the temperature limit for life to 113 °C. *Extremophiles* 1:14–21.

Block, S. M. 1997. Real engines of creation. *Nature* 386:217–219.

Boddey, R. M., et al. 1995. Biological nitrogen fixation associated with sugar cane and rice: Contributions and prospects for improvement. *Plant Soil* 174:195–209.

Bogorov, V. G. 1969. *Zhizn' okeana* (*Life of the Ocean*). Moscow: Znaniye.

Bolin, B. 1970. The carbon cycle. *Scientific American* 223(3):124–132.

Bolin B., E. T. Degens, P. Duvigneaud, and S. Kempe. 1979. The global biogeochemical carbon cycle. In *The Global Carbon Cycle,* ed. B. Bolin, E. T. Degens, S. Kempe, and P. Ketner, 1–56. New York: John Wiley.

Bosecker, K. 1997. Bioleaching: Metal solubilization by microorganisms. *FEMS Microbiology Review* 20:591–604.

Boss, A. 1998. *Looking for Earths: The Race to Find New Solar Systems.* New York: John Wiley.

Botsford, L. W., J. C. Castilla, and C. H. Peterson. 1997. The management of fisheries and marine ecosystems. *Science* 277:509–515.

Bousquet, P., et al. 2000. Regional changes in carbon dioxide fluxes of land and oceans since 1980. *Science* 290:1342–1346.

Bowen, H. J. M. 1966. *Trace Elements in Biochemistry.* New York: Academic.

Bowring, S. A., et al. 1993. Calibrating rates of early Cambrian evolution. *Science* 261:1293–1298.

Box, E. O. 1988. Estimating the seasonal carbon source-sink geography of a naturally steady-state terrestrial biosphere. *Journal of Applied Meteorology* 27:1109–1124.

Boyd, P. W., et al. 2000. A mesoscale phytoplankton bloom in the polar Southern Ocean stimulated by iron fertilization. *Nature* 407:695–702.

BP. 2000. Statistical review of world energy 1999. <http://www.bp.com/worldenergy>.

Bradley, J. F. N. 1975. *Civil War in Russia, 1917–1920.* New York: St. Martin's.

Branden, C., and J. Tooze. 1991. *Introduction to Protein Structure.* New York: Garland.

Breuer, D., and T. Spohn. 1995. Possible flush instability in the mantle convection at the Archean-Proterozoic transition. *Nature* 378:608–610.

Breymeyer, A. I., and G. M. Van Dyne, eds. 1980. *Grasslands, Systems Analysis and Man*. Cambridge: Cambridge University Press.

Brian, M. V., ed. 1978. *Production Ecology of Ants and Termites*. Cambridge: Cambridge University Press.

Brimblecombe, P. 1987. *The Big Smoke*. London: Routledge.

Brocks, J. J., et al. 1999. Archean molecular fossils and the early rise of eukaryotes. *Science* 285:1033–1036.

Broda, E. 1975. *The Evolution of Bioenergetic Processes*. New York: Pergamon.

Broecker, W. S. 1995. Chaotic climate. *Scientific American* 273(5):62–68.

Broecker, W. S. 1997. Thermohaline circulation, the Achilles heel of our climate system: Will man-made CO_2 upset the current balance? *Science* 278:1582–1588.

Brooks, R. 2001. The relationship between matter and life. *Nature* 409:409–411.

Broughton, W. J., and S. Puhler, eds. 1986. *Nitrogen Fixation*. Oxford: Clarendon.

Brovkin, V. N. 1994. *Behind the Front Lines of the Civil War: Political Parties and Social Movements in Russia, 1918–1922*. Princeton, NJ: Princeton University Press.

Brown, G. 2000. *The Energy of Life*. New York: Free Press.

Brown, G. G. 1995. How do earthworms affect microfloral and faunal community diversity? *Plant & Soil* 170:209–231.

Brown, J. H., and G. B. West, eds. 2000. *Scaling in Biology*. Oxford: Oxford University Press.

Brown, S., and A. E. Lugo. 1982. The storage and production of organic matter in tropical forests and their role in the global carbon cycle. *Biotropica* 14:161–187.

Bruner, A. G., et al. 2001. Effectiveness of parks in protecting tropical biodiversity. *Science* 291:125–127.

Brunig, E. F. 1983. Vegetation structure and growth. In *Tropical Rain Forest Ecosystems*, ed. F. B. Golley, pp. 49–75. Amsterdam: Elsevier.

Bryant, D. A., ed. 1994. *The Molecular Biology of Cyanobacteria*. Dordrecht, the Netherlands: Kluwer Academic.

Bryant, D., et al. 1997. *The Last Frontier Forests*. Washington, DC: World Resources Institute.

Budyko, M. 1999. Climate catastrophes. *Global and Planetary Change* 20:281–288.

Buermann, W., and R. Myneni. 2000. Data sets of global vegetation (LAI and FPAR) now available. <http://earthobservatory.nasa.gov/Newsroom/Stories/LAI-FPAR-data.html>.

Bult, C. J., et al. 1996. Complete genome sequence of the methanogenic archaeon, *Methanococcus jannaschii*. *Science* 273: 1058–1073.

Burrow, A. 2000. Supernova explosions in the universe. *Nature* 403:727–733.

Burrows, A., and R. Angel. 1999. Direct detection at last. *Nature* 402:732–733.

Butcher, S. S., et al., eds. 1992. *Global Biogeochemical Cycles*. London: Academic.

Butler, A. 1998. Acquisition and utilization of transition metal ions by marine organisms. *Science* 282:207–210.

Butler, J. H., et al. 1999. A record of atmospheric halocarbons during the twentieth century from polar firn air. *Nature* 399:749–755.

Cagin, S., and P. Dray. 1993. *Between Earth and Sky: How CFCs Changed Our World and Endangered the Ozone Layer.* New York: Pantheon.

Calder, W. A. 2000. Diversity and convergence: Scaling for conservation. In *Scaling in Biology*, ed. J. Brown and G. West, 297–323. New York: Oxford University Press.

Calvin, M. 1989. Forty years of photosynthesis and related activities. *Photosynthesis Research* 211:3–16.

Canfield, D. E., K. S. Habicht, and B. Thamdrup. 2000. The Archean sulfur cycle and the early history of atmospheric oxygen. *Science* 288:658–661.

Canfield, D. E., and A. Teske. 1996. Late Proterozoic rise in atmospheric oxygen concentration inferred from phylogenetic and sulphur-isotope studies. *Nature* 382:127–132.

Capone, D. G., et al. 1997. *Trichodesmium,* a globally significant marine cyanobacterium. *Science* 276:1221–1229.

Carbon Dioxide Assessment Committee. 1983. *Changing Climate.* Washington, DC: National Academy Press.

Carson, R. 1962. *Silent Spring.* Boston: Houghton Mifflin.

Cartledge, T. G., et al. 1992. *Biosynthesis and the Integration of Cell Metabolism.* Oxford: Butterworth-Heinemann.

Cary, S. C., et al. 1998. Worms bask in extreme temperatures. *Nature* 391:545–546.

Cech, T. 1986. RNA as an enzyme. *Scientific American* 255(5):64–75.

Center for Atmospheric Science. 2000. The ozone hole tour. <www.atm.ch.cam.uk/tour>.

Cerling, T. E., et al. 1997. Global vegetation change through the Miocene/Pliocene boundary. *Nature* 389:153–158.

Cess, R. D., et al. 1995. Absorption of solar radiation by clouds: Observations versus models. *Science* 267:496–499.

Cess, R. D., and M. H. Zhang. 1996. How much solar radiation do clouds absorb? *Science* 271:1133–1134.

Chao, L. 2000. The meaning of life. *BioScience* 50:245–250.

Chapman, J. L., and M. R. Reiss. 1999. *Ecology.* Cambridge: Cambridge University Press.

Chapman, S. 1959. *IGY: Year of Discovery.* Ann Arbor: University of Michigan Press.

Charlson, R. J., et al. 1987. Oceanic phytoplankton, atmospheric sulfur, cloud albedo and climate. *Nature* 326:655–661.

Chatterjee, S. 1998. *The Rise of Birds: 225 Million Years of Evolution.* Baltimore: Johns Hopkins University Press.

Chatton, E. 1937. *Titres et travaux scientifiques.* Sottano, Italy: Setes Place.

Chisholm, S. W., et al. 1988. A novel free-living prochlorophyte abundant in the oceanic euphotic zone. *Nature* 334:340–343.

Church, T. M. 1996. An underground route for the water cycle. *Nature* 380:579–580.

Chyba, C. F., and McDonald G. D. 1995. The origin of life in the solar system: Current issues. *Annual Review of Earth and Planetary Sciences* 23:215–249.

Clark, S. 1999. Polarized starlight and the handedness of life. *American Scientist* 87:336–343.

Clarke, A. C. 1999. Improving the neighborhood. *Nature* 402:19.

Clarke, M. R., H. R. Martins, and P. Pascoe. 1993. The diet of sperm whales (*Physeter macrocepalus* Linnaeus 1758) off the Azores. *Philosophical Transactions of the Royal Society London* B 339:67–82.

Cleveland C. C., et al. 1999. Global patterns of terrestrial biological (N_2) fixation in natural ecosystems. *Global Biogeochemical Cycles* 13:623–645.

Climatic Environment Monitoring with GPS Atmospheric Profiling (CLIMAP). 1976. The surface of the ice-age Earth. *Science* 191:1131–1137.

Cloud, P. 1968. Atmospheric and hydrospheric evolution on the primitive earth. *Science* 160:729–736.

Coale, K. H., et al. 1996. A massive phytoplankton bloom induced by an ecosystem-scale iron fertilization in the equatorial Pacific Ocean. *Nature* 383:495–501.

Cody, G. D., et al. 2000. Primordial carbonylated iron-sulfur compounds and the synthesis of pyruvate. *Science* 289:1337–1339.

Cohen, J., and I. Stewart. 2001. Where are the dolphins? *Nature* 409:1119–1122.

Cohen, J. E., and D. Tilman. 1996. Biosphere 2 and biodiversity: The lessons so far. *Science* 274:1150–1151.

Colbeck, I., and A. R. MacKenzie. 1994. *Air Pollution by Photochemical Oxidants.* Amsterdam: Elsevier.

Coleman, D.C., and D. A. Crossley. 1996. *Fundamentals of Soil Ecology.* San Diego: Academic.

Coles, P., and F. Lucchin. 1995. *Cosmology: The Origin and Evolution of Cosmos Structure.* New York: John Wiley.

Commoner, B. 1971. *Closing Circle.* New York: Knopf.

Condie, K. C. 1989. Origin of the Earth's crust. *Palaeogeography, Palaeoclimatology, Palaeoecology* 75:57–81.

Condie, K. C. 1998. Episodic growth of juvenile crust and catastrophic events in the mantle. In *Origin and Evolution of Continents,* ed. Y. Motoyoshi and K. Shiraishi, 1–7. Tokyo: National Institute of Polar Research.

Connell, J. H., and M. D. Lowman. 1989. Low-diversity tropical rain forests: Some possible mechanisms for their existence. *American Naturalist* 134:88–119.

Cooper, A., and D. Penny. 1997. Mass survival of birds across the Cretaceous-Tertiary boundary: Molecular evidence. *Science* 275:1109–1113.

Corliss, J. B., et al. 1979. Submarine thermal springs on the Galápagos Rift. *Science* 203:1073–1083.

Corliss, J. B., J. A. Baros, and S. E. Hoffman.1981. An hypothesis concerning the relationship between submarine hot springs and the origin of life on Earth. *Oceanologica Acta* (special issue):59–69.

Costanza, R., et al. 1997. The value of the world's ecosystem services and natural capital. *Nature* 387:253-260.

Courtillot, V. 1999. *Evolutionary Catastrophes: The Science of Mass Extinctions.* Cambridge: Cambridge University Press.

Courtillot, V., and Y. Gaudemer. 1996. Effects of mass extinctions on biodiversity. *Nature* 381:146–148.

Cowan, D. A. 1992. Biochemistry and molecular biology of the extremely thermophilic bacteria. In *Molecular Biology and Biotechnology of Extremophiles,* ed. R. A. Herbert and R. J. Sharp, 1–43. Glasgow: Blackie.

Cox, A. N., W. C. Livingston, and M. S. Matthews. 1991. *Solar Interior and Atmosphere.* Tucson, AZ: University of Arizona Press.

Cox, C. S., and C. M. Wathes, eds. 1995. *Bioaerosols Handbook.* Boca Raton, FL: Lewis.

Cox, P. M., et al. 2000. Acceleration of global warming due to carbon-cycle feedbacks in a coupled climate model. *Nature* 408:184–187.

Cramer, W., et al. 1999. Ecosystem composition and structure. In *The Terrestrial Biosphere and Global Change,* ed. B. Walker et al., 190–228. New York: Cambridge University Press.

Cramer, W., and C. B. Field. 1999. Comparing global models of terrestrial net primary productivity (NPP): Overview and key results. *Global Change Biology* 5:1–15.

Crawford, I. 2000. Where are they? *Scientific American* 283(1):38–43.

Crawley, M. J. 1983. *Herbivory The Dynamics of Animal-Plant Interactions.* Berkeley: University of California Press.

Crick, F. 1981. *Life Itself: Its Origin and Nature.* New York: Simon & Schuster.

Crick, F. 1966. *Of Molecules and Men.* Seattle: University of Washington Press.

Crick, F., and L. E. Orgel. 1973. Directed panspermia. *Icarus* 19:341–346.

Crick, F., and L. E. Orgel. 1981. *Life Itself.* New York: Simon and Schuster.

Crowley, T. J. 1983. The geological record of climatic change. Review of Geophysics and Space Physics 21:828–877.

Crowley, T. J. 2000. Causes of climate change over the past 1000 years. Science 289:270–276.

Crutzen, P. 1970. The influence of nitrogen oxides on the atmospheric ozone content. *Quarterly Journal of the Royal Meteorological Society* 96:320–325.

Cullmann, W., E. Götz, and G. Gröner. 1986. *The Encyclopedia of Cacti.* Portland, OR: Timber.

Culver, S. J., and P. F. Rawson. 2000. *Biotic Response to Global Change: The Last 145 Million Years.* New York: Cambridge University Press.

Cumberland, J. H., J. R. Hibbs, and I. Hoch, eds. 1982. *The Economics of Managing Chlorofluorocarbons.* Washington, DC: Resources for the Future.

Cummins, C. S. 1989. Bacterial cell wall. In *Practical Handbook of Microbiology,* ed. W. M. O'Leary, 349–379. Boca Raton, FL: Chemical Rubber Company.

Cyr, H. 2000. Individual energy use and the allometry of population density. In *Scaling in Biology,* ed. J. H. Brown and G. B. West, 269–283. Oxford: Oxford University Press.

Daily, G. C., ed. 1997. *Nature's Services Societal Dependence on Natural Ecosystems.* Washington, DC: Island Press.

Daly, H. 1999. The lurking inconsistency. *Conservation Biology* 13:693-694.

Damuth, J. 1981. Population density and body size in mammals. *Nature* 290:699–700.

Damuth, J. 1991. Of size and abundance. *Nature* 351:268–269.

Dana, J. D. 1863. On cephalization. *New Englander* 22:495–506.

Dansgaard, W., et al. 1993. Evidence for general instability of past climate from a 250-kyr ice-core record. *Nature* 364:218–220.

Danson, F. M., and S. E. Plummer. 1995. *Advances in Environmental Remote Sensing.* New York: John Wiley.

Darwin, C. 1881. *The Formation of Vegetable Mould: Through the Action of Worms, with Observations on Their Habits.* London: John Murray.

Darwin, C. 1897. *Journal of Researches into the Natural History and Geology of the Countries Visited during the Voyage of H.M.S. Beagle round the World, under the Command of Capt. Fitz Roy, R.N.* New York: D. Appleton.

Davies, G. F. 1980. Review of oceanic and global heat flow estimates. *Reviews of Geophysics and Space Physics* 18:718–722.

Davies, G. F. 1999. *Dynamic Earth: Plates, Plumes and Mantle Convection.* Cambridge: Cambridge University Press.

Davies, P. 1998. *The Fifth Miracle: The Search for the Origin of Life.* London: Allen Lane.

Dawkins, R. 1995. God's utility function. *Scientific American* 273(5):80–85.

Deamer, D. W., and G. R. Fleischacker. 1994. *Origins of Life: The Central Concepts.* Boston: Jones and Bartlett.

Decker R., and B. Decker. 1981. The eruptions of Mount St. Helens. *Scientific American* 244(3):68–80.

Deckert, G., et al. 1998. The complete genome of the hyperthermophilic bacterium *Aquifex aeolicus. Nature* 392:353–358.

de Duve, C. 1984. *A Guided Tour of the Living Cell.* New York: Scientific American Library.

de Duve, C. 1991. *Blueprint for a Cell: The Nature and Origin of Life.* Burlington, NC: Neil Patterson Publishers / Carolina Biological Supply Company.

de Duve, C. 1995a. *Vital Dust: Life as a Cosmic Imperative.* New York: Basic Books.

de Duve, C. 1995b. The beginning of life on Earth. *American Scientist* 83:428–437.

de Duve, C. 1996. The birth of complex cells. *Scientific American* 274(4):50–57.

Deevey, E. S. 1960. The human population. *Scientific American* 203(3):194–204.

Deevey, E. S. 1970. Mineral cycles. *Scientific American* 223(3):149–158.

Delmas, R. J. 1992. Environmental information from ice cores. *Reviews of Geophysics* 30:1–21.

DeLong, E. F. 1998. Archaeal means and extremes. *Science* 280:542–543.

DeLong, E. F., et al. 1994. High abundance of Archaea in Antarctic marine picoplankton. *Nature* 371:695–697.

Delsemme, A. H. 2001. An argument for the cometary origin of the biosphere. *American Scientist* 89:432–442.

Delwiche, C. F. 1999. Tracing the thread of plastid diversity through the tapestry of life. *The American Naturalist* 154:S164–S177.

Demers, S., S. de Mora, and M. Vernet, eds. 2000. *Effects of UV Radiation on Marine Ecosystems.* Cambridge: Cambridge University Press.

Denny, M. W. 1993. *Air and Water The Biology and Physics of Life's Media.* Princeton, NJ: Princeton University Press.

Desbruyères, D., and L. Laubier. 1980. *Alvinella pompejana* gen. sp. nov., aberrant Ampharetidae from East Pacific Rise hydrothermal vents. *Oceanologica Acta* 3:267–274.

Des Marais, D. J. 2000. When did photosynthesis emerge on Earth? *Science* 289:1703–1705.

Des Marais, D. J., et al. 1992. Carbon isotope evidence for the stepwise oxidation of the Proterozoic environment. *Nature* 359:605–609.

De Vooys, C. G. N. 1979. Primary production in aquatic systems. In *The Global Carbon Cycle,* ed. B. Bolin et al., 259–292. New York: John Wiley.

Dhand, R., ed. 2000. Functional genomics. *Nature* 405:819–865.

Dick, S. J. 1996. *The Biological Universe.* Cambridge: Cambridge University Press.

Di Giulio, M. 2000. The universal ancestor lived in a thermophilic or hyperthermophilic environment. *Journal of Theoretical Biology* 203:203–213.

Dines, W. H. 1917. The heat balance of the atmosphere. *Quarterly Journal of Royal Meteorological Society* 43:151–158.

Dismukes, G. C., V. V. Klimov, and Y. Baranov. 2001. The origin of atmospheric oxygen on Earth: The innovation of oxygenic photosynthesis. *Proceedings of the National Academy of Sciences USA* 98:2170–2175.

Dobrovolsky, V. V. 1994. *Biogeochemistry of the World's Land.* Moscow: Mir Publishers.

Dobzhansky, T. 1972. Darwinian evolution and the problem of extraterrestrial life. *Perspectives in Biology and Medicine* 15:157–175.

Dole, L. R., ed. 1996. *Circumstellar Habitable Zones.* Menlo Park, CA: Travis House.

Dole, S. H. 1964. *Habitable Planets for Man.* Waltham, MA: Blaisdell.

Donnison, J. R., and D. F. Mikulskis. 1992. Three-body orbital stability criteria for circular orbits. *Monthly Notes of the Royal Astronomical Society* 254:21–26.

Donovan, S. K., and C. R. C. Paul, eds. 1998. *The Adequacy of the Fossil Record.* New York: John Wiley.

Doolittle, R. F. 1998. Microbial genomes open up. *Nature* 392:339–342.

Doolittle, R. F., et al. 1996. Determining divergence times of the major kingdoms of living organisms with a protein clock. *Science* 271:470–477.

Doolittle, W. F. 2000. Uprooting the tree of life. *Scientific American* 282(2):9095.

Doolittle, W. F., and J. R. Brown. 1995. Tempo, mode, the progenote, and the universal root. In *Tempo and Mode in Evolution,* ed. W. M. Fitch and F. J. Ayala, 3–25. Washington, DC: National Academy Press.

Douglas, A. E. 1995. The ecology of symbiotic micro-organisms. *Advances in Ecological Research* 26:69–103.

Douglas, A. E. 1998. Nutritional interactions in insect-microbial symbioses: Aphids and their symbiotic bacteria *Buchnera. Annual Review of Entomology* 43:17–37.

Doyle, L. R., H. J. Deeg, and T. M. Brown. 2000. Searching for shadows of other Earths. *Scientific American* 283(3):58–65.

Drake, F. 1980. N is neither very small not very large. In *Strategies for the Search for Life in the Universe,* ed. M.D. Papagiannis, 27–34. Dordrecht, the Netherlands: Kluwer.

Drake, B. G., M. A. Gonzalez-Meler, and S. P. Long. 1996. More efficient plants: A consequence of rising atmospheric CO_2? *Annual Review of Plant Physiology and Molecular Biology* 48:609–639.

Drapeau, M. D. 1999. Further origin of species. *American Scientist* 87:277.

Dressler, B. O., R. A. F. Grieve, and V. C. Sharpton. 1994. *Large Meteorite Impacts and Planetary Evolution.* Washington, DC: Geological Society of America.

Dressler, D., and H. Potter. 1991. *Discovering Enzymes.* New York: Scientific American Library.

Driscoll, C. T., et al. 2001. Acidic deposition in the Northeastern United States: Sources and inputs, ecosystem effects, and management strategies. *BioScience* 51:180–198.

D'Souza, G., et al. 1996. *Advances in the Use of NOAA AVHRR Data for Land Applications.* Dordrecht, the Netherlands: Kluwer.

Duchesne, L. C., and D. W. Larson. 1989. Cellulose and the evolution of plant life. *BioScience* 39:238–241.

Ducklow, H. W., and C. A. Carlson. 1992. Oceanic bacterial production. In *Advances in Microbial Ecology,* ed. K. C. Marshall, 113–182. New York: Plenum.

Dyson, F. 1979. *Disturbing the Universe.* New York: Harper & Row.

Dyson, F. 1999a. *Origins of Life.* Cambridge: Cambridge University Press.

Dyson, F. 1999b. *The Sun, the Genome and the Internet, Tools of Scientific Revolutions.* Oxford: Oxford University Press.

Eastman, J. T., and A. L. DeVries. 1986. Antarctic fishes. *Scientific American* 255(5):106–114.

Ebermayer, E. W. F. 1882. *Naturgesetzliche Grundlagen des Wald- und Ackerbaues (Natural Foundations of Forestry and Agriculture).* Vol. I. *Die Bestandteile der Pflanzen (The Components of Plants).* Berlin: Springer-Verlag.

Eckley, N. 2001. Traveling toxics. *Environment* 43(7):24–36.

Eddy, J. A. 1977. The case of the missing sunspots. *Scientific American* 236(5):80–92.

Edmonds, R. L., ed. 1982. *An Analysis of Coniferous Forest Ecosystems in the Western United States.* New York: Van Nostrand Reinhold.

Edwards, C. A., and J. R. Lofty. 1972. *Biology of Earthworms.* London: Chapman & Hall.

Ehleringer, J. R., and C. B. Field. 1993. *Scaling Physiological Processes.* San Diego: Academic.

Ehrenfeld, D. 1986. Why put value on biodiversity? In *Biodiversity,* ed. E. O. Wilson, 212–216. Washington, DC: National Academy Press.

Ehrlich, P. 2000. *Human Natures.* Washington, DC: Island Press.

Ehrlich, P., A. H. Ehrlich, and J. P. Holdren. 1970. *Ecoscience.* San Francisco: W. H. Freeman.

Ehrlich, P. R., and P. H. Raven 1964. Butterflies and plants: A study of coevolution. *Evolution* 18:586–608.

Eisenberg J.F. 1981. *The Mammalian Radiations.* Chicago: University of Chicago Press.

Elderfield, H., and A. Schultz. 1996. Mid-ocean ridge hydrothermal fluxes and the chemical composition of the ocean. *Annual Review of Earth and Planetary Sciences* 24:191–224.

Elphick, J., ed. 1995. *The Atlas of Bird Migration.* New York: Random House.

Elton, C. 1927. *Animal Ecology.* New York: Macmillan.

Emanuel, W. R., H. H. Shugart, and M. D. Stevenson. 1985. Climatic change and the broad-scale distribution of ecosystem complexes. *Climatic Change* 7:29–43.

Embley, T. M., and W. Martin. 1998. A hydrogen-producing mitochondrion. *Nature* 396:517–519.

Enquist, B. J., J. H. Brown, and G. B. West. 1998. Allometric scaling of plant energetics and population density. *Nature* 395:163–165.

Enquist, B. J., et al. 1999. Allometric scaling of production and life-history variation in vascular plants. *Nature* 401:907–911.

Enquist, B. J., et al. 2000. Quarter-power allometric scaling in vascular plants: Functional basis and ecological consequences. In *Scaling in Biology,* ed. J. H. Brown and G. B. West, 167–198. Oxford: Oxford University Press.

Ensenrink, M. 1999. Biological invaders sweep in. *Science* 285:1834–1836.

EROS (Earth Resources Observation Systems) Data Center. 1978–1986. *LANDSAT Data Users Notes.* Sioux Falls, SD: EROS Data Center.

Erwin, D., J. Valentine, and D. Jablonski. 1997. The origin of animal body plans. *American Scientist* 85:126–137.

Erwin, D. H. 1993. *The Great Paleozoic Crisis.* New York: Columbia University Press.

Erwin, T. L. 1982. Tropical forests: Their richness in Coleoptera and other Arthropod species. *Coleopterists Bulletin* 36:74–75.

Esser, G. 1987. Sensitivity of global carbon pools and fluxes to human and potential climatic impacts. *Tellus* 39B:245–260.

Etheridge, D. M., et al. 1998. Atmospheric methane between 1000 A.D. and present: Evidence of anthropogenic emissions and climatic variability. *Journal of Geophysical Research* 103:15979–15993.

Evenari, M., et al., eds. 1985. *Hot Deserts and Arid Shrublands.* New York: Elsevier.

Falkowski, P., et al. 1998. Biogeochemical controls and feedbacks on ocean primary production. *Science* 281:200–205.

Falkowski, P., et al. 2000. The global carbon cycle: A test of our knowledge of Earth as a system. *Science* 290:291–296.

Famintsyn, A. S. 1891. Concerning the symbiosis of algae with animals. *Transactions of the Botanical Laboratory of the Academy of Sciences* 1:1–22.

FAO (Food and Agriculture Organization). 2000. *FAOSTAT Statistics Database.* <http://apps.fao.org>.

Farley, J. 1977. *The Spontaneous Generation Controversy from Descartes to Oparin.* Baltimore: Johns Hopkins University Press.

Farman, J. C., B. G. Gardiner, and J. D. Shanklin. 1985. Large losses of total ozone in Antarctica reveal seasonal ClO_x/NO_x interaction. *Nature* 315:207–210.

Farrell, B. D. 1998. "Inordinate fondness" explained: Why are there so many beetles? *Science* 281:555–559.

Ferris, M. J., and B. Palenik. 1998. Niche adaptation in ocean cyanobacteria. *Nature* 396:226–228.

Ferry, J. G., ed. 1993. *Methanogenesis: Ecology, Physiology, Biochemistry & Genetics.* New York: Chapman & Hall.

Field, C. B., M. J. Behrenfeld, J. T. Randerson, and P. Falkowski. 1998. Primary production of the biosphere: Integrating terrestrial and oceanic components. *Science* 281:237–240.

Field, C. B., J. T. Randerson, and C. M. Malmstrom.1995. Global net primary production: Combining ecology and remote sensing. *Remote Sensing of the Environment* 51:74–88.

Fisk M. R., S. J. Giovanoni, and I. H. Thorseth. 1998. Alteration of oceanic volcanic glass: Textural evidence of microbial activity. *Science* 281:978–980.

Fitch, W. M., and F. J. Ayala, eds. 1995. *Tempo and Mode in Evolution.* Washington, DC: National Academy Press.

Fitch, W. M., and K. Upper. 1987. The phylogeny of tRNA sequences provides evidence for ambiguity reduction in the origin of the genetic code. *Cold Spring Harbor Symposium on Quantitative Biology* 52:759–767.

Fleck, B., V. Domingo, and A. Poland, eds. 1995. The SOHO mission. *Solar Physics* 162 (special section).

Fleischmann, R. D., et al. 1995. Whole-genome random sequencing and assembly of *Haemophilus influenzae. Science* 269:496–512.

Fletcher, M., and G. D. Floodgate, eds. 1985. *Bacteria in Their Natural Environments.* London: Academic.

Fliermans, C. B., and D. L. Balkwill. 1989. Microbial life in deep terrestrial subsurfaces. *BioScience* 39:370–377.

Fogg, G. E. 1958. Actual and potential yield in photosynthesis. *The Advancement of Science* 14:359–400.

Foley, J. A. 1994. Net primary productivity in the terrestrial biosphere: The application of a global model. *Journal of Geophysical Research* 99:20773–20783.

Forey, P., and P. Janvier. 1994. Evolution of early vertebrates. *American Scientists* 82:554–565.

Forterre, P. 1996. A hot topic: The origin of hyperthermophiles. *Cell* 85:789–792.

Fortey, R. 1997. *Life: An Unauthorized Biography.* New York: HarperCollins.

Fortey, R. 1998. *Life: A Natural History of the First Four Billion Years.* New York: Knopf.

Foukal, P. 1987. Physical interpretation of variations in total solar irradiance. *Journal of Geophysical Research* 92:801–807.

Foukal, P. 1990. The variable sun. *Scientific American* 262(2):34–41.

Fowler T. K. 1997. *The Fusion Quest.* Baltimore: Johns Hopkins University Press.

Frängsmyr, T., ed. 1983. *Linneaus: The Man and His World.* Berkeley: University of California Press.

Frank, J. 1998. How the ribosome works. *American Scientist* 86:428–439.

Fraser, C. M., J. A. Eisen, and S. L. Salzberg. 2000. Microbial genome sequencing. *Nature* 406:799–803.

Frederickson, J. K., and T. C. Onstott. 1996. Microbes dep inside the Earth. *Scientific American* 275(4):68–93.

Friedman, H. 1986. *Sun and Earth.* New York: Scientific American Library.

Friis-Christensen, E., and K. Lassen. 1991. Length of the solar cycle: An indicator of solar activity closely associated with climate. *Science* 254:698–700.

Frisch, P. C. 2000. The galactic environment of the Sun. *American Scientist* 88:52–59.

Froehlich, P. N., et al. 1982. The marine phosphorus cycle. *American Journal of Science* 282:474–511.

Fröhlich, C. 1987. Variability of the solar "constant" on time scales of minutes to years. *Journal of Geophysical Research* 92:796–800.

Froude, D. O., et al. 1983. Ion microprobe identification of 4,100–4,200 Myr-old terrestrial zircons. *Nature* 304:616–618.

Fuhrman, J. A. 1999. Marine viruses and their biogeochemical and ecological effects. *Nature* 399:541–548.

Fung, I. Y., et al. 1987. Application of advanced very high resolution radiometer vegetation index to study atmosphere-biosphere exchange of CO_2. *Journal of Geophysical Research* 92: 2999–3015.

Fung, I. Y., et al. 2000. Iron supply and demand in the upper ocean. *Global Biogeochemical Cycles* 14:281–295.

Galtier, N., N. Tourasse, and M. Gouy. 1999. A nonhyperthermophilic common ancestor to extant life forms. *Science* 283:220–221.

Ganachaud, A., and C. Wunsch. 2000. Improved estimates of global ocean circulation, heat transport and mixing from hydrographic data. *Nature* 408:453–457.

Garrels, R. M., and F. T. Mackenzie. 1971. *Evolution of Sedimentary Rocks.* New York: W. W. Norton.

Gartner B. L., ed. 1995. *Plant Stems: Physiology and Functional Morphology.* San Diego: Academic.

Gaston, K. J. 1991. The magnitude of global insect species richness. *Conservation Biology* 5:283–296.

Gates, D. M. 1993. *Climate Change and Its Biological Consequences.* Sunderland, MA: Sinauer Associates.

Gattuso, J.-P., D. Allemand, and M. Frankignoulle. 1999. Photosynthesis and calcification at cellular, organismal and community levels in coral reefs: A review on interactions and control by carbonate chemistry. *American Zoologist* 39:160–183.

Gattuso, J.-P., and R. W. Buddemeier. 2000. Calcification and CO_2. *Nature* 407:311–313.

Gee, H. 1999. *In Search of Deep Time: Beyond the Fossil Record to a New History of Life.* New York: Free Press.

Gehring, W. J., and R. Wehner. 1995. Heat shock protein synthesis and thermotolerance in *Cataglyphis,* an ant from the Sahara desert. *Proceedings of the National Academy of Sciences USA* 92:2994–2998.

Gensel, P. G., and H. N. Andrews. 1987. The evolution of early land plants. *American Scientists* 75:478–489.

Gentry, A. H. 1988. Species richness of upper Amazonia. *Proceedings of the National Academy of Sciences USA* 85:156–159.

Gesteland, R. F., and J. F. Atkins, eds. 1993. *The RNA World.* Cold Spring Harbor NY: Cold Spring Harbor Laboratory Press.

Giardina, C. P., and M. G. Ryan. 2000. Evidence that decomposition rates of organic carbon in mineral soil do not vary with temperature. *Nature* 404:858–861.

Gierasch, L. M., and J. King, eds. 1990. *Protein Folding: Deciphering the Second Half of the Genetic Code.* Washington, DC: American Association for the Advancement of Science.

Gilbert, D. L., et al., eds. 1994. *Groundwater Ecology.* San Diego: Academic.

Gilliard, E. T. 1958. *Living Birds of the World.* New York: Doubleday.

Gillon, J. 2000. Feedback on Gaia. *Nature* 406:685–686.

Gillooly, J. F., et al. 2001. Effects of size and temperature on metabolic rate. *Science* 293:2248–2251.

Glanz, J. 1997. Bold prediction downplays the Sun's next peak. *Science* 275:927.

Glen, W. 1990. What killed dinosaurs? *American Scientist* 78:354–370.

Glynn, P. W., et al. 1994. Reef coral reproduction in the eastern Pacific: Costa Rica, Panama, and Galapagos Islands (Ecuador). *Marine Biology* 118:191–208.

Godbold, D. L., and A. Hutterman. 1994 *Effects of Acid Precipitation on Forest Processes.* New York: Wiley-Liss.

Goffeau, A., et al. 1996. Life with 6000 genes. *Science* 274:546–567.

Gold, T. 1992. The deep, hot biosphere. *Proceedings of the National Academy of Sciences USA* 89:6045–6049.

Gold, T. 1999. *The Deep Hot Biosphere.* New York: Copernicus.

Golley, F. B. 1968. Secondary productivity in terrestrial communities. *American Zoologist* 8:53–59.

Golley, F. B. 1972. Energy flux in ecosystems. In *Ecosystem Structure and Function,* ed. J. A. Wiens, 69–88. Corvallis: Oregon State University Press.

Golley, F. B., ed. 1983. *Tropical Rain Forest Ecosystems.* Amsterdam: Elsevier.

Golley, F. B., K. Petrusewicz, and L. Ryszkowski, eds. 1975. *Small Mammals: Their Productivity and Population Dynamics.* Cambridge: Cambridge University Press.

Good, N., and D. H. Bell. 1980. Photosynthesis, plant productivity and crop yield. In *The Biology of Crop Productivity,* ed. P. S. Carlson, 3–51. New York: Academic.

Gordon, A. L, and Fine R. A. 1996. Pathways of water between the Pacific and Indian Oceans in the Indonesian Sea. *Nature* 379:146–149.

Gordon, H. R., and A. Morel. 1983. *Remote Sensing of Ocean Color for Interpretation of Satellite Visible Imagery.* New York: Springer-Verlag.

Gottschalk, G. 1986. *Bacterial Metabolism.* New York: Springer-Verlag.

Goudriaan, G., and P. Ketner. 1984. Are land biota a source or sink for CO_2? In *Interactions between Climate and Biosphere,* ed. H. Lieth, R. Fantechi, and H. Schnitzler, 247–252. Lisse, Switzerland: Swets and Zeitlinger.

Gould, S. J. 1989. *Wonderful Life: The Burgess Shale and the Nature of History.* New York: Norton.

Grace, J., and M. Rayment. 2000. Respiration in the balance. *Nature* 404:819–820.

Graether, S. P., et al. 2000. Beta-helix structure and ice-binding properties of a hyperactive antifreeze protein from an insect. *Nature* 406:325–328.

Graf, H. F. J. Feichter, and B. Longmann. 1997. Volcanic sulfur emissions: Estimates of source strength and its contribution to the global sulfate distribution. *Journal of Geophysical Research* 102:10727–10738.

Grant, W. D., and K. Horikoshi. 1992. Alkaliphiles: Ecology and biotechnology applications. In *Molecular Biology and Biotechnology of Extremophiles,* ed. R. A. Herbert and R. J. Sharp, 143–162. Glasgow: Blackie.

Grant, W. D., R. T. Gemmell, and T. J. McGenity. 1998. Halophiles. In *Extremophiles Microbial Life in Extreme Environments,* ed. K. Horikoshi and W. D. Grant, 93–132. New York: Wiley-Liss.

Gray, J., and W. Shear. 1992. Early life on land. *American Scientist* 80:444–456.

Gray, J. E., et al. 2000. The HIC signalling pathway links CO_2. *Nature* 408:713–716.

Gray, M. W., and W. F. Doolittle. 1982. Has the endosymbiont hypothesis been proven? *Microbiological Reviews* 46:1–42.

Gregor, C. B., et al., eds. 1988. *Chemical Cycles in the Evolution of the Earth.* New York: John Wiley.

Greven, H. 1980. *Die Bärtierchen: Tardigrada.* Wittenberg: A. Ziemsen.

Griffin, W. T., et al. 1997. Methods for obtaining deep subsurface microbiological samples by drilling. In *The Microbiology of the Terrestrial Deep Subsurface,* ed. P. S. Amy and D. L. Haldeman, 23–44. Boca Raton, FL: Lewis.

Griffin, D. W., et al. 2001. African desert dust in the Caribbean atmosphere: microbiology and public health. *Aerobiologia* 17:203–213.

Grimm, V., and C. Wissel. 1997. Babel, or the ecological stability discussions: An inventory and analysis of terminology and a guide for avoiding confusion. *Oecologia* 109:323–334.

Grotzinger, J. P., and A. H. Knoll. 1999. Stromatolites in Precambrian carbonates: Evolutionary mileposts or environemntal dipsticks? *Annual Review of Earth and Planetary Science* 27:313–358.

Grotzinger, J. P., and D. H. Rothman. 1996. An abiotic model for stromatolite morphogenesis. *Nature* 383:423–425.

Grotzinger, J. P. et al . 1995. Biostratigraphic and geochronological constraints on early animal evolution. *Science* 270:598–604.

Guidry, M. W., F. T. Mackenzie, and R. S. Arvidson. 2000. Role of tectonic in phosphorus distribution and cycling. In *Marine Authigenesis: From Global to Microbial,* ed. C. R. Glenn, L. Prévôt-Lucas, and J. Lucas, 35–51. Tulsa, OK: Society for Sedimentary Geology.

Gutman, G., and A. Ignatov. 1995. Global land monitoring from AVHRR: Potential and limitations. *International Journal of Remote Sensing* 16:2301–2309.

Gutman, G., et al. 1995. The enhanced NOAA global data set from advanced very high resolution radiometer. *Bulletin of the American Meteorological Society* 76:1141–1156.

Guzik, J. A., et al. 1987. A comparison between mass-losing and standard solar models. *Astrophysical Journal* 319:957–965.

Haeckel, E. H. 1894. *Systematische Phylogenie: Entwurf eines natürlichen Systems der Organismen auf Grund ihrer Stammengeschichte (Systematic phylogeny: The outline of natural systems of organisms based on their evolutionary history).* Reimer: Berlin.

Haigh, J. D. 1996. The impact of solar variability on climate. *Science* 272:981–984.

Hall, D. O., et al., eds. 1993. *Photosynthesis and Production in Changing Environment.* Amsterdam: Elsevier.

Hall, D. O., and K. K. Rao. 1999. *Photosynthesis.* New York: Cambridge University Press.

Hall, D. O., J. M.O. Scurlock. 1993. Biomass production and data. In *Photosynthesis and Production in Changing Environment,* ed. D. O. Hall et al., 659–678. Amsterdam: Elsevier.

Hall, B. D., et al. 2001. Halocarbons and other atmospheric trace species. In *Climate Monitoring and Diagnostics Laboratory Summary Report No. 25 1998–1999,* ed. R. C. Schnell, D. B. King, and R. M. Rosson, 91–112. Washington, DC: U.S. Department of Commerce.

Hanna, J. M., and D. E. Brown. 1983. Human heat tolerance: An anthropological perspective. *Annual Review of Anthropology* 12:259–284.

Hansen, J., et al. 1999. GISS analysis of surface temperature change. *Journal of Geophysical Research* 104:30997–31022.

Hansen, J., et al. 2000. Global warming in the twenty-first century: An alternative scenario. *Proceedings of the National Academy of Sciences USA* 97:9875–9880.

Harashima, K., et al., eds. 1989. *Aerobic Photosynthetic Bacteria.* Berlin: Springer-Verlag.

Harcourt, A. H. 1996. Is the gorilla a threatened species? How should we judge? *Biological Conservation* 75:165–176.

Harland, W. B., A. G. Smith, and D. G. Smith. 1989. *A Geologic Time Scale.* Cambridge: Cambridge University Press.

Harper, C. L., and S. B. Jacobsen. 1996. Noble gases and Earth's accretion. *Science* 273:1814–1818.

Harper, J. L. 1977. *Population Biology of Plants.* London: Academic.

Harrad, S., ed. 2001. *Persistent Organic Pollutants: Environmental Behaviors and Pathways of Human Exposure.* Boston: Kluwer Academic.

Harris, W. V. 1971. *Termites.* London: Longman.

Hartenstein, R. 1986. Earthworm biotechnology and global biogeochemistry. *Advances in Ecological Research* 15:379–409.

Hatch, M. D. 1992. C_4 photosynthesis: An unlikely process full of surprises. *Plant Cell Physiology* 4:333–342.

Hawken, P., A. Lovins, and A. H. Lovins. 1999. *Natural Capitalism.* Boston: Little, Brown.

Hay, W. W. 1998. Detrital sediment flues from continents to oceans. *Chemical Geology* 145:287–323.

Hays, J. D., J. Imbrie, and N. J. Shackleton. 1976. Variations in the Earth's orbit: Pacemaker of the ice ages. *Science* 194:1121–1132.

Hazen, R. M., and E. Roedder. 2001. How old are bacteria from the Permian age? *Nature* 411:155.

Heckman, D. S., et al. 2001. Molecular evidence for the early colonization of land by fungi and plants. *Science* 293:1129–1133.

Hector, A. et al. 1999. Plant diversity and productivity experiments in European grasslands. *Nature* 286:1123–1127.

Hedges, S. B., et al. 1996. Continental breakup and the ordinal diversification of birds and mammals. *Nature* 381:226–229.

Henderson, G. M., and N. C. Slowey. 2000. Evidence from U-Th dating against Northern Hemisphere forcing of the penultimate deglaciation. *Nature* 404:61–66.

Henderson-Sellers, A., and V. Gornitz. 1984. Possible climatic impacts of land covert transformations, with particular emphasis on tropical deforestation. *Climatic Change* 6:231–257.

Hengartner, M. O. 2000. The biochemistry of apoptosis. *Nature* 407:770–776.

Herbert, R. A., and R. J. Sharp, eds. 1992. *Molecular Biology and Biotechnology of Extremophiles.* Glasgow: Blackie.

Hershey, A. D. 1970. Genes and hereditary characteristics. *Nature* 226:697–700.

Hewitt, G. 2000. The genetic legacy of the Quaternary ice ages. *Nature* 405:907–913.

Heywood, V. H., and R. T. Watson, eds. 1995. *Global Biodiversity Assessment.* Cambridge: Cambridge University Press.

Hibbett, D. S., L.-B. Gilbert, and M. J. Donoghue. 2000. Evolutionary instability of ectomycorrhizal symbioses in basidiomycetes. *Nature* 407:506–508.

Hill, W. E., et al., eds. 1990. *The Ribosome Structure, Function & Evolution.* Washington, DC: American Society for Microbiology.

Hillis, D. M. 1997. Biology recapitulates phylogeny. *Science* 276:218–219.

Hinga, K. R. 1979. The food requirement of whales in the southern hemisphere. *Deep Sea Research* 26:569–577.

Hirota, Y. 1995. The Kuroshio. Part II. Zooplankton. *Oceanography and Marine Biology: An Annual Review* 33:151–220.

Hobbs, P. V. 1999. Effects of aerosols on clouds and radiation. In *Global Energy and Water Cycles,* ed. K. A. Browning and R. J. Gurney, 91–99. New York: Cambridge University Press.

Hodge, P. 1994. *Meteorite Craters and Impact Structures of the Earth.* Cambridge: Cambridge University Press.

Hoffman, P. F., et al. 1998. A Neoproterozoic snowball Earth. *Science* 281:1342–1346.

Hofman, H. J., et al. 1999. Origin of 3.45 Ga confirm stromatolites in Warrawoona Group, Western Australia. *Geological Society of America Bulletin* 111:1256–1262.

Holeman, J. N. 1968. The sediment yield of major rivers of the world. *Water Resources Research* 4:737–747.

Holland, K. T., J. S. Knapp, and J. G. Shoesmith. 1987. *Anaerobic Bacteria.* Blackie: Glasgow.

Hölldobler, B., and E. O. Wilson. 1990. *The Ants.* Cambridge, MA: Harvard University Press.

Holm, N. G., ed. 1992. *Marine Hydrothermal Systems and the Origin of Life.* Norwell, MA: Kluwer.

Holmén, K. 1992. The global carbon cycle. In *Global Biogeochemical Cycles,* ed. S. S. Butcher, 259. London: Academic.

Holsinger, J. R. 1988. Troglobites: The evolution of cave-dwelling organisms. *American Scientist* 76:147–153.

Hoppert, M., and F. Mayer. 1999. Prokaryotes. *American Scientists* 87:518–525.

Horgan, J. 1991. In the beginning. *Scientific American* 264(2):117–125.

Horikoshi, K. 1998. Alkaliphiles. In *Extremophiles Microbial Life in Extreme Environments,* ed. K. Horikoshi and W. D. Grant, 155–179. New York: Wiley-Liss.

Horikoshi, K., and W. D. Grant, eds. 1998. *Extremophiles Microbial Life in Extreme Environments.* New York: Wiley-Liss.

Horikoshi, K., et al. 1997. Microbial flora in the deepest sea mud of Mariana Trench. *FEMS Microbiology Letters* 152:279–285.

Horowitz, N. H. 1986. *To Utopia and Back: The Search for Life in the Solar System.* New York: W. H. Freeman.

Houde, E. D., and E. S. Rutherford. 1993. Recent trends in estuarine fisheries: Predictions of fish production and yield. *Estuaries* 16:161–176.

Houghton, J. T., et al., eds. 1990. *Climate Change: The IPCC Scientific Assessment.* New York: Cambridge University Press.

Houghton, J. T., G. J. Jenkins, and J. J. Ephraims, eds. 1996. *Climate Change 1995 The Science of Climate Change.* New York: Cambridge University Press.

Houghton, J. T., et al., eds. 2001. *Climate Change 2001: The Scientific Basis.* New York: Cambridge University Press.

Houghton, R. A. 1995. Land-use change and the carbon cycle. *Global Change Biology* 1:275–287.

Houghton, R. A. 1999. The annual net flux of carbon to the atmosphere from changes in land use 1850–1990. *Tellus* 51B:298–313.

Hoyle, F. 1980. *Steady-State Cosmology Revisited.* Cardiff, Wales: University College Cardiff Press.

Hoyle, F. 1982. The universe: Past and present reflections. *Annual Review of Astronomy and Astrophysics* 20:1–35.

Hoyle, F., and F. Wickramasinghe. 1978. *Lifecloud: The Origin of Life in the Universe.* New York: Harper & Row.

Huelsenbeck, J. P., and B. Rannala. 1997. Phylogenetic methods come of age: Testing hypotheses in an evolutionary context. *Science* 276:227–232.

Hughes, G. L. Daily, and P. R. Ehrlich. 1997. Population diversity: Its extent and extinction. *Science* 278:689–692.

Hulme, M. 1995. Estimating global changes in precipitation. *Weather* 50:34–42.

Humphreys, W. F. 1979. Production and respiration in animal communities. *Journal of Animal Ecology* 48:427–453.

Humphries, C. J., P. H. Williams, and R. I. Vane-Wright. 1995. Measuring biodiversity value for conservation. *Annual Review of Ecology and Systematics* 26:93–111.

Hunter, G. K. 2000. *Vital Forces: The Discovery of the Molecular Basis of Life.* San Diego: Academic.

Huston, M. 1993. Biological diversity, soils, and economics. *Science* 262:1676–1679.

Huston, M. 1994. *Biological Diversity: The Coexistence of Species in Changing Landscapes.* New York: Cambridge University Press.

Huston, M. 1997. Hidden treatments in ecological experiments: Re-evaluating the ecosystem function if biodiversity. *Oecologia* 110:449–460.

Hutchings, J. A. 2000. Collapse and recovery of marine fishes. *Nature* 406:882–885.

Hutchinson, G. E. 1948. On living in the biosphere. *The Scientific Monthly* 67:393–397.

Hutchinson, G. E. 1970. The Biosphere. *Scientific American* 223(3):45–53.

Hyde, W. T., et al. 2000. Neoproterozoic "snowball Earth" simulations with a coupled climate/ice-sheet model. *Nature* 405:4250–29.

ICOLD (International Commission on Large Dams). 1994. *Dams and the Environment: Water Quality and Climate.* Paris: ICOLD.

Ingledew, W. J. 1990. Acidophiles. In *Microbiology of Extreme Environments,* ed. C. Edwards, 33–54. New York: McGraw-Hill.

International Human Genome Sequencing Consortium (IHGSC). 2001. Initial sequencing and analysis of the human genome. *Nature* 409:860–921.

IPCC (Intergovernmental Panel on Climatic Change). 1994. *Radiative Forcing of Climate Change.* Cambridge: Cambridge University Press.

Irving, P. M., ed. 1991 *Acidic Deposition: State of Science and Technology.* Washington, DC: U.S. National Acid Precipitation Assessment Program.

Isaac, J. D. 1969. The nature of oceanic life. *Scientific American* 221(3):147–162.

Isenberg, A. C. 2000. *The Destruction of the Bison.* New York: Cambridge University Press.

Ittekkot, V., et al. 2000. Hydrological alterations and marine biogeochemistry: A silicate issue? *BioScience* 50:776–782.

IUCN (International Union for the Conservation of Nature). 2001. Continental overview of elephant estimates. <http://www.iucn.org/themes/ssc/aed/products/continen/contable.htm>.

IWC (International Whaling Commission). 2001. Whale population estimates. <http://ourworld.compuserve.com/homepages/iwcoffice/Estimate.htm>.

Jackson, I., ed. 1998. *The Earth's Mantle.* New York: Cambridge University Press.

Jackson, R. B., et al. 1997. A global budget for fine root biomass, surface area, and nutrient contents. *Proceedings of the National Academy of Sciences USA* 94:7362–7366.

Jacobs, J. A. 1992. *Deep Interior of the Earth.* London: Chapman & Hall.

Jacobson, M. Z. 2001. Strong radiative heating due to the mixing state of black carbon in atmospheric aerosols. *Nature* 409:695–697.

Jahnke, R. A. 1992. The phosphorus cycle. In *Global Biogeochemical Cycles,* ed. S. S. Butcher et al., 301–315. London: Academic.

James, A. N., K. J. Gaston, and A. Balmford. 1999. Balancing the Earth's accounts. *Nature* 401:323-324.

Jannasch, H. W. 1989. Chemosynthetically sustained ecosystems in the deep sea. In *Autotrophic Bacteria,* ed. H. G. Schlegel and B. Bowien, 147–166. Berlin, Springer-Verlag.

Jannasch, H. W., and M. J. Mottl. 1985. Geomicrobiology of deep-sea hydrothermal vents. *Science* 229:717–725.

Janvier, P. 1996. *Early Vertebrates.* New York: Oxford University Press.

Janzen, D. H. 1966. Coevolution of mutualism between ants and acacias in Central America. *Evolution* 20:249–275.

Janzen, D. H. 1985. The natural history of mutualisms. In *The Biology of Mutualism,* ed. D. H. Boucher, 40–49. New York: Oxford University Press.

Jarrell, K. F., et al. 1999. Recent excitement about the Archaea. *BioScience* 49:530–541.

Javaux, E. J., A. H. Knoll, and M. R. Walter. 2001. Morphological and ecological complexity in early eukaryotic ecosystems. *Nature* 412:66–69.

Jewitt, D. 2000. Eyes wide shut. *Nature* 403:145–148.

Johnson, K. S., F. P. Chavez, and G. E. Friederich. 1999. Continental-shelf sediment as a primary sources of iron for coastal phytoplankton. *Nature* 398:697–700.

Jokipii, J. R., and F. B. McDonald. 1995. Quest for the limits of the heliosphere. *Scientific American* 272(4):58–63.

Jonas, P. 1999. Precipitating cloud systems. In *Global Energy and Water Cycles,* ed. K. A. Browning and R. J. Gurney, 109–118. New York: Cambridge University Press.

Jones, I. S. F, and D. Otaegui. 1997. Photosynthetic greenhouse gas mitigation by ocean nourishment. *Energy Conversion & Management* 38S:367–372.

Jousson, O., et al. 2000. Invasive alga reached California. *Nature* 408:157.

Jouzel, J., et al. 1987. Vostok ice core: A continuous isotope temperature record over the last climatic cycle (160,000 years). *Nature* 329:403–409.

Kahle, C. E., ed. 1974. *Plate tectonics: Assessments and Reassessments.* Tulsa, OK: American Association of Petroleum Geologists.

Kaiser, D., and R. Losick. 1993. How and why bacteria talk to each other. *Cell* 73:873–885.

Kaiser, J. 2000. Rift over biodiversity divides ecologists. *Science* 289:1282–1283.

Kajander, E. O., et al. 1997. Nanobacteria from blood, the smallest culturable autonomously replicating agent on Earth. *Proceedings of Society of Photo-optical Instrumentation Engineers (SPIE)* 3111:420–428.

Kaler, J. B. 1992. *Stars.* New York: Scientific American Library.

Karner, M. B., E. F. DeLong, and D. M. Karl. 2001. Archaeal dominance in the mesopelagic zone of the Pacific Ocean. *Nature* 409:507–510.

Kasting, J. F. 1993. Earth's early atmosphere. *Science* 259:920–925.

Kasting, J. F. 1997. Habitable zones around low mass stars and the search for extraterrestrial life. *Origins of Life* 27:291–307.

Kasting, J. F., and D. H. Grinspoon. 1991. The faint young sun problem. In *The Sun in Time,* ed. C. P. Sonett, M. S. Giampapa, and M. S. Matthews, 447–462. Tucson: University of Arizona Press.

Kasting, J. F., D. P. Whitmire, and R. T. Reynolds. 1993. Habitable zones around main sequence stars. *Icarus* 101:108–128.

Katz, L. A. 1999. The tangled web: Gene genealogies and the origin of eukaryotes. *The American Naturalist* 154:S137-S145.

Katz, M. E., et al. 1999. The source and fate of massive carbon input during the latest Paleocene thermal maximum. *Science* 286:1531–1533.

Kauffman, J. B., et al. 1995. Fire in the Brazilian Amazon. 1. Biomass, nutrient pools, and losses in slashed primary forest. *Oecologia* 104:397–408.

Kaula, W. M. 1968. *An Introduction to Planetary Physics.* New York: John Wiley.

Keeling, C. D. 1998. Reward and penalties of monitoring the Earth. *Annual Review of Energy and the Environment* 23:25–82.

Keith, D. 2000. The Earth is not yet an artifact. *IEEE Technology and Society Magazine* 19(4):25–28.

Kellogg, L. H. 1993. Chaotic mixing in the Earth's mantle. *Advances in Geophysics* 34:1–33.

Kellogg, L. H., B. H. Hager, and R. D. van der Hilst. 1999. Compositional stratification in the deep mantle. *Science* 283:1881–1884.

Kendrew, J. C., et al. 1958. Structure of myoglobin. *Nature* 185:422–427.

Kenrick, P. 1999. The family tree flowers. *Nature* 420:358–359.

Kenrick, P., and P. R. Crane. 1997. *The Origin and Early Diversification of Land Plants.* Washington, DC: Smithsonian Institution.

Kerr, R. A. 1996. A new dawn for Sun-climate links? *Science* 271:1360–1361.

Kerr, R. A. 1997. Putative Martian microbes called microscopy artifacts. *Science* 278:1706–1707.

Kerr, R. A. 1998. Requiem for life on Mars? Support for microbes fades. *Science* 282:1398–1400.

Kessler, A. 1985. *Heat Balance Climatology.* Amsterdam: Elsevier.

Khakhina, L. N. 1992. *Concepts of Symbiogenesis: A Historical Stduy of the Research of Russian Botanists.* New Háven: Yale University Press.

Khasawneh, F. E., E. C. Sample, and E. J. Kamprath, eds. 1980. *The Role of Phosphorus in Agriculture.* Madison, WI: American Society of Agronomy.

Kieft, T. L., and T. J. Phelps. 1997. Life in the slow lane: Activities of microorganisms in the subsurface. In *The Microbiology of the Terrestrial Deep Subsurface.* ed. P. S. Amy and D. L. Haldeman, 137–157. Boca Raton, FL: Lewis.

Kiehl, J. T., and Trenberth K. E. 1997. Earth's annual global mean energy budget. *Bulletin of the American Meteorological Society* 78:197–208.

King, M. D., and D. D. Herring. 2000. Monitoring Earth's vital signs. *Scientific American* 282(4):92–97.

Kirchner, J. W., and A. Weil. 2000. Delayed biological recovery from extinctions throughout the fossil record. *Nature* 404:177–180.

Kleiber, M. 1932. Body size and metabolism. *Hilgardia* 6:315–353.

Kleiber, M. 1961. *The Fire of Life.* New York: John Wiley.

Klenk, H.-P., et al. 1997. The complete genome sequence of the hyperthermophilic, sulphate-reducing archaeon *Archaeoglobus fulgidus*. *Nature* 390:364–370.

Klinge, H., et al. 1975. Biomass and structure in a Central Amazonian rain forest. In *Tropical Ecological Systems,* ed. F. Golley and E. Medina, 115–122. New York: Springer-Verlag.

Kluyver, A. J. 1956. Microbial metabolism and the energetic basis of life. In *The Microbe's Contribution to Biology,* ed. A. J. Kluyver and C. B. Van Niel, 3. Cambridge, MA: Harvard University Press.

Knoll, A. H., 1999. A new molecular window on early life. *Science* 285:1025–1026.

Knoll, A. H., et al. 1996. Comparative Earth history and late Permian mass extinction. *Science* 273:452–457.

Ko, M. K. W., N. D. Sze, and M. Prather. 1994. Better protection of the ozone layer. *Nature* 367:505–508.

Koblents-Mishke, O. I., et al. 1968. New data on the magnitude of primary production in the oceans. *Doklady Akademii Nauk SSSR Seria Biologicheskaya (Transactions of the Academy of Sciences of the USSR Biological Series)* 183:1189–1192.

Koerner, L. 1999. *Linnaeus: Nature and Nation.* Cambridge, MA: Harvard University Press.

Kolber, Z. S., et al. 2000. Bacterial photosynthesis in surface waters of the open ocean. *Nature* 407:177–179.

Kolber, Z. S., et al. 2001. Contribution of aerobic photoheterotrophic bacteria to the carbon cycle in the ocean. *Science* 292:2492–2495.

Kondratyev, K. I. 1988. *Climate Shocks: Natural and Anthropogenic.* New York: John Wiley.

Kooyman, G. L., and Ponganis P. J. 1997. The challenges of diving to depth. *American Scientist* 85:530–539.

Kovda V. A. 1971. The problem of biological and economic productivity of the earth's land areas. *Soviet Geography* 12:6–23.

Kovda V. A., et al. 1968. Contemporary scientific concepts relating to the biosphere. In *Use and Conservation of the Biosphere,* 13–29. Paris: UNESCO.

Kozaki, A., and G. Takeba. 1996. Photorespiration protects C_3 plants from photooxidation. *Nature* 384:557–560.

Krall, S., R. Peveling, and B. D. Diallo. 1997. *New Strategies in Locust Control.* Boston: Birkenhauser Verlag.

Kress, V. 1997. Magma mixing as a source for Pinatubo sulphur. *Nature* 389:591–593.

Kroll, R. G. 1990. Alkalophiles. In *Microbiology of Extreme Environments.* ed. C. Edwards, 55–92. New York: McGraw-Hill.

Kuenen, J. G., and P. Bos. 1989. Habitats and ecological niches of chemolitho(auto)trophic bacteria. In *Autotrophic Bacteria,* ed. H. G. Schlegel and B. Bowien, 53–80. Madison, WI: Science Tech.

Kuhn, J. R., et al. 1998. The Sun's shape and brightness. *Nature* 392:155–157.

Kukla, G. 2000. The last interglacial. *Science* 287:987–989.

Kumar, S., and S. B. Hedges. 1998. A molecular timescale for vertebrate evolution. *Nature* 392:917–920.

Kumar, S., ed. 1998. *Apoptosis: Biology and Mechanisms.* Berlin: Springer-Verlag.

Kump, L. R. 2000. What drives climate? *Nature* 408:651–651.

Kump, L. R., and F. T. Mackenzie. 1996. Regulation of atmospheric O_2: Feedback in the microbial feedbag. *Science* 271:459–460.

Kurzweil, R. 1990. *The Age of Intelligent Machines.* Cambridge, MA: MIT Press.

Kurzweil, R. 1999. *The Age of Spiritual Machines.* New York: Viking.

Kvenvolden, K. A. 1993. Gas hydrates—Geological perspective and global change. *Reviews of Geophysics* 31:173–187.

Labandeira, C. C., and J. J. Sepkoski. 1993. Insect diversity in the fossil record. *Science* 262:310–315.

Lal, R. 1995. Erosion-crop productivity relationships for soils in Africa. *Soil Science Society of America Journal* 59:661–667.

Lamond, A. I., and W. C. Earnshaw. 1998. Structure and function in the nucleus. *Science* 280:547–553.

Landsberg, J. J. 1986. *Physiological Ecology of Forest Production.* New York: Academic.

Larralde, R., M. P. Robertson, and S. L. Miller. 1995. Rates of decomposition of ribose and other sugars: Implications for chemical evolution. *Proceedings of the National Academy of Sciences USA* 92:8158–8160.

Lassen, K., and E. Friis-Christensen. 1995. Variability of the solar cycle during the past five centuries and the apparent association with terrestrial climate. *Journal of Atmospheric and Terrestrial Physics* 57:835–846.

Lawlor, D. W. 1993. *Photosynthesis: Molecular, Physiological and Environmental Processes.* Longman Scientific: Essex.

Layborne, R. C. 1974. Collision between a vulture and an aircraft at an altitude of 37,000 feet. *Wilson Bulletin* 86:461–462.

Laybourn-Parry, J. 1992. *Protozoan Plankton Ecology.* London: Chapman & Hall.

Lazcano, A., and S. L. Miller. 1994. How long did it take for life to begin and evolve to cyanobacteria? *Journal of Molecular Evolution* 39:546–554.

Lazcano, A., and S. L. Miller. 1996. The origin and early evolution of life: Prebiotic chemistry, the pre-RNA world, and time. *Cell* 85:793–798.

Leader, D., and J. Blahnik. 1993. *Bread Alone.* New York: William Morrow and Company.

Lean, J., and D. Rind. 1999. Evaluating Sun-climate relationships since the Little Ice Age. *Journal of Atmospheric and Solar-terrestrial Physics* 61:25–36.

LeConte, J. 1891. *Evolution.* New York: D. Appleton.

Lee, L. K. 1990. The dynamics of declining soil erosion rates. *Journal of Soil and Water Conservation* 45:622–624.

Lefohn, A. S., J. D. Husar, and R. B. Husar. 1999. Estimating historical anthropogenic global sulfur emission patterns for the period 1850–1990. *Atmospheric Environment* 33:3425–3444.

Leidy, J. 1881. The parasites of the termites. *Journal of the Academy of Natural Sciences of Philadelphia.* 3:425–447.

Lepš, J., et al. 1982. Community, stability, complexity, and species life history strategies. *Vegetatio* 50:53–63.

Le Roy, E. 1927. *L'exigence idéaliste et la fait de l'évolution (Idealistic Exigency and the Feat of Evolution).* Paris: Boivin.

Levi, P. 1984. *The Periodic Table.* New York: Schocken.

Levine, A. J. 1992. *Viruses.* New York: Scientific American Library.

Levine, J. S., ed. 1991. *Global Biomass Burning.* Cambridge, MA: MIT Press.

Levings, C. S. III, and I. K. Vasil, eds. 1995. *The Molecular Biology of Plant Mitochondria.* Dordrecht, the Netherlands: Kluwer Academic.

Lewis, J. 1999. *Worlds without End: The Exploration of Planets Known and Unknown.* Reading, MA: Perseus.

L'Haridon, S., et al. 1995. Hot subterranean biosphere in a continental oil reservoir. *Nature* 377:223–224.

Li, C., J. Chen, and T. Hua. 1998. Precambrian sponges with cellular structures. *Science* 279:879–882

Li, J. K. 2000. Scaling and invariants in cardiovascular biology. In *Scaling in Biology,* ed. J. H. Brown and G. B. West, 113–128. Oxford: Oxford University Press.

Li, X., et al. 2000. A pigment-binding protein essential for regulation of photosynthetic light harvesting. *Nature* 403:391–395.

Liebig, J., von. 1862. *Die Naturgesetze des Feldbaues (Natural Laws of Agriculture).* Braunschweig, Germany: Vieweg.

Lieth, H. 1964. Versuch einer kartographischen Darstellung der Produktivität der Pflanzendecke auf der Erde. In *Geographisches Taschenbuch 1964/65,* 72–80. Wiesbaden, Germany: Max Steiner Verlag.

Lieth, H. 1973. Primary production: Terrestrial ecosystems. *Human Ecology* 1:3003–332.

Lighthart, B., and A. J. Mohr. 1994. *Atmospheric Microbial Aerosols.* New York: Chapman & Hall.

Lindeman, R. 1942. The trophic-dynamic aspect of ecology. *Ecology* 23:399–418.

Lindert, P. 2000. *Shifting Ground: The Changing Agricultural Soils of China and Indonesia.* Cambridge, MA: MIT Press.

Linnaeus, C. 1735. *Systema Naturae.* Leyden, the Netherlands: J. F. Gronovius.

Linnaeus, C. 1758. *Systema Naturae.* Stockholm: Laurentius Salvius.

Linstedt, S. L., and W. A. Calder. 1981. Body size, physiological time, and longevity of homeothermic animals. *Quarterly Review of Biology* 56:1–16.

Lissauer, J. J. 1999. How common are habitable planets? *Nature* 402 (Supp. 2):C11–C14.

Lockwood, M., R. Stamper, and M. N. Wild. 1999. A doubling of the Sun's coronal magnetic field during the past 100 years. *Nature* 399:437–439.

Long, S. P., M. J. Roberts, and M. B. Jones, eds. 1992. *Primary Productivity if Grass Ecosystems of the Tropics and Sub-tropics.* London: Chapman & Hall.

Loomis, S. H. 1995. Freezing tolerance of marine invertebrates. *Oceanography and Marine Biology: An Annual Review* 33:373–350.

Lorenz, E. N. 1976. *The Nature and Theory of the General Circulation of the Atmosphere.* Geneva: WMO.

Loubere, P. 2000. Marine control of biological production in the eastern equatorial Pacific Ocean. *Nature* 406:497–500.

Lovelock, J. E. 1972. Gaia as seen through the atmosphere. *Atmospheric Environment* 6:579–580.

Lovelock, J. E. 1979. *Gaia: A New Look at Life on Earth.* Oxford: Oxford University Press.

Lovelock, J. E. 1988. *The Ages of Gaia.* New York: W. W. Norton.

Lovelock, J. E., R. J. Maggs, and R. A. Rasmussen. 1972. Atmospheric dimethylsulphide and the natural sulphur cycle. *Nature* 237:452–453.

Lowe, D. R. 1994. Early environments: Constraints and opportunities for early evolution. In *Early Life on Earth,* ed. S. Bengston, 24–35. New York: Columbia University Press.

Lowman, M. D., and N. M. Nadkarni, eds. 1995. *Forest Canopies.* San Diego: Academic.

Lüttge, U. 1998. Crassulacean acid metabolism. In *Photosynthesis: A Comprehensive Treatise,* ed. A. S. Raghavendra, 136–149. New York: Cambridge University Press.

Lutz, R. A., et al. 1994. Rapid growth at deep-sea vents. *Nature* 371:663–664.

Lutz, W., W. Sanderson, and S. Scherbov. 2001. The end of world population growth. *Nature* 412:543–545.

Lutzoni, F., M. Pagel, and V. Reeb. 2001. Major fungal lineages are derived from lichen symbiotic ancestors. *Nature* 411:937–940.

L'vovich, M. I., et al. 1990. Use and transformation of terrestrial water systems. In *The Earth as Transformed by Human Action,* ed. B. L. Turner, 235–252. Cambridge: Cambridge University Press.

Lwoff, A., et al. 1946. Nomenclature of nutritional types of microorganisms. *Cold Spring Harbor Symposia on Quantitative Biology* 11:302–303.

Macdonald, A. M., and C. Wunsch. 1996. An estimate of global ocean circulation and heat fluxes. *Nature* 382:436–439.

Mackenzie, F. T., and J. A. Mackenzie. 1995. *Our Changing Planet.* Englewood Cliffs, NJ: Prentice-Hall.

MacLeod, N., ed. 1996. *Cretaceous-Tertiary Mass Extinctions: Biotic and Environmental Change.* New York: W. W. Norton.

Madigan, M. T., and B. L. Marrs. 1997. Extremophiles. *Scientific American* 276(4):82–87.

Mage, D., et al. 1996 Urban air pollution in megacities of the world. *Atmospheric Environment* 30:681–686.

Magurran, A. E. 1988. *Ecological Diversity and its Measurement.* Princeton, NJ: Princeton University Press.

Mann, K. H. 1984. Fish population in open oceans ecosystems. In *Flows of Energy and Materials in Marine Ecosystems,* ed. M. J. R. Fasham, 435–458. New York: Plenum.

Mann, M. E., R. S. Bradley, and M. K. Hughes. 1998. Global-scale temperature patterns and climate forcing over the past six centuries. *Nature* 392:779–787.

Mann, M. E., R. S. Bradley, and M. K. Hughes. 1999. Northern hemisphere temperatures during the past millennium: Inferences, uncertainties, and limitations. *Geophysical Research Letters* 26: 759–762.

Mann, M. E., R. S. Bradley, and M. K. Hughes. 2000. Global-scale temperature patterns and climate forcing over the past six centuries. *Nature* 392:779–787.

Manzer, L. E. 1990. The CFC-ozone issue: Progress on the development of alternatives to CFCs. *Science* 249:31–35.

Margulis, L. 1970. *Origin of Eukaryotic Cells.* New Haven: Yale University Press.

Margulis, L. 1993. *Symbiosis in Cell Evolution.* New York: W. H. Freeman.

Margulis, L. 1998. *Symbiotic Planet: A New Look at Evolution.* New York: Basic.

Margulis, L., and R. Fester, eds. 1991. *Symbiosis as a Source of Evolutionary Innovation.* Cambridge, MA: MIT Press.

Margulis, L., and D. Sagan. 2000. *What Is Life?* Berkeley: University of California Press.

Margulis, L., and K. V. Schwartz. 1982. *Five Kingdoms.* New York: W. H. Freeman.

Marieb, E. N. 1998. *Human Anatomy and Physiology.* Menlo Park, CA: Benjamin/Cummings.

Marland, G., T. A. Boden, R. J. Andres, A. L. Brenkert, and C. A. Johnson. 2000. Global, regional, and national fossil fuel CO_2 emissions. <http://cdiac.esd.ornl.gov/trends/emis/em_cont.htm>.

Marquet, P. A. 2000. Invariants, scaling laws, and ecological complexity. *Science* 289:1487–1488.

Marshall, A. T. 1996. Calcification in hermatypic and ahermatypic corals. *Science* 271:637–642.

Martin, J. H., and S. E. Fitzwater. 1988. Iron deficiency limits phytoplankton growth in the north-east Pacific subarctic. *Nature* 331:341–343.

Martin, W., and M. Müller. 1998. The hydrogen hypothesis for the first eukaryote. *Nature* 392:37–41.

Matthews, E. 1983. Global vegetation and land use: New high-resolution data bases for climatic studies. *Journal of Climate and Applied Meteorology* 22:474–487.

Matthews, E. 1984. Global inventory of the pre-agricultural and present biomass. In *Interactions between Climate and Biosphere,* ed. H. Lieth, R. Fantechi, and H. Schnitzler, 237–246. Lisse, Switzerland: Swets and Zeitlinger.

Matthews, E., et al. 2000a. *The Weight of Nations.* Washington, DC: World Resources Institute.

Matthews, E., et al. 2000b. *Pilot Analysis of of Global Ecosystems: Forest Ecosystems.* Washington, DC: World Resources Institute.

Maxam, A. M., and W. Gilbert. 1977. A new method of sequencing DNA. *Proceedings of the National Academy of Sciences USA* 74:560–564.

May, R. M. 1988. How many species are there on Earth? *Science* 241:1441–1449.

May, R. M. 1990. How many species? *Philosophical Transactions of the Royal Society London* 330B:293–304.

Mayr, E., ed. 1974. *The Species Problem.* New York: Arno.

Mayr, E. 1982. *The Growth of Biological Thought: Diversity, Evolution and Inheritance.* Cambridge, MA: Harvard University Press.

Mayr, E. 1998. Two empires or three? *Proceedings of the National Academy of Sciences USA* 95:9720–9723.

McCloskey, J. M., and H. Spalding. 1989. A reconnaissance-level inventory of the amount of wilderness remaining in the world. *Ambio* 18:221–227.

McDonald, A. 1999. Combating acid deposition and climate change. *Environment* 41(3):4–11, 43–41.

McElwain, J. C., D. J. Beerling, and F. I. Woodward. 1999. Fossil plants and global warming at the Triassic-Jurassic boundary. *Science* 285:1386–1390.

McGrady-Steed, J., P. M. Harris, and P. J. Morin. 1997. Biodiversity regulates ecosystem predictability. *Nature* 390:162–165.

McKay, C. P., and W. J. Borucki. 1997. Organic synthesis in experimental impact shocks. *Science* 276:390–391.

McKay, C. P., et al. 1996. Search for past life on Mars: Possible relic biogenic activity in Martian meteorite ALH84001. *Science* 273:924–930.

McMahon, T. 1973. Size and shape in biology. *Science* 179:1201–1204.

McMahon, T., and J. T. Bonner. 1983. *On Size and Life.* New York: Scientific American Library.

McMenamin, M. A. S., and D. L. S. McMenamin. 1990. *The Emergence of Animals: The Cambrian Breakthrough.* New York: Columbia University Press.

McNaughton, S. J., and F. F. Banyikwa. 1995. Plant communities and herbivory. In *Serengeti: Research, Management, and Conservation of an Ecosystem,* ed. A. R. E. Sinclair and P. Arcese, 49–70. Chicago: University of Chicago Press.

Melillo, J. M., et al. 1993. Global climate change and terrestrial net primary production. *Nature* 363:234–240.

Melville, H. 1851. *Moby Dick.* New York: Harper & Brothers.

Menzies, R. J., R. Y. George, and G. T. Rowe. 1973. *Abyssal Environments and Ecology of the World Ocean.* New York: John Wiley.

Merezhkovskii, K. S. 1909. *Teoriya dvukh plazm kak osnova symbiogenezisa, novogo ucheniya o proiskhozhdenii organizmov* (Theory of two plasms as the basis of symbiogenesis, a new understanding of the origin of organisms). Kazan, Russia: Kazan University.

Meyer, D. A. 1998. Asian longhorned beetle. *Hardwood Research Bulletin* 500.

Meyer, O. 1985. Metabolism of aerobic carbon monoxide-utilizing bacteria. In *Microbial Gas Metabolism: Mechanistic, Metabolic and Biotechnological Aspects,* ed. R. K. Poole and C. S. Dow, 131–151. London: Academic.

Milanković, M. 1941. *Kanon der Erdbestrahlung und seine Anwendung auf das Eiszeitenproblem.* Belgrade: Srpska kral'evska akdemija.

Miles, E. W., and D. R. Davies. 2000. On the ancestry of barrels. *Science* 289:1490.

Miller, S. L. 1953. A production of amino acids under possible primitive earth conditions. *Science* 117:528–529.

Miller, S. L., and J. L. Bada. 1988. Submarine hot springs and the origin of life. *Nature* 334:609–611.

Miller, S. L., and L. E. Orgel. 1974. *The Origins of Life on the Earth.* Englewood Cliffs NJ: Prentice-Hall.

Milliman, J. D., and J. P. Syvitski. 1992. Geomorphic/tectonic control of sediment discharge to the ocean: The importance of small mountainous rivers. *Journal of Geology* 100:525–544.

Mittermeier, R. A., et al. 2000. *Hotspots: Earth's Biologically Richest and Most Endangered Terrestrial Ecoregions.* Chicago: University of Chicago Press.

Mojzis, S. J., et al. 1996. Evidence for life on Earth before 3,800 million years ago. *Nature* 384:55–59

Molina, M. J., and F. S. Rowland. 1974. Stratospheric sink for chlorofluoromethanes: chlorine atom catalyzed destruction of ozone. *Nature* 249:810–812.

Monod, J. 1971. *Chance and Necessity.* New York: Knopf.

Moore, P. D., B. Chaloner, and P. Stott. 1996. *Global Environmental Change.* Oxford: Blackwell Science.

Moore, W. S. 1996. Large groundwater inputs to coastal waters revealed by ^{226}Ra enrichments. *Nature* 380:612–614.

Moravec, H. 1988. *Mind Children: The Future of Robot and Human Intelligence.* Cambridge, MA: Harvard University Press.

Moravec, H. 1998. *Robot: Mere Machine to Transcend Mind.* New York: Oxford University Press.

Moreira, D., H. Le Guyader, and H. Phillippe. 2000. The origin of red algae and the evolution of chloroplasts. *Nature* 405:69–72.

Morell, V. 1996. Life's last domain. *Science* 273:1043–1045.

Morell, V. 1997. Microbiology's scarred revolutionary. *Science* 276:699–702.

Morgan J., et al. 1997. Size and morphology of the Chicxulub impact crater. *Nature* 390:472–476.

Morowitz, H. J. 1968. *Energy Flow in Biology.* New York: Academic.

Morris, S. C. 1998. *The Crucible of Creation: The Burgess Shale and the Rise of Animals.* Oxford: Oxford University Press.

Morse, J. W., and Mackenzie F. T. 1998. Hadean ocean carbonate chemistry. *Aquatic Geochemistry* 4:301–319.

Motz, L, ed. 1979. *Rediscovery of the Earth.* New York: Van Nostrand Reinhold.

Müller, D. 1960. Kreislauf des Kohlenstoffs. *Handbuch der Pflanzenphysiologie* 12:934–948.

Mullis, K. B., et al. 1986. Specific enzymatic amplification of DNA in vitro: The polymerase chain reaction. *Cold Spring Harbor Symposia on Quantitative Biology* 51:263–273.

Myers, J. 1974. Conceptual development in photosynthesis, 1924–1974. *Plant Physiology* 54:420–426.

Myers, N., et al. 2000. Biodiversity hotspots for conservation priorities. *Nature* 403:853–858.

Myneni, R. B., and D. L. Williams. 1994. On the relationship between FAPAR and NDVI. *Remote Sensing of the Environment* 49:200–211.

Myneni, R. B., et al. 1997. Estimation of global leaf area index and absorbed PAR using radiative transfer models. *IEEE Transactions on Geoscience and Remote Sensing* 4:1380–1393.

Nabhan, G. P., and S. L. Buchmann. 1997. Services provided by pollinators. In *Nature's Services Societal Dependence on Natural Ecosystems,* ed. G. C. Daily, 133–150. Washington, DC: Island.

Naeem, S., et al. 1994. Declining biodiversity can alter the performance of ecosystems. *Nature* 368:734–736.

Naeem, S., D. R. Hahn, and G. Schuurman. 2000. Producer-consumer codependency modulates biodiversity effects. *Nature* 403:762–764.

Nagy, K.A., I.A. Girard, and T.K. Brown. 1999. Energetics of free-ranging mammals, reptiles and birds. *Annual Review of Nutrition* 19:247–277.

Nance, R. D., T. R. Worsley, and J. B. Moody. 1988. The supercontinent cycle. *Scientific American* 259(1):72–79.

Naqvi, S. W. A., et al. 2000. Increased marine production of N_2O due to intensifying anoxia on the Indian continental shelf. *Nature* 408:346–349.

NASA (National Aeronautics and Space Administration). 1999. *Aerosols.* Greenbelt, MD: NASA. <http://eospso.gsfc.nasa.gov/ftp_docs/Aerosols.pdf>.

NASA (National Aeronautics and Space Administration). 2000. Capabilities of various planet detection methods. <http://www.kepler.arc.nasa.gov/Capabilities.html>.

NASA (National Aeronautics and Space Administration). 2001. Triana mission. <http://triana.gsfc.nasa.gov/home>.

Needham, A. E. 1965. *Uniqueness of Biological Materials.* Elmsford, NY: Pergamon.

Nelson, K. E., et al. 1999. Evidence for lateral gene transfer between Archaea and bacteria from genome sequence of Thermotoga maritima. *Nature* 399:323–329.

Nepstad, D.C., et al. 1999. Large-scale impoverishment of Amazonian forests by logging and fire. *Nature* 398:505–508.

Nerem, R. S. 1995. Global mean sea level variations from TOPEX/POSEIDON altimeter data. *Science* 268:708–710.

Nesis, K. N. 1997. Gonatid squids in the subarctic North Pacific: Ecology, biogeography, niche diversity and role in the ecosystem. *Advances in Marine Biology* 32:243–315.

Newell, R. E., et al. 1989. Carbon monoxide and the burning Earth. *Scientific American* 261(4):82–88.

Newman, M. 2001. A new picture of life's history on Earth. *Proceedings of the National Academy of Sciences USA* 98:5955–5956.

Nicol, S., and W. de la Mare. 1993. Ecosystem management and the Antarctic krill. *American Scientist* 81:36–47.

Noble, I. R., and R. Dirzo. 1997. Forests as human-dominated ecosystems. *Science* 277:522–525.

Nobre, C. A., P. J. Sellers, and J. Shukla. 1991. Amazon deforestation and regional climate change. *Journal of Climate* 4:957–988.

Noddack, W. 1937. Der Kohlenstoff im Haushalt der Natur (Carbon in nature's economy). *Angewandte Chemie (Applied Chemistry)* 50:505–510.

Norberg, U. M. 1990. *Vertebrate Flight.* Berlin: Springer-Verlag.

Norris, P. R., and W. J. Ingledew. 1992. Acidophilic bacteria: Adaptation and applications. In *Molecular Biology and Biotechnology of Extremophiles,* ed. R. A. Herbert and R. J. Sharp, 115–142. Glasgow: Blackie.

Norris, P. R., and D. B. Johnson. 1998. Acidophilic microorganisms. In *Extremophiles Microbial Life in Extreme Environments,* ed. K. Horikoshi and W. D. Grant, 133–153. New York: Wiley-Liss.

NRC (National Research Council). 1994. *Solar Influences on Global Change.* Washington, DC: National Academy Press.

NRC (National Research Council). 2000. *Reconciling Observations of Global Temperature Change.* Washington, DC: National Academy Press.

Nriagu, J. O., R. D. Coker, and L. A. Barrie. 1991. Origin of sulfur in Canadian Arctic haze from isotope measurements. *Nature* 349:142–145.

Nurse, P. 2000. The incredible life and times of biological cells. *Science* 289:1711–1716.

Ochman, H., J. G. Lawrence, and E. A. Groisman. 2000. Lateral gene transfer and the nature of bacterial innovation. *Nature* 405:299–304.

Odum, H. T. 1957. Trophic structure and productivity of Silver Springs, Florida. *Ecological Monographs* 27:55–112.

OECD. 1994. *Global Change of Planet Earth.* Paris: OECD.

Oechel, W. C., et al. 2000. Acclimation of ecosystem CO_2 exchange in the Alaskan Arctic in response to decadal climate warming. *Nature* 406:978–981.

Officer, C., and J. Page. 1996. *The Great Dinosaur Extinction Controversy.* Reading, MA: Addison-Wesley.

Ohmoto, H. 1996. Evidence in pre-2.2 Ga paleosols for the early evolution of atmospheric oxygen and terrestrial biota. *Geology* 24:1135–1138.

Ohmoto, H., T. Kakegawa, and D. R. Lowe. 1993. 3.4-billion-year-old biogenic pyrites from Barberton, South Africa: Sulfur isotope evidence. *Science* 262:555–557.

Oki, T. 1999. The global water cycle. In *Global Energy and Water Cycles,* ed. K. A. Browning and R. J. Gurney, 10–29. New York: Cambridge University Press.

Oldeman, I. R., et al. 1990. *World Map of the Status of Human-Induced Soil Degradation: An Explanatory Note.* Wageningen, the Netherlands: International Soil Information and Reference Center and Nairobi: UNEP.

Olson, J. S., J. A. Watts, and L. J. Allison.1983. *Carbon in Live Vegetation of Major World Ecosystems.* Oak Ridge, TN: Oak Ridge National Laboratory.

Oltmanns, F. 1904. *Morphologie und Biologie der Algen (Morphology and Biology of Algae).* Jena, Germany: Gustav Fischer.

Omar, G. I., and M. S. Steckler. 1995. Fission track evidence on the initial rifting of the Red Sea: Two pulses, no propagation. *Science* 270:1341–1344.

Oparin, A. I. 1924. *Proiskhozhdenie zhizni (The Origin of Life).* Moscow: Moskovskii rabochii.

Oparin, A. I. 1938. *The Origin of Life.* New York: McMillan Publishing.

Oreskes, N. 1999. *The Rejection of Continental Drift.* New York: Oxford University Press.

ORNL (Oak Ridge National Laboratory). 2000. NPP Database Holdings at the ORNL DAAC. <http://www-esdis.ornl.gov/NPP>.

Orphan, V. J., et al. 2001. Methane-consuming archaea revealed by directly coupled isotopic and phylogenetic analysis. *Science* 293:484–487.

Pace, N. R. 1997. A molecular view of microbial diversity and the biosphere. *Science* 276:734–740.

Pagel, M. 1999. Inferring the historical patterns of biological evolution. *Nature* 401:877–884.

Palkova, Z., et al. 1997. Ammonia mediates communications between yeast colonies. *Nature* 390:532–546.

Palumbi, S. R. 2001. Humans as the world's greatest evolutionary force. *Science* 293:1786–1790.

Pandey, A., and M. Mann. 2000. Proteomics to study genes and genomes. *Nature* 405:837–846.

Parkes, R. J. 2000. A case of bacterial immortality? *Nature* 407:844–845.

Parkes, R. J., et al. 1994. Deep bacterial biospere in Pacific Ocean sediments. *Nature* 371:410–413.

Patterson, C. C. 1965. Contaminated and natural lead environments. *Archives of Environmental Health* 11:344–360.

Paul, E. A., and F. E. Clark. 1989. *Soil Microbiology and Biochemistry.* New York: Academic.

Pauling, L., R. B. Corey, and H. R. Branson. 1951. The structure of proteins: Two hydrogen-bonded helical configurations of the polypeptide chain. *Proceedings of the National Academy of Sciences USA* 37:205–211.

Pavlov, A. A., et al. 2000. Greenhouse warming by CH_4 in the atmosphere of early Earth. *Journal of Geophysical Research* 105:11981–11983.

Pearson, P. N., and M. R. Palmer. 2000. Atmospheric carbon dioxide concentrations over the past 60 million years. *Nature* 406:695–699.

Pedersen, K. 1993. The deep subterranean biosphere. *Earth-Science Reviews* 34:243–260.

Peebles, P. J. E. 1993. *Principles of Physical Cosmology.* Princeton, NJ: Princeton University Press.

Pelizzari, U. 1998. Umberto Pelizzari's last attempt. <http://www.softntt.it>.

Pennisi, E. 1998. Genome data shake tree of life. *Science* 280:672–674.

Pennycuick, C. P. 1972. *Animal Flight.* London: Edward Arnold.

Perry, D. A. 1994. *Forest Ecosystems.* Baltimore: Johns Hopkins University Press.

Persson, T., ed. 1980. *Structure and Function of Northern Coniferous Forests.* Stockholm: Ecological Bulletins.

Peschek, G. A., ed. 1999. *The Phototrophic Prokaryotes.* New York: Kluwer Academic/Plenum.

Peters, R. H. 1983. *The Ecological Implications of Body Size.* New York: Cambridge University Press.

Petit, J. R., et al. 1999. Climate and atmospheric history of the past 420,000 years from the Vostok ice core, Antarctica. *Nature* 399:429–426.

Pfeffer, W. 1881. *Pflanzenphysiologie* (*Physiology of Plants*). Leipzig, Germany: W. Engelmann.

Pfeffer, W. 1897. *Pflanzenphysiologie.* Leipzig, Germany: Wilhelm Engelmann Verlag.

Philipson, W. R., J. M. Ward, and B. G. Butterfield.1971. *The Vascular Cambium.* London: Chapman & Hall.

Phillips, J. G. P. J. Butler, and P. J. Sharp, eds. 1985. *Physiological Strategies in Avian Biology.* Glasgow: Blackie.

Phillips, K. J. H. 1992. *Guide to the Sun.* Cambridge: Cambridge University Press.

Pickard, B. G. and R. N. Beachy. 1999. Intercellular connections are developmentally controlled to help move molecules through the plant. *Cell* 98:5–8.

Pierson, B. K. 1994. The emergence, diversification, and role of photosynthetic eubacteria. In *Early Life on Earth,* ed. S. Bengston, 161–180. New York: Columbia University Press.

Pilkington, M., and R. A. F. Grieve. 1992. The geophysical signature of terrestrial impact craters. *Review of Geophysics* 30:161–181.

Pimentel, D., ed. 1993. *World Soil Erosion and Conservation.* New York: Cambridge University Press.

Pimentel, D., et al. 1997. Economic and environmental benefits of biodiversity. *BioScience* 47:747-757.

Pimm, S. L., et al. 1995. The future of biodiversity. Science 269:347–350.

Pimm, S. L. et al. 2001. Can we defy nature's end? *Science* 293:2207–2208.

Pirozynski, K.A., and D. W. Malloch. 1975. The origin of land plants: A matter of mycotrophism. *Biosystems* 6:153–164.

Pittock, A. B. 1978. A critical look at long-term sun-weather relationships. *Review of Geophysics and Space Physics* 16:400–420.

Pittock, A. B. 1983. Solar variability, weather and climate: An update. *Quarterly Journal of the Royal Meteorological Society* 109:23–55.

Plumptree, A. J., and S. Harris. 1995. Estimating the biomass of large mammalian herbivores in a tropical montane forest: A method of faecal counting that avoids assuming a "steady state" system. *Journal of Applied Ecology* 32:111–120.

Polis, G. A., ed. 1991. *The Ecology of Desert Communities.* Tucson: University of Arizona Press.

Pollack, H. N., and D. S. Chapman. 1977. The flow of heat from the Earth's interior. *Scientific American* 237(2):60–76.

Pollitz, F. E. 1999. From drifting to rifting. *Nature* 398:21–22.

Ponnamperuma, F. N. 1984. Straw as a source of nutrients for wetland rice. In *Organic Matter and Rice,* 117–136. Los Banos, the Philippines: International Rice Research Institute.

Post, W., A. W. King, and S. D. Wullschleger. 1997. Historical variations in terrestrial biospheric carbon storage. *Global Biogeochemical Cycles* 11:99–109.

Potter, C. S. 1999. Terrestrial biomass and the effects of deforestation on the global carbon cycle. *BioScience* 49:769–778.

Potter, C. S., et al. 1993. Terrestrial ecosystem production: A process model based on global satellite and surface data. *Global Biogeochemical Cycles* 7:811–841.

Powell, J. L. 1998. *Night Comes to the Cretaceous: Dinosaur Extinction and the Transformation of Modern Geology.* New York: W. H. Freeman.

Price, C. 2000. Evidence for a link between global lightning activity and upper tropospheric water vapour. *Nature* 406:290–293.

Prins, H. H. T., and J. Reitsma. 1989. Mammalian biomass in an African equatorial forest. *Journal of Animal Ecology* 58:851–861.

Priscu, J. C., et al. 1998. Perennial Antarctic lake ice: An oasis for life in a polar desert. *Science* 280:2095–2098.

Priscu, J. C., et al. 1999. Geomicrobiology of subglacial ice above Lake Vostok, Antarctica. *Science* 286:2141–2144.

Proctor, M., P. Yeo, and A. Lack. 1996. *The Natural History of Pollination.* Portland, OR: Timber.

Psenner, R., and B. Sattler. 1998. Life at the freezing point. *Science* 280:2073–2074.

Purvis, A., and A. Hector. 2000. Getting the measure of biodiversity. *Nature* 405:212–219.

Qiu, Y.-L., et al. 1998. The gain of three mitochondrial introns identifies liverworts as the earliest land plants. *Nature* 394:671–674.

Qiu, Y.-L., et al. 1999. The earliest angiosperms: Evidence from mitochondrial, plastid and nuclear genes. *Nature* 402:404–407.

Quinn, T. J., and C. Fröhlich. 1999. Accurate radiometers should measure the output of the Sun. *Nature* 401:841.

Rabinowitch, E. I. 1945. *Photosynthesis and Related Processes.* New York: Interscience.

Rabinowitz, D., et al. 2000. A reduced estimate of the number of kilometre-sized near-Earth asteroids. Nature 403:165–166.

Raczyński, J. 1978. *Zubr.* Cracow: Państwowe Wydawnictwo Rołnicze a Lesne.

Radick, R. R. 1991. The luminosity variability of solar-type stars. In *The Sun in Time,* ed. C. P. Sonett, M. S. Giampapa, and M. S. Matthews, 787–808. Tucson: University of Arizona Press.

Raghavendra, A. S., ed. 1998. *Photosynthesis: A Comprehensive Treatise.* New York: Cambridge University Press.

Rai, A. N., ed. 1990. *CRC Handbook of Symbiotic Cyanobacteria.* Boca Raton, FL: Chemical Rubber Company.

Raich, J. W., and C. S. Potter. 1995. Global patterns of carbon dioxide emissions from soils. *Global Biogeochemical Cycles* 9:23–26.

Ramanathan, V. 1998. Trace-gas greenhouse effect and global warming. *Ambio* 27:187–197.

Ramanathan, V., B. R. Bakkstrom, and E. F. Harrison. 1989. Climate and the Earth's radiation budget. *Physics Today* 42(5):22–32.

Ramankutty, N., and J. A. Foley. 1999. Estimating historical changes in global land cover: Croplands from 1700 to 1992. *Global Biogeochemical Cycles* 13:997–1027.

Rampino, M. R., and S. Self. 1992. Volcanic winter and accelerated glaciation following the Toba super-eruption. *Nature* 359:50–52.

Rao, P. K., et al. 1990. *Weather Satellites: Systems, Data, and Environmental Applications.* Boston, MA: American Meteorological Society.

Rasmussen, B. 2000. Filamentous microfossils in a 3,235-million-year-old volcanogenic massive sulphide deposit. *Nature* 405:676–679.

Ray, J. 1686–1704. *Historia Plantarum* (History of Plants). London: M. Clark.

Raymo, M. E., W. F. Ruddiman, and P. N. Froelich. 1988. The influence of late Cenozoic mountain building on ocean geochemical cycles. *Geology* 16:649–653

Raynaud, D., et al. 1993. The ice record of greenhouse gases. *Science* 259:926–934.

Reagan, D. P., and R. B. Waide. 1996. *The Food Web of a Tropical Rain Forest.* Chicago: University of Chicago Press.

Redecker, D., R. Kodner, and L. E. Graham. 2000. Glomalean fungi from the Ordovician. *Science* 289:1920–1921.

Reichle, D. E., ed. 1981. *Dynamic Properties of Forest Ecosystems.* Cambridge: Cambridge University Press.

Reid, R. P., et al. 2000. The role of microbes in accretion, lamination and early lithification of modern marine stromatolites. *Nature* 406:989–992.

Reinfelder, J. R., M. L. Krapiel, and F. M. M. Morel. 2000. Unicellular C_4 photosynthesis in a marine diatom. *Nature* 407:996–999.

Reith, M. 1995. Molecular biology of rhodophyte and chromophyte plastids. *Annual Review of Plant Physiology and Plant Molecular Biology* 46:549–575.

Renne, P. R., and A. R. Basu.1991. Rapid eruption of the Siberian Traps flood basalts at the Permo-Triassic boundary. *Science* 253:176–179.

Rensberger, B. 1996. *Life Itself: Exploring the Realm of the Living Cell.* New York: Oxford University Press.

Revelle, R., and H. E. Suess. 1957. Carbon dioxide exchange between atmosphere and ocean and the question of an increase of atmospheric CO_2 during the past decades. *Tellus* 9:18–27.

Rial, J. A. 1999. Pacemaking the ice ages by frequency modulation of Earth's orbital eccentricity. *Science* 285:564–568.

Richards, J. F. 1990. Land transformation. In *The Earth as Transformed by Human Action,* ed. B. L. Turner et al., 161–178. Cambridge: Cambridge University Press.

Richardson, J. S. 1985. Schematic drawings of protein structures. *Methods in Enzymology* 115:359–380.

Richelson, J. T. 1998. Scientists in black. *Scientific American* 278(2):48–55.

Richmond, R., R. Sridhar, and M. J. Daly. 1999. Physicochemical survival pattern for the radiophile *Deinococcus radiodurans. Proceedings of Society of Photo-optical Instrumentation Engineers (SPIE)* 3755:210–222.

Richter, D. D., and D. Markewitz. 1995. How deep is soil? *BioScience* 45:600–609.

Riebesell, U., et al. 2000. Reduced calcification of marine plankton in response to increased atmospheric CO_2. *Nature* 407:364–367.

Riley, G. A. 1944. The carbon metabolism and photosynthetic efficiency of the Earth as a whole. *American Scientist* 32:129–134.

Robinson, B. H. 1995. Light in the ocean's midwaters. *Scientific American* 273(1):59–64.

Robinson, J. M. 1996. Atmospheric bulk chemistry and evolutionary megasymbiosis. *Chemosphere* 33:1641–1653.

Roger, A. J. 1999. Reconstructing early events in eukaryotic evolution. *American Naturalist* 154:S146-S163.

Rogers, D. J., and S. E. Randolph. 2000. The global spread of malaria in a future, warmer world. *Science* 289:1763–1765.

Rohring, E., and B. Ulrich, eds. 1991. *Temperate Deciduous Forests.* Amsterdam: Elsevier.

Romankevich, E. A. 1988. Living matter of the Earth. *Geokhimiya* 2:292–306.

Rood, R. T., and J. S. Trefil. 1981. *Are We Alone?* New York: Scribner.

Roper, C. F. E., and K. J. Boss. 1982. The giant squid. *Scientific American* 246(4):96–105.

Rosenberg, D. M., P. McCully, and C. M. Pringle. 2000. Global-scale environmental effects of hydrological alterations: Introduction. *BioScience* 50:746–751.

Rosenzweig, M. L. 1995. *Species Diversity in Space and Time.* New York: Cambridge University Press.

Rosenzweig, M. L. 1997. Tempo and mode of speciation. *Science* 277:1622–1623.

Rosenzweig, M. L., and Z. Abramsky. 1993. How are diversity and productivity related? In *Species Diversity in Ecological Communities,* ed. R. E. Ricklefs and D. Schlutter, 52–65. Chicago: University of Chicago Press.

Rossman, A. Y. 2001. A special issue on global movement of invasive plants and fungi. *BioScience* 51:93–94.

Rothschild, L. J., and R. L. Mancinelli. 2001. Life in extreme environments. *Nature* 409:1092–1101.

Roubik, D. W. 2000. Pollination system stability in tropical America. *Conservation Biology* 14:1235–1236.

Rowan, R., and D. A. Powers. 1991. A molecular classification of zooxanthellae and the evolution of animal-algal symbioses. *Science* 251:1348–1351.

Rowland, F. S. 1989. Chlorofluorocarbons and the depletion of stratospheric ozone. *American Scientist* 77:36–45.

Royal Ministry of Foreign Affairs. 1971. *Air Pollution across National Boundaries.* Stockholm: Royal Ministry of Foreign Affairs.

Ruddiman, W. F., ed. 1997. *Tectonic Uplift and Climate Change.* New York: Plenum.

Rudnick, R. L., W. F. McDonough, and R. J. O'Connell.1998. Thermal structure, thickness and composition of continental lithosphere. *Chemical Geology* 145:395–411.

Ruepp, A., et al. 2000. The genome sequence of the thermoacidophilic scavenger *Thermoplasma acidophilum. Nature* 407: 508–513.

Ruimy, A., G. Dedieu, and B. Saugier. 1996. TURC: A diagnostic model of continental gross primary productivity and net primary productivity. *Global Biogeochemical Cycles* 10:269–285.

Ruiz, G. M., et al. 2000. Global spread of microorganisms by ships. *Nature* 408:49.

Runnegar, B. 1994. Proterozoic eukaryotes: Evidence from biology and geology. In *Early Life on Earth,* ed. S. Bengston, 287–297. New York: Columbia University Press.

Russell, N. J. 1992 Physiology and molecular biology of psychrophilic micro-organisms. In *Molecular Biology and Biotechnology of Extremophiles,* ed. R. A. Herbert and R. J. Sharp, 203–224. Glasgow: Blackie.

Russell, N. J., and T. Hamamoto. 1998. Psychrophiles. In *Extremophiles Microbial Life in Extreme Environments,* ed. K. Horikoshi and W. D. Grant, 25–45. New York: Wiley-Liss.

Rye, B., and H. D. Holland. 1998. Paleosols and the evolution of atmospheric oxygen: A critical review. *American Journal of Science* 298:621–672.

Ryther, J. H. 1969. Photosynthesis and fish production in the sea. *Science* 166:72–76.

Sagan, C., and C. Chyba. 1997. The early faint sun paradox: Organic shielding of ultraviolet-labile greenhouse gases. *Science* 276:1217–1221.

Sagan, C., and G. Mullen. 1972. Earth and Mars: Evolution of atmospheres and surface temperatures. *Science* 177:52–56.

Sagan, C., et al. 1993. A search for life on Earth from the Galileo spacecraft. *Nature* 365:715–721.

Sage, R. F., and R. K. Monson. 1999. *C_4 Plant Biology.* San Diego: Academic.

Sahagian, D., R. W. Schwartz, and D. K. Jacobs. 1994. Direct anthropogenic contributions to sea level rise in the twentieth century. *Nature* 367:54–56.

Saikkonen, K., et al. 1998. Fungal endophytes: A continuum of interactions with host plants. *Annual Review of Ecology and Systematics* 29:319–343.

Sandbrook, R. 1999. New hopes for the United Nations Environment Programme (UNEP)? *Global Environmental Change* 9:171–174.

Sanders, I. R. 1999. No sex please, we're fungi. *Nature* 399:737–739.

Sanger, F., S. Nicklen, and R. E. Coulson.1977. DNA sequencing with chain terminating inhibitors. *Proceedings of the National Academy of Sciences USA* 74: 5463–5467.

Sankaran, M., and S. J. McNaughton. 1999. Determinants of biodiversity regulate compositional stability of communities. *Nature* 401:691–693.

Santer, B. D., et al. 2000. Interpreting differential at the surface and in the lower troposphere. *Science* 287:1227–1232.

Sarbu, S. M., T. C. Kane, and B. K. Kinkle.1996. A chemoautotrophically based cave ecosystem. *Science* 272:1953–1955.

Satheesh, S. K., and V. Ramanathan. 2000. Large differences in tropical aerosol frocing at the top of the atmosphere and Earth's surface. *Nature* 405:60–63.

SCEP (Study of Critical Environmental Problems). 1970. *Man's Impact on the Global Environment.* Cambridge, MA: MIT Press.

Schaefer, M. B. 1965. The potential harvest of the sea. *Transaction of the American Fisheries Society* 94:123–128.

Schaller, G. B. 1972. *The Serengeti Lion.* Chicago: University of Chicago Press.

Schidlowski, M. 1991. Quantitative evolution of global biomass through time: Biological nad geochemical constraints. In *Scientists on Gaia,* ed. S. H. Schneider and P. J. Boston, 211–222. Cambridge, MA: MIT Press.

Schidlowski, M., and P. Aharon. 1992. Carbon cycle and carbon isotope record: Geochemical impact of life over 3.8 Ga of earth history. In *Early Organic Evolution: Implications for Mineral and Energy Resources,* ed. M Schidlowski et al., 147–175. Berlin: Springer Verlag.

Schieber, J., D. Krinsley, and L. Riciputi. 2000. Diagenetic origin of quartz silt in mudstones and implications for silica cycling. *Nature* 406:981–985.

Schlesinger, W. H. 1977. Carbon balance in terrestrial detritus. *Annual Review of Ecology and Systematics* 8:51–81.

Schlesinger, W. H. 1991. *Biogeochemistry: An Analysis of Global Change.* San Diego: Academic.

Schlesinger, W. H., J. P. Winkler, and J. P. Megonigal. 2000. Soils and the global carbon cycle. In *The Carbon Cycle,* ed. T. M. L. Wigley and D. S. Schimel. Cambridge: Cambridge University Press.

Schmid, P. E., M. Tokeshi, and J. M. Schmid-Araya. 2000. Relation between population density and body size in stream communities. *Science* 289:1557–1559.

Schmidt-Nielsen, K. 1964. *Desert Animals.* Oxford: Oxford University Press.

Schmidt-Nielsen, K. 1984. *Scaling: Why is Animal Size So Important.* Cambridge: Cambridge University Press.

Schmitt, R. W. 1999. The ocean's response to the freshwater cycle. In *Global Energy and Water Cycles,* ed. K. A. Browning and R. J. Gurney, 144–154. New York: Cambridge University Press.

Schneider, J. 2000. *The Extrasolar Planet Encyclopedia.* <www.obspm.fr/encycl/encycl.html>.

Schneider, S. H. 2001. Earth systems engineering and management. *Nature* 409:417–421.

Schneider, S. H., and P. J. Boston, eds. 1991. *Scientists on Gaia.* Cambridge, MA: MIT Press.

Schopf, J. W., ed. 1983. *Earth's Earliest Biosphere: Its Origin and Evolution.* Princeton, NJ: Princeton University Press.

Schopf, J. W. 1993. Microfossils of the early Archean Apex chert: New evidence of the antiquity of life. *Science* 260:640–646.

Schopf, J. W. 1994. The oldest known records of life: Early Archaean stromatolites, microfossils, and organic matter. In *Early Life on Earth,* ed. S. Bengston, 193–206. New York: Columbia University Press.

Schopf, J. W. 1999. *Cradle of Life: The Discovery of Earth's Earliest Fossils.* Princeton, NJ: Princeton University Press.

Schopf, J. W., and Klein, C., eds. 1992. *The Proterozoic Biosphere: A Multidisciplinary Study.* New York: Cambridge University Press.

Schrödinger, E. 1944. *What Is Life? Physical Aspect of the Living Cell.* Cambridge: Cambridge University Press.

Schroeder, H. 1919. Die jährliche Gesamtproduktion der grünen Pflanzdecke der Erde (Annual production of the Earth's green plant cover). *Naturwissenschaften* 7:8–12.

Schwartz, J. H. 1999. *Sudden Origins: Fossils, Genes, and the Emergence of Species.* New York: John Wiley.

Schwartzman, D. 1999. *Life, Temperature, and the Earth.* New York: Columbia University Press.

Sclater, G. J., C. Jaupart, and D. Galson. 1980. The heat flow through oceanic and continental crust and the heat loss of the Earth. *Reviews of Geophysics and Space Physics* 18:269–311.

Scott, J. A. 1986. *The Butterflies of North America.* Stanford, CA: Stanford University Press.

Seeley, T. D. 1995. *The Wisdom of the Hive.* Cambridge, MA: Harvard University Press.

Seilacher, A., P. K. Bose, and F. Pflüger. 1998. Triploblastic animals more than 1 billion years ago: Trace fossil evidence from India. *Science* 282:80–83.

Selby, M. J. 1985. *Earth's Changing Surface.* Oxford: Clarendon.

Shackleton, N. J. 2000. The 100,000-year ice-age cycle identified and found to lag temperature, carbon dioxide, and orbital eccentricity. *Science* 289:1897–1902.

Shapiro, J. A., and Dworkin M, eds. 1997. *Bacteria as Multicellular Organisms.* New York: Oxford University Press.

Shapiro, R. 1988. Prebiotic ribose synthesis: A critical analysis. *Origins of Life* 18:71–85.

Sharpton, V. L., and P. D. Ward, eds. 1991. *Global Catastrophes in Earth History.* Boulder, CO: Geological Society of America.

Sheffield C. 1981. *Earth Watch: A Survey of the World from Space.* New York: Macmillan.

Shen, Y., et al. 2001. Isotopic evidence for microbial sulphate reduction in the Archaean era. *Nature* 410:77–81.

Shigo, A. 1986. *A New Tree Biology.* Durham, NC: Shigo and Tress.

Shiklomanov, I. A. 1997. *Comprehensive Assessment of the Freshwater Resources of the World.* Stockholm: World Meteorological Organization and Stockholm Environment Institute.

Shiklomanov, I. A. 1999. *World Water Resources and Water Use.* St. Petersburg: State Hydrological Institute.

Shklovskii, I. S., and C. Sagan. 1966. *Intelligent Life in the Universe.* San Francisco: Holden-Day.

Short, N. M., et al. 1976. *Mission to Earth: LANDSAT Views the World.* Washington, DC: NASA.

Shu, G., et al. 1999. Lower Cambrian vertebrates from South China. *Nature* 402:42–46.

Sidebottom, C., et al. 2000. Heat-stable antifreeze protein from grass. *Nature* 406:256.

Siegenthaler, U., and J. L. Sarmiento. Atmospheric carbon dioxide and the ocean. *Nature* 365:119–125.

Sigman, D. M., and E. A. Boyle. 2000. Glacial/interglacial variations in atmospheric carbon dioxide. *Nature* 407:859–869.

Simonson, R. W. 1968. Concept of soil. *Advances in Agronomy* 20:1–47.

Simpson, G. G. 1964. The non-prevalence of humanoids. *Science* 143:769–75

Skole, D., and C. Tucker. 1993. Tropical deforestation and habitat fragmentation in the Amazon: Satellite data from 1978 to 1988. *Science* 260:1905–1910.

Sladen, W. J. L., C. M Menzies, and W. L. Reichel. 1966. DDT residues in Adelie penguins and a crab-eater seal from Antarctica: Ecological implications. *Nature* 210:670–673.

Slater, P., ed. 1986. *Earth Remote Sensing Using the LANDSAT Thematic Mapper and SPOT Sensor Systems.* Bellingham, WA: The International Society for Optical Engineering.

Sleep, N. H., et al. 1989. Annihilation of ecosystems by large asteroid impacts on the early Earth. *Nature* 342:139–142.

Smetacek, V. 2000. The giant diatom pump. *Nature* 406:574–575.

Smil, V. 1984. *The Bad Earth.* Armonk, NY: M.E. Sharpe.

Smil, V. 1985. *Carbon Nitrogen Sulfur.* New York: Plenum.

Smil, V. 1987. *Energy Food Environment.* Oxford: Oxford University Press.

Smil, V. 1991. *General Energetics.* New York: John Wiley.

Smil, V. 1993. *Global Ecology.* London: Routledge.

Smil, V. 1999a. China's agricultural land. *China Quarterly* 158:414–429.

Smil, V. 1999b. Nitrogen in crop production: An account of global flows. *Global Biogeochemical Cycles* 13:647–662.

Smil, V. 1999c. Crop residues: Agriculture's largest harvest. *BioScience* 49:299–308.

Smil, V. 1999d. *Energies.* Cambridge, MA: MIT Press.

Smil, V. 2000a. *Cycles of Life: Civilization and the Biosphere.* New York: Scientific American Library.

Smil, V. 2000b. *Feeding the World.* Cambridge, MA: MIT Press.

Smil, V. 2000c. Phosphorus in the environment: Natural flows and human interferences. *Annual Review of Energy and Environment* 25:53–88.

Smil, V. 2000d. Energy in the 20th century: Resources, conversions, costs, uses, and consequences. *Annual Review of Energy and the Environment* 25:21–51.

Smil, V. 2001. *Enriching the Earth: Fritz Haber, Carl Bosch, and the Transformation of World Food Production.* Cambridge: MIT Press.

Smith, A. B., and C. H. Jeffery. 1998. Selectivity of extinction among sea urchins at the end of the Cretaceous period. *Nature* 392:69–71.

Smith, C. A., and E. J. Wood. 1992. *Biosynthesis.* London: Chapman & Hall.

Smith, C. R., et al. 1998. Faunal community structure around a whale skeleton in the deep northeast Pacific Ocean: Macrofaunal, microbial and bioturbation effects. *Deep-Sea Research II* 45:335–364.

Smith, J. B., and D. A. Tirpak, eds. 1990. *The Potential Effects of Global Climate Change on the United States.* New York: Hemisphere.

Smith, J. M., and E. Szathmary. 1999. *The Origins of Life: From the Birth of Life to the Origin of Language.* Oxford: Oxford University Press.

Smith, M. D., et al. 1992. Molecular biology of radiation-resistant bacteria. In *Molecular Biology and Biotechnology of Extremophiles,* ed. R. A. Herbert and R. J. Sharp, 258–280. Glasgow: Blackie.

Smith, R. C., et al. 1992. Ozone depletion: Ultraviolet radiation and phytoplankton biology in Antarctic waters. *Science* 255:952–957.

Smith, T. M., et al. 1992. The response of terrestrial C storage to climate change: Modeling C dynamics at varying temporal and spatial scales. *Water Air and Soil Pollution* 64:307–326.

Snelgrove, P. V. R. 1999. Getting to the bottom of marine biodiversity: Sedimentary habitats. *BioScience* 49:129–138.

Soil and Water Conservation Society. 1999. Erosion control stalls: threatens food production, air, water. <http://www.swcs.org/t_publicaffairs_sspressrelease.html>.

Sokolov, B. S., and A. L. Ianshin, eds. 1986. *V.I. Vernadskii i sovremennost' (V. I. Vernadsky and the Comtemporaneity).* Moscow: Nauka.

Sokolova, M. N. 1997. Trophic structure of abyssal macrobenthos. *Advances in Marine Biology* 32:427–525.

Solomon, A. M., et al. 1993. The interaction of climate and land use in future terrestrial carbon storage and release. *Water Air and Soil Pollution* 70:595–614.

Soltis, P. S., D. E. Soltis, and M. W. Chase.1999. Angiosperm phylogeny inferred from multiple genes as a tool for comparative biology. *Nature* 402:402–404.

Somerville, C. 2000. The twentieth century trajectory of plant biology. *Cell* 100:13–25.

Sorokin, Y. I. 1993. *Coral Reef Ecology.* New York: Springer-Verlag.

Spaink, H. P. 2000. Root nodulation and infection factors produced by rhizobial bacteria. *Annual Review of Microbiology* 54:257–288.

Spotila, J. R., and D. M. Gates. 1975. Body size, insulation and optimum body temperatures of homeotherms. In *Perspectives of Biophysical Ecology,* ed. D. M. Gates and R. B. Schmerl, 291–301. New York: Springer-Verlag.

Stearns, S. C., ed. 1987. *The Evolution of Sex and Its Consequences.* Basel, Switzerland: Birkhäuser.

Steering Group for the Workshop on Size Limits of Very Small Microorganisms. 1999. *Size Limits of Very Small Microorganisms.* Washington, DC: National Research Council.

Stein, C. A., and S. Stein. 1994. Constraints on hydrothermal heat flux through the oceanic lithosphere from global heat flow. *Journal of Geophysical Research* 99:3081–3095.

Stenseth, N. C., and R. A. Ims, eds. 1993. *The Biology of Lemmings.* London: Academic.

Stern, R. A., and W. Bleeker. 1998. Age of the world's oldest rocks refined using Canada's SHRIMP: The Acasta Gneiss Complex, Northwest Territories, Canada. *Geoscience Canada* 25:27–32.

Stetter, K. O. 1996. Hyperthermophilic prokaryotes. *FEMS Microbiological Reviews* 18:2918–2926.

Stetter, K. O. 1998. Hyperthermophiles: Isolation, classification, and properties. In *Extremophiles Microbial Life in Extreme Environments,* ed. K. Horikoshi and W. D. Grant, 1–24. New York: Wiley-Liss.

Stevens, T. O., and J. P. McKinley. 1995. Lithoautrophic microbial ecosystems in deep basalt aquifers. *Science* 270:450–454.

Stocker, T. F., and A. Schmittner. 1997. Influence of carbon dioxide emission rates on the stability of the thermohaline circulation. *Nature* 388:862–865.

Storey, K. B., and J. M. Storey. 1988. Freeze tolerance in animals. *Physiological Reviews* 68:27–84.

Struhl, K. 1999. Fundamentally different logic of gene regulation in eukaryotes and prokaryotes. *Cell* 98:1–4.

Sturges, W. T., et al. 2000. A potent greenhouse gas identified in the atmosphere: SF_5CF_3. *Science* 289:611–613.

Sturkie, P. D., ed. 1986. *Avian Physiology.* Berlin: Springer-Verlag.

Sturrock, P. A. 1986. *Physics of the Sun.* Dordrecht, the Netherlands: D. Reidel.

Suess, E. 1875. *Die Entstehung der Alpen (The Origin of the Alps).* Vienna: W. Braunmüller.

Suess, E. 1883–1901. *Das Antlitz der Erde (The Face of the Earth).* Prague: F. Tempsky.

Suess, E. 1904–1924. *The Face of the Earth,* trans. H. B. C. Solas. Oxford: Clarendon.

Suhai, S., ed. 2000. *Genomics and Proteomics: Functional and Computational Aspects.* New York: Kluwer Academic/Plenum.

Summons, R. E., et al. 1999. 2-methylhopanoids as biomarkers for cyanobacterial oxygenic photosynthesis. *Nature* 400:554.

Sun, G., et al. 1998. In search of the first flower: A Jurassic angiosperm, *Archaefructus,* from Northeast China. *Science* 282:1692–1694.

Swan, L. W. 1961. The ecology of the high Himalayas. *Scientific American* 205(4):68–78.

Szalai, V. A., and G. W. Brudvig. 1998. How plants produce dioxygen. *American Scientist* 86:542–551.

Szathmary, E. 1997. The first two billion years. *Nature* 387:662–663.

Szekielda, K.-H. 1988. *Satellite Monitoring of the Earth.* New York: John Wiley.

Szewzyk, U., R. Szewzyk, and T. A. Stevenson. 1994. Thermophilic, anaerobic bacteria isolated from a deep borehole in granite in Sweden. *Proceedings of the National Academy of Sciences USA* 91:1810–1813.

Szostak, J. W., D. P. Bartel, and L. Luisi.2001. Synthesizing life. *Nature* 409:387–390.

Taintner, J. A. 1988. *The Collapse of Complex Societies.* Cambridge: Cambridge University Press.

Taylor, F. 1974. Implications and extensions of the serial endosymbiosis theory of the origin of eukaryotes. *Taxon* 23:229–258.

Taylor, S. R. 1992. *Solar System Evolution*. Cambridge: Cambridge University Press.

Taylor, S. R., and S. M. McLennan. 1985. *The Continental Crust: Its Composition and Evolution*. London: Blackwell Scientific.

Taylor, S. R., and S. M. McLennan. 1995. The geochemical evolution of the continental crust. *Reviews of Geophysics* 33:241–265.

Teal, J. M. 1962. Energy flow in the salt marsh ecosystem of Georgia. *Ecology* 43:614–624.

Teilhard de Chardin P. 1956. *The Appearance of Man*, trans. J. M. Cohen. New York: Harper.

Teilhard de Chardin, P. 1966. *Man's Place in Nature*. New York: Harper & Row.

Terres, J. K. 1991. *The Audubon Society Encyclopedia of North American Birds*. New York: Wings.

Tevini, M., ed. 1993. *UV-B Radiation and Ozone Depletion: Effects on Humans, Animals, Plants, Microorganisms, and Materials*. Boca Raton, FL: Lewis.

Thompson, D. W. 1917. *On Growth and Form*. Cambridge: Cambridge University Press.

Thornton, I. 1996. *Krakatau The Destruction and Reassembly of an Island Ecosystem*. Cambridge, MA: Harvard University Press.

TIGR (The Institute of Genomic Research). 2000. TIGR Microbial Database: A listing of published genomes and chromosomes and those in progress. <www.tigr.org/tdb/mdb/mdbcomplete.html>.

Tilman, D. 1999. Diversity and production in European grasslands. *Science* 286:1099–100.

Tilman, D., and J. A. Downing. 1994. Biodiversity and stability in grasslands. *Nature* 367:363–365.

Tilman, D., et al. 1997. The influence of functional diversity and composition on ecosystem processes. *Nature* 277:1300–1302.

Tobin, J. E. 1995. Ecology and diversity of tropical forest canopy ants. In *Forest Canopies*, ed. M. D. Lowman and N. M. Nadkarni, 129–147. San Diego: Academic.

Toggweiler, J. R. 1994. The ocean's overturning circulation. *Physics Today* 47(11):45–50.

Tolbert, N. E. 1997. The C_2 oxidative photosynthetic carbon cycle. *Annual Review of Plant Physiology* 48:1–25.

Tortell, P. D., M. Maldonado, and N. Price. 1996. The role of heterotrophic bacteria in iron-limited ocean ecosystems. *Nature* 383:330–332.

Towe, K. M. 1994. Earth's early atmosphere: Constraints and opportunities for early evolution. In *Early Life on Earth*, ed. S. Bengston, 36–47. New York: Columbia University Press.

Transeau, E. N. 1926. The accumulation of energy by plants. *Ohio Journal of Science* 26:1–10.

Tréguer, P., and P. Pondaven. 2000. Silica control of carbon dioxide. *Nature* 406:358–359.

Tréguer, P., et al. 1995. The silica balance in the world ocean: A reestimate. *Science* 268:375–379.

Trenberth, K. E., and C. J. Guillemot. 1995. Evaluation of the global atmospheric moisture budget as seen from analyses. *Journal of Climate* 8:2255–2272.

Trimble, S. W. 1999. Decreased rates of alluvial sediment storage in the Coon Creek Basin, Wisconsin, 1975–93. *Science* 285:1244–1246.

Trüper, H. G. 1999. How to name a prokaryote? Etymological considerations, proposals and practical advice in prokaryote nomenclature. *FEMS Microbiology Review* 23:231–249.

Tucker, C. J., and S. E. Nicholson. 1999. Variations in the size of the Sahara Desert from 1980 to 1997. *Ambio* 28:587–591.

Tucker, C. J., and J. R.G. Townshend. 2000. Strategies for monitoring tropical deforestation using satellite data. *International Journal of Remote Sensing* 21:1461–1472.

Tumilevskii, L. 1967. *Vernadskii.* Moscow: Molodaya Gvardiya.

Tunnicliffe, V. 1992. Hydrothermal-vent communities of the deep sea. *American Scientist* 80:336–349.

Tunnicliffe, V., A. G. McArthur, and D. McHugh. 1998. A biogeographical perspective of the deep-sea hydrothermal vent fauna. *Advances in Marine Biology* 34:355–442.

Turner, B. L., et al., eds. 1990. *The Earth as Transformed by Human Action.* Cambridge: Cambridge University Press.

Turner, J. S. 2000. *The Extended Organism.* Cambridge, MA: Harvard University Press.

Tyler, P. A. 1995. Conditions for the existence of life at the deep-sea floor: An update. *Oceanography and Marine Biology: An Annual Review* 33:221–244.

Tyrrell, T. 1999. The relative influences of nitrogen and phosphorus on oceanic primary production. *Nature* 400:525–531.

U.N. Conference on the Human Environment. 1972. *Conference on the Environment.* New York: United Nations Organisation.

UNEP (United Nations Environmental Programme). 1995. *Montreal Protocol on Substances that Deplete the Ozone Layer.* Nairobi: UNEP.

UNESCO (United Nations Educational, Scientific, and Cultural Organisation). 1968. *Use and Conservation of the Biosphere.* Paris: UNESCO.

UNESCO (United Nations Educational, Scientific, and Cultural Organisation). 2000. The Man and the Biosphere Programme. <http://www.unesco.org/mab>.

UNO (United Nations Organisation). 1976. *World Energy Supplies 1950–1974.* New York: United Nations.

UNO (United Nations Organisation). 2001. *World Population Prospects 2000.* New York: United Nations Organisation.

U.S. Environmental Protection Agency. 2001. *Methane in the Atmosphere.* <http://www.epa.gov/ghginfo/topic1.htm>.

USGS (U.S. Geological Survey). 2000. Zebra mussel information. <http://nas.er.usgs.gov/zebra.mussel>.

Uvarov, B. 1977. *Grasshoppers and Locusts.* Cambridge: Cambridge University Press.

Valencia R. M., H. Balslev, and G. Paz y Miño. 1994. High tree alpha-diversity in Amazonian Ecuador. *Biodiversity and Conservation* 3:21–28.

Valiela, I. 1984. *Marine Ecological Processes.* New York: Springer-Verlag.

Van Cappellen, P., and E. D. Ingall. 1996. Redox stabilization of the atmosphere and oceans by phosphorus-limited marine production. *Science* 271:493–496.

Van Dover, C. L. 2000. *The Ecology of Deep-Sea Hydrothermal Vents.* Princeton, NJ: Princeton University Press.

Van Etten, J. L., and R. H. Meints. 1999. Giant viruses infecting algae. *Annual Review of Microbiology* 53:447–494.

van Gardingen, P. R., G. M. Foody, and P. J. Curran, eds. 1997. *Scaling-up: From Cell to Landscape.* Cambridge: Cambridge University Press.

Veizer, J., Y. Godderis, and L. M. François. 2000. Evidence for decoupling of atmospheric CO_2 and global climate during the Phanerozoic eon. *Nature* 408:698–701.

Verhoogen, J. 1980. *Energetics of the Earth.* Washington, DC: National Research Council.

Vernadskii, V. I. 1904. *Osnovi kristalografii (Fundamentals of Crystalography).* Moscow: Izdatel'stvo Moskovskogo geologicheskogo instituta (Moscow Geological Institute).

Vernadskii, V. I. 1926. *Biosfera (Biosphere).* Leningrad: Nauchnoe khimiko-tekhnicheskoye izdatel'stvo (Scientific Chemico-Technical Publishing).

Vernadskii, V. I. 1931. O biogeokhimicheskom izuchenii iavleniia zhizni (Biogeochemical study of life). *Izvestiia AN SSSR* 6:633–653.

Vernadskii, V. I. 1939. *Problemy biogeokhimii (Problems of biogeochemistry).* Moscow-Leningrad: Izdatel'stvo Akademii Nauk SSSR (Academy of Sciences of the USSR).

Vernadskii, V. I. 1940. *Biogeokhimicheskie ocherki, 1922–1932 (Biogeochemical Studies, 1922–1932).* Moscow: Izdatel'stvo Akademii Nauk SSSR (Academy of Sciences of the USSR).

Vernadskii, V. I. 1954–1960. *Izbrannye sochineniia (Selected Works).* Moscow: Izdatel'stvo Akademii Nauk SSSR (Academy of Sciences of the USSR).

Vernadskii, V. I. 1995. *Publitsisticheskie stat'i (Publicistic Writings).* Moscow: Nauka.

Vernadsky, V. I. 1923. A plea for the establishment of a biogeochemical laboratory. *Transactions of the Liverpool Biological Society* 37:38–43.

Vernadsky, V. I. 1924. *La géochimie (Geochemistry).* Paris: Félix Alcan.

Vernadsky, V. I. 1929. *La Biosphère.* Paris: Félix Alcan.

Vernadsky, V. I. 1944. Problems of biogeochemistry, II. *Transactions of the Connecticut Academy of Arts & Sciences* 35:483–517.

Vernadsky, V. I. 1945. The biosphere and the noosphere. *American Scientist* 33:1–12.

Vernadsky, V. I. 1986. *The Biosphere.* Oracle, AZ: Synergetic.

Vernadsky, V. I. 1998. *The Biosphere,* trans. D. B. Langmuir. New York: Copernicus.

Vincent, W. F., and C. Howard-Williams. 2000. Life on snowball Earth. *Science* 287:2421.

Vinogradov, M. E. 1997. Some problems of vertical distribution of meso- and macroplankton in the ocean. *Advances in Marine Biology* 32:1–85.

Vinogradova, N. G. 1997. Zoogeography of the abyssal and hadal zones. *Advances in Marine Biology* 32:325–375.

Visick, K. V., and M. J. McFall-Ngai. 2000. An exclusive contract: Specificity in the *Vibrio fischeri-Euprymna scolopes* relationship. *Journal of Bacteriology* 182:1779–1787.

Vitorello, I., and H. N. Pollack. 1980. On the variation of continental heat flow with age and the thermal evolution of the continents. *Journal of Geophysical Research* 85:983–995.

Vitousek, P. M., et al. 1986. Human appropriation of the products of photosynthesis. *BioScience* 36:368–373.

Vitousek, P. M., et al. 1997a. Human domination of Earth's ecosystems. *Science* 277:494–499.

Vitousek, P. M., et al. 1997b. Human alteration of the global nitrogen cycle: Sources and consequences. *Ecological Applications* 7:737–750.

Volk, T. 1998. *Gaia's Body: Toward a Physiology of Earth.* New York: Copernicus.

von Bertalanffy, L. 1940. Der Organismus als physikalisches System betrachtet (The organism considered as physical system). *Naturwissenschaften (Natural Sciences)* 28:521–531.

von Willert, D. J. 1994. *Welwitschia mirabilis* Hook fil.—das Überlebenswunder der Namibwüste (The survival miracle of the Namib desert). *Naturwissenschaften* (Natural Sciences) 81:430–442.

Vörösmarty, C. J., and D. Sahagian. 2000. Anthropogenic distrubance of the terrestrial water cycle. *BioScience* 50:753–765.

Vreeland, R. H., W. D. Rosenzweig, and D. W. Powers.2000. Isolation of a 250 million-year-old halotolerant bacterium from a primary salt crystal. *Nature* 407:897–900.

Wächtershäuser, G. 1992. Groundworks for an evolutionary biochemistry: The iron-sulphur world. *Progress in Biophysics and Molecular Biology* 58:85–201.

Waide, R. B., et al. 1999. The relationship between productivity and species richness. *Annual Review of Ecology and Systematics* 30:257–300.

Waisel, Y., A. Eshel, and U. Kafkafi, eds. 1996. *Plant Roots The Hidden Half.* New York: Marcel Dekker.

Walker, J. C. G., P. B. Hays, and J. F. Kasting. 1981. A negative feedback mechanism for the long-term stabilization of the Earth's surface temperature. *Journal of Geophysical Research* 86:9776–9782.

Walsh, W. B. 1968. *Russia and the Soviet Union: A Modern History.* Ann Arbor, MI: University of Michigan Press.

Walter, M. 1996. Old fossils could be fractal frauds. *Science* 383:385–386.

Walters, W., and H. Fierstine. 1964. Measurements of swimming speeds of yellowfin tuna and wahoo. *Nature* 202:208–209.

Ward, P. D., and D. Brownlee. 2000. *Rare Earth: Why Complex Life Is Uncommon the Universe.* New York: Springer-Verlag.

Ward, P. D., D. R. Montgomery, and R. Smith R. 2000. Altered River Morphology in South Africa Related to the Permian-Triassic Extinction. *Science* 289:1740–1743.

Wardle, D. A., et al. 1997. The influence of island area on ecosystem properties. *Nature* 277:1296–1299.

Waring, R. H., and S. W. Running. 1998. *Forest Ecosystems: Analysis at Multiple Scales.* San Diego: Academic.

Warnant, P., et al. 1994. CARAIB: A global model of terrestrial biological productivity. *Global Biogeochemical Cycles* 8:255–270.

Warneck, P. 2000. *Chemistry of the Natural Atmosphere.* San Diego: Academic.

Watanabe, Y., J. E. Martini, and H. Ohmoto. 2000. Geochemical evidence for terrestrial ecosystems 2.6 billion years ago. *Nature* 408:574–577.

Watson, A. J., and P. S. Liss. 1998. Marine biological controls on climate via the carbon and sulphur geochemical cycles. *Philosophical Transactions of the Royal Society London* 353:41–51.

Watson, J. D., and F. H. Crick. 1953. Molecular structure of nucleic acids. *Nature* 171:737–738.

Watson, R. T., et al., eds. 1996. *Climate Change 1995 Impacts, Adaptations and Mitigation of Climate Change: Scientific Analysis.* Cambridge: Cambridge University Press.

Wayne, R. K., J. Leonard, and A. Cooper.1999. Full of sound and fury: The recent history of ancient DNA. *Annual Review of Ecology and Systematics* 30:457–477.

Webster, P. J. 1981. Monsoons. *Scientific American* 245(2):108–118.

Weiss, B. P., et al. 2000. A low temperature transfer of ALH84001 from Mars to Earth. *Science* 290:791–795.

Weitnauer, E. 1956. On the question of the nocturnal behavior of the swift. *Der ornithologische Beobachter (Ornithological Observer)* 53(3):74–79.

Wellsbury, P., et al. 1997. Deep marine biosphere fuelled by increasing organic matter availability during burial and heating. *Nature* 388:573–576.

Wesley, J. P. 1974. *Ecophysics.* Springfield, IL: Charles C. Thomas.

West, G. B., et al. 1997. A general model for the origin of allometric scaling laws in biology. *Science* 276:122–126.

West, G. B., et al. 2000. The origin of universal scaling laws in biology. In *Scaling in Biology,* ed. J. H. Brown and G. B. West, 87–112. Oxford: Oxford University Press.

West, J. B., and L. Sukhamay, eds. 1984. *High Altitude and Man.* Bethesda, MD: American Physiological Society.

Westheimer, F. H. 1987. Why nature chose phosphates. *Science* 235:1173–1177.

Westman, W. E. 1977. How much are nature's services worth? *Science* 197:960-964.

Wetherill, G. W. 1990. Formation of the Earth. *Annual Review of Earth and Planetary Sciences* 18:205–256.

Wheeler, J. C. 2000. *Cosmic Catastrophes: Supernovae, Gamma-Ray Bursts, and Adventures in Hyperspace.* Cambridge: Cambridge University Press.

White, O., et al. 1999. Genome sequence of the radioresistant bacterium *Deinococcus radiodurans* R1. *Science* 286:1571–1577.

Whitehead, J. A. 1989. Giant ocean cataracts. *Scientific American* 260(2):50–57.

Whitman, W. B., D.C. Coleman, and W. J. Wiebe. 1998. Prokaryotes: The unseen majority. *Proceedings of the National Academy of Sciences USA* 95:6578–6583.

Whittaker, R. H. 1959. On the broad classification of organisms. *Quarterly Review of Biology* 34:210–226.

Whittaker, R. H. 1972. Evolution and measurement of species diversity. *Taxon* 21:231–251.

Whittaker, R. H. 1975. *Communities and Ecosystems.* New York: Macmillan.

Whittaker, R. H., and G. E. Likens. 1973. Carbon in the biota. In *Carbon and the Biosphere,* ed. G. M. Woodwell and E. V. Pecan, 281–300. Washington, DC: U.S. Atomic Energy Commission.

Whittaker, R. H., and G. E. Likens. 1975. The biosphere and man. In *Primary Productivity of the Biosphere,* ed. H. Lieth and R. H. Whittaker, 305–328. New York: Springer-Verlag.

Whitton, B. A., and M. Potts. 2000. *The Ecology of Cyanobacteria: Their Diversity in Time and Space.* Boston: Kluwer Academic.

Wielgolaski, F. E., ed. 1997. *Polar and Alpine Tundra.* New York: Elsevier.

Wigley, T. M. L., and D. S. Schimel, eds. 2000. *The Carbon Cycle.* Cambridge: Cambridge University Press.

Wignall, P. B., and R. J. Twitchett. 1996. Oceanic anoxia and the end Permian mass extinction. *Science* 272:1155–1158.

Wilde, S. A., et al. 2001. Evidence from detrital zircons for the existence of continental crust and oceans on the Earth 4.4 Gyr ago. *Nature* 409:175–178.

Wilhelm, S. W., and C. A. Suttle. 1999. Viruses and nutrient cycles in the sea. *BioScience* 49:781–788.

Williams, D. M., J. F. Kasting, and R. A. Wade.1997. Habitable moons around extrasolar giant planets. *Nature* 385:234–236.

Williams, G. C. 1992. *Gaia,* nature worship and biocentric falllacies. *Quarterly Review of Biology* 67:479–486.

Williams, G. R. 1996. *The Molecular Biology of Gaia.* New York: Columbia University Press.

Williams, T. C., and J. M. Williams. 1978. An oceanic mass migration of land birds. *Scientific American* 239(4):166–176.

Wilson, E. B. 1925. *The Cell in Development and Heredity.* New York: Macmillan.

Wilson, E. O., ed. 1986. *Biodiversity.* Washington, DC: National Academy Press.

Wilson, E. O. 1992. *The Diversity of Life.* Cambridge, MA: Belknap.

Wilson, E. O., and D. I. Perlman. 2000. *Conserving Earth's Biodiversity.* Covelo, CA: Island.

Wilson, J. T., ed. 1972. *Continents Adrift: Readings from Scientific American.* San Francisco: W. H. Freeman.

Wilson, R. C. 1997. Total solar irradiance trend during solar cycles 21 and 22. *Science* 277:1963–1965.

Wilson, R. C., S. Drury, and J. L. Chapman.2000. *The Great Ice Age.* London: Routledge.

Wilson, T. L. 2001. The search for extraterrestrial intelligence. *Nature* 409:1110–1114.

Winogradsky, S. 1887. Über Schwefelbacterien (Sulfur bacteria). *Botanische Zeitung (Botanical Journal)* 45:489–523.

Winogradsky, S. 1890. Recherches sur les organismes de la nitrification. *Annales de l'Institut Pasteur* 4:213–231.

Woese, C. R. 1987. Bacterial evolution. *Microbiological Reviews* 51:221–271.

Woese, C. R. 1998a. The universal ancestor. *Proceedings of the National Academy of Sciences USA* 95:6854–6859.

Woese, C. R. 1998b. Default taxonomy: Ernst Mayr's view of the microbial world. *Proceedings of the National Academy of Sciences USA* 95:11043–11046.

Woese, C. R., and G. E. Fox. 1977. Phylogenetic structure of the prokaryotic domain: The primary kingdoms. *Proceedings of the National Academy of Sciences of Sciences USA* 74:5088–5090.

Woese, C. R., D. Kandler, and M. L. Wheelis.1990. Towards a natural system of organisms: Proposal for the domains Archaea, Bacteria, and Eucarya. *Proceedings of the National Academy of Sciences USA* 87:4576–4579.

Wolery T. J., and N. H. Sleep. 1988. Interactions of geochemical cycles with the mantle. In *Chemical Cycles in the Evolution of the Earth,* ed. C. B. Gregor et al., 77–103. New York: John Wiley.

Wollast, R., F. T. Mackenzie, and L. Chou, eds. 1993. *Interactions of C, N, P and S Biogeochemical Cycles and Global Change.* Berlin: Springer-Verlag.

Wolszczan, A. 1994. Confirmation of Earth-mass planets orbiting the millisecond pulsar PSR B1257+12. *Science* 264:538–542.

Woodward, F. I., 1987, Stomatal numbers are sensitive to increases in CO_2 from pre-industrial levels. *Nature* 327:617–618.

Woolfson, A. 2000. *Life without Genes.* New York: HarperCollins.

World Commission on Environment and Development. 1987. *Our Common Future.* New York: Oxford University Press.

World Resources Institute. 2000. *World Resources 2000–2001.* Amsterdam: Elsevier.

Worthington, E. B. 1975. *The Evolution of IBP.* Cambridge: Cambridge University Press.

Wray, G. A., J. S. Levinton, and L. Shapiro. 1996. Molecular evidence for deep Precambrian divergences among Metazoan phyla. *Science* 274:568–573.

Wright, D. D., et al. 1997. Tree and liana enumeration and diversity on a one-hectare plot in Papua New Guinea. *Biotropica* 29:250–260.

Würsig, B. 1988. The behavior of baleen whales. *Scientific American* 258(4):102–107.

Xiao, S., et al. 1998. Three-dimensional preservation of algae and animal embryos in a Neoproterozoic phosphorite. *Nature* 391:553–558.

Xie, O., and P. A. Arkin. 1996. Analyses of global monthly precipitation using gauge observations, satellite estimates, and numerical model predictions. *Journal of Climate* 9:840–858.

Xiong, J., et al. 2000. Molecular evidence for the early evolution of photosynthesis. *Science* 289:1724–1729.

Yachandra, V. K., et al. 1993. Where plants make oxygen: A structural model for the photosynthetic oxygen-evolving manganese cluster. *Science* 260:675–679.

Yayanos, A. A. 1998. Empirical and theoretical aspects of life at high pressure in the deep sea. In *Extremophiles Microbial Life in Extreme Environments,* ed. K. Horikoshi and W. D. Grant, 47–92. New York: Wiley-Liss.

Yockey, H. P. 1995. Information in bits and bytes: Reply to Lifson's review of *Information Theory and Molecular Biology. BioEssays* 17:85–88.

Yoder, J. A., et al. 1998. Primary production in the global ocean. <http://dixon.gso.uri.edu/ppmeth.htm>.

York, D. 1993. The earliest history of the Earth. *Scientific American* 268(1):90–96.

Young, T. P., C. H. Stubblefield, and L. A. Isbell.1997. Ants on swollen-thorn acacias: Species coexistence in a simple system. *Oecologia* 109:98–107.

Zapol, W. M. 1987. Diving adaptations of the Weddel seal. *Scientific American* 256(6):100–105.

Zehr, J. P., et al. 2001. Unicellular cuanobacteria fix N_2 in the subtropical North Pacific Ocean. *Nature* 412:635–638.

Zengler, K., et al. 1999. Methane formation from long-chain alkanes by anaerobic microorganisms. *Nature* 401:266–269.

Zierenberg, R. A., M. W. W. Adams, and A. J. Arp. 2000. Life in extreme environments: Hydrothermal vents. *Proceedings of the National Academy of Sciences USA* 97:12961–12962.

Zimmer, C. 1998. *At the Water's Edge: Macroevolution and the Transformation of Life.* New York: Free Press.

Zimmerman, P. R., et al. 1982. Termites: A potentially large source of atmospheric methane, carbon dioxide and molecular hydrogen. *Science* 218:563–565.

Zinke, P. J., et al. 1986. *Worldwide Organic Soil Carbon and Nitrogen Data.* Oak Ridge, TN: Oak Ridge National Laboratory.

ZoBell, C.E. 1952. Bacterial life at the bottom of the Philippine Trench. *Science* 115:507–508.

Scientific Name Index

Name Index

Subject Index